Computational
Methods
in Chemistry

Computational Methods in Chemistry

Edited by

Joachim Bargon

IBM Research Laboratory
San Jose, California

SPRINGER SCIENCE+BUSINESS MEDIA, LLC

Library of Congress Cataloging in Publication Data

Symposium on Computational Methods in Chemistry, Bad Neuenahr, Ger., 1979.
  Computational methods in chemistry.

  (The IBM research symposia series)
  "Proceedings of the Symposium on Computational Methods in Chemistry, held in Bad
Neuenahr, German Federal Republic, September 17—19, 1979."
  Sponsored by IBM Germany.
  Includes index.
  1. Spectrum analysis—Data processing—Congresses. 2. Quantum chemistry—Data process-
ing—Congresses. I. Bargon, Jaochim. II. IBM Deutschland. III. Title. IV. Series: International
Business Machines Corporation. IBM research symposia series.
QD95.S93 1979   542'.8   80-14881

ISBN 978-1-4684-3730-0   ISBN 978-1-4684-3728-7 (eBook)
DOI 10.1007/978-1-4684-3728-7

Proceedings of the International Symposium on Computational
Methods in Chemistry, sponsored by IBM Germany and held in
Bad Neuenahr, German Federal Republic, September 17—19, 1979.

PREFACE

The papers collected in this volume were presented at an international symposium on Computational Methods in Chemistry. This symposium was sponsored by IBM Germany and was held September 17-19, 1979, in Bad Neuenahr, West Germany.

According to Graham Richards [Nature 278, 507 (1979)] the "Third Age of Quantum Chemistry" has started, where the results of quantum chemical calculations have become so accurate and reliable that they can guide the experimentalists in their search for the unknown.

The particular example highlighted by Richards was the successful prediction and subsequent identification of the relative energies, transition probabilities and geometries of the lowest triplet states of acetylene. The theoretical predictions were based chiefly upon the work of three groups: Kammer [Chem. Phys. Lett. 6, 529 (1970)] had made qualitatively correct predictions; Demoulin [Chem. Phys. 11, 329 (1975)] had calculated the potential energy curves for the two lowest triplet states (^3B and ^3A) of acetylene; and Wetmore and Schaefer III [J. Chem. Phys. 69, 1648 (1978)] had determined the geometries of the cis (^3B_u and ^3A_u) and the trans ($^3B_2$ and $^3A_2$) isomers of these two states. In a guided search, Wendt, Hunziker and Hippler [J. Chem. PHys. 70, 4044 (1979)] succeeded in finding the predicted near infrared absorption of the cis triplet acetylene (no corresponding absorption for the trans form was found, which is in agreement with theory), and the resolved structure of the spectrum confirmed the predicted geometries conclusively.

This convincing success of quantum chemical predictions triggered our curiosity to assess the extent to which experimentalists, working in different fields of chemistry, could benefit from recent achievements of theoretical methods. At the same time, we wanted to inform the theoreticians about the current needs of the experimentalists.

The focus of this symposium was chiefly on two areas of chemistry, namely the computational progress in various kinds of spectroscopy (NMR, IR, ESR, PES, and Mass Spectrometry), and the recent achievements of quantum chemistry, sometimes called "Computer Chemistry" by the lectureres in this volume. Considerable time was spent during discussion periods to compare the strengths and weaknesses of semi-empirical versus ab initio methods when applied to problems of varying size and to properties of a different nature. In the chairman's opinion, it seems that both approaches will continue to co-exist, each taking their specific role. In the case of theoretical studies of chemical reactions, for example semi-empirical methods can provide an economical way to map out potential energy surfaces in a preliminary and crude form, perhaps by automatically forming energy gradients combined with automated geometry optimization, followed by a refinement of all parameters with sophisticated ab initio methods in the areas of highest interst.

A great number of theoretical and computational methods, customary in chemistry nowadays, had to be omitted or seriously neglected. Thus no Monte Carlo calculations were included not as the consequence of parochial thinking due to the casino next door, but rather because of the hopeless task to consider all of the known computational methods in chemistry within the framework of a coherent but limited symposium. The topics chosen here were considered to be of interest to a group of scientists whose bandwidth would overlap in the area of physical organic chemistry.

Joachim Bargon   Paul Schweitzer
IBM San Jose Research Laboratory   IBM Germany, Sindelfingen
Symposium Chairman   Symposium Manager

CONTENTS

SETTING UP, USING, AND MAINTAINING COMPUTER-READABLE SPECTRA COMPILATIONS

J.T. Clerc

University of Berne

Switzerland

H. Könitzer

Swiss Federal Institute of Technology

Zürich, Switzerland

INTRODUCTION

Identification and structure elucidation of organic compounds
is today mostly done with spectroscopic methods. Even though the
theoretical basis of most methods in molecular spectroscopy is quite
advanced, the analyst still has to rely heavily on the use of various semi-empirical interpretation methods based on previously collected reference data. Traditionally, these reference data are compiled into spectra collections or digested and condensed into correlation tables familiar to every analytical chemist. There are
three basic limitations with this traditional approach. Firstly,
only those features initially selected for indexing can be used for
retrieval. Secondly, multidimensional searches using logical combinations of several descriptors are hardly possible. Thirdly,
manipulation of the data becomes tedious even with data collections
of moderate size. If modern data processing equipment and methods
are used to manipulate traditionally organized data collections, the
afore-mentioned restrictions are somewhat relaxed but by no means
completely removed. To index the data with additional features becomes possible but is still difficult and limited to predefined
features. The same holds true for the multidimensional search capabilities. Handling of the data, however, is definitely improved by
the use of computers.

Transforming conventional and traditional spectra collections into computer-readable form brings some advantages for the user. However, philosophy and concepts, often originating from the "data processing stone age", are thereby perpetuated in disguised form preventing optimal use of today's powerful hardware. Thus, it seems appropriate to reconsider the various concepts, methods and goals in the context of the possibilities offered by modern computers and analytical instruments. It is worth-while to do this very carefully. In setting up computer readable spectral data collections test studies quite often lead to solutions that, even though they exhibit grave and obvious shortcomings, will completely block any new development and/or approach. Once the data types and their computer representation are fixed and an experimental collection of some thousand spectra is set up, the activation energy for a fundamental change becomes very high. Thus, the experimental collection starts to grow despite its inherent limitations, becomes established and sets the standard for other compilations of spectral data. However, even well designed data collections tend to become obsolete due to technical advancements.

WHICH DATA SHOULD BE INCLUDED?

Collections of spectral data of organic compounds are expected to satisfy various needs. The practical analyst wants to confirm the identity of a tentatively identified compound by comparing the measured spectra with reference spectra from the data collection. This implies a unique and unambiguous identification of chemical compounds suitable for computer storage and manipulation. Using this canonical identification he retrieves the respective data set from the collection. He expects the result to be ready within a reasonable time and presented in a form suitable for direct use. If the chemical compound he searches for is not represented in the data collection, the system should offer him those entries most similar to the requested one. In the general case these will be those compounds having a high number of substructural elements in common with the structure in question. Thus, substructure search capabilities are called for. If the analyst deals with a compound on which only little and/or conflicting prior information is available, he may try to find a matching spectrum in the library. If this approach is not successful, he may wish to retrieve reference compounds with spectral characteristics similar to the recorded one. The structures of the compounds retrieved may then give valuable hints regarding the structure of the unknown. This operation mode requires access to the library through spectral data. Furthermore, routines for comparing spectra and assessing a degree of similarity have to be available. For easy interpretation of the results the structures of the retrieved reference compounds should be put out in a form easily perceived by the chemist. As spectral data do not depend only on the structure but are also influenced by the sampling technique and various recording parameters, the data sets have to be complemented with all parameters relevant in this context.

No matter how the reference data collection is accessed, the data are used mainly to resolve ambiguities in spectral interpretation. It is therefore of utmost importance that they are reliable. Users generally consult a reference data collection when in doubt about values and significance of spectroscopic parameters, and so will normally not be in a position to judge the quality of the retrieved reference data sets directly. Thus, the data sets collected in the spectra library should be verified with greatest care.

An important field of application of spectroscopic reference data collections is the investigation of new correlations between structural and spectral features and the development of automated classification routines. Here again a high degree of reliability is of paramount importance. Many of the powerful new algorithms for uncovering hidden correlations between structure and spectra are extremely sensitive to outlayers. Thus, it is not uncommon that the parameters of a classification procedure are primarily determined by a few erroneous spectra in the training set, rather than by the bulk of correct spectra. Furthermore, hitherto unknown correlations will involve spectral and/or structural features not yet in common usage. Thus, the data collection must not be restricted to include features of known significance only. It rather should contain all available structural and spectral data items, stored in such a way that extraction of new features, possibly relevant to the solution of future, still undefined problems becomes possible.

In summary, a reference spectra library should therefore include the following pieces of information. Firstly, the complete structure of every compound has to be recorded in computer readable form. The structure representation should be unique and unambiguous, it should give unrestricted substructure search capabilities and allow for structure output easy to read by the chemist. Secondly, the spectral data have to be stored with adequate precision, accuracy and resolution. The curve trace reconstructed from the stored data should be virtually identical to the original curve trace, as judged by an experienced analyst. Then the requirement for a potentially unlimited set of spectral features is best fulfilled. Thirdly, supplemental information should include all parameters necessary to rerun the spectrum. Above all, the data should be of the highest reliability attainable at reasonable costs.

WHICH REFERENCE COMPOUNDS SHOULD BE INCLUDED?

The answers obtainable from a reference data collection are
obviously limited to data sets included in the collection. A universal spectra library including every known chemical compound would be
optimal in this context. However, it is obvious that such a collection can never be realized. Furthermore, a data collection of excessive size becomes very expensive to use and to maintain. Thus,
a careful analysis of the problem of which compounds to include is
worth-while.

The analyst expects useful results from the reference data
collection. Results will be useful if they answer the user's queries
at least partially. As the set of answers the system can provide is
given by the contents of the data collection, it is the range of
user queries that defines the optimum contents of a spectroscopic
reference data collection. Since universality can never be attained,
the user will, in general, not be presented with a reference compound identical to the unknown at hand and he will have to be satisfied with a set of compounds similar to his unknown. We should therefore concentrate on including carefully chosen simple model compounds rather than highly complex molecules. Furthermore, large
series of homologous compounds should not be included; a few examples will do. A spectra compilation of limited size, if modelled
along these lines, can provide useful reference data for a broad
range of problems at acceptable costs. The variety of the model
compounds to be included in the compilation depends on the user's
field of interest. The compound classes on which a user group does
research should certainly be represented by many reference compounds
of closely similar structure. Other compound classes of lesser significance to the users may be typified by a smaller number of models.
Consequently, different user groups will need different reference
libraries. In the general case, it will thus hardly be possible to
acquire a ready-made data collection that optimally suits one's own
needs. However, a base collection covering a wide range of chemical
classes with a limited number of model compounds per class is expected to be common to most applications. This base collection will
then have to be enriched with compounds from the user's field of
interest. The person best qualified for the selection of these compounds and having easiest access to them is the user himself. Consequently, the user of a spectroscopic reference data collection will
have to acquire his own reference spectra and incorporate them into
the base collection. Convenient procedures for this operation have
to be available. As a further consequence, optimal results cannot
in general be expected from general-purpose data collections. Either
they are (implicitly or explicitly) specialized to a given field or
they contain a great many reference compounds of low utility. Most

currently available data collections belong to the second type.
They tend to include all spectra available without proper regard
to a well balanced content. The performance of the data bank system
is affected in two ways. First, the costs for using the collection
are too high, and second, the large number of homologous compounds
present often results in a hit list that is monopolized by one
compound class. Thus, the user loses an important part of the information the system could supply and, even worse, he is lured into
false confidence by the seemingly consistent answers. It is, however, not primarily the data bank designers that are to be blamed
for this. As long as a potential user's first question is "how
many spectra do you have?", quantity will rank before quality.

MAN-MACHINE COMMUNICATIONS

Even today a surprisingly high percentage of analytical chemists
have little or no experience with computers. The "activation energy"
for getting started with computers and becoming moderately efficient
proves to be quite high. There are several reasons for this.

The language gap between computer people and chemists has become so wide that communication is severely restricted due to an
entirely different terminology (computer-Chinese). Furthermore, the
various rules and conventions for using many of the larger computer
installations are true folklore in the sense that they rely on oral
transmission from generation to generation. In order to overcome
these and other barriers, the language in which the user has to
formulate his queries should be simple and modelled along the patterns and concepts of the analyst's natural language. Our own experience with remote batch systems, however, shows that the real
difficulties are not primarily arising from the command language
of the spectroscopic data bank, but rather from the computing
centre's operating system language and its associated terminology.
This problem could probably be overcome with smaller dedicated
systems that are optimized to one single task or to a few closely
related applications.

Another equally important point is concerned with the presentation of the results. Here again the user expects output formats
compatible with his line of thinking. In particular, to make visual
comparison easy, spectra should be displayed as curve traces in a
format very closely matching the standard output format of the
spectra recorded in the laboratory. The output of a spectroscopic
information system is often put to use in two rather different ways.
In a first step, the analyst wants to check quickly on various preliminary hypotheses. Most of them can be discarded as useless at

once after a quick look at a reference data set. Quite often a new
or modified hypothesis is formulated which the analyst also wants
to test on the spot. For this application, an interactive system
is best suited. Virtually immediate response to the user's queries
is most important at this stage. The output of the system is preferably displayed on a video screen. The graphic quality of the
output has not to meet highest aesthetic standards; heavy emphasis
should rather be placed on a format easy to perceive. In a later
stage, when (hopefully) only a few tentative solutions survive,
these will have to be checked more carefully and in detail. For this,
data have to be available as hard copy. Finally, to document his
findings and conclusions, the analyst will probably include various
reference data sets into his report. Here, beside a clear and con-
cise presentation, aesthetic aspects become important.

As shown in previous sections, a spectroscopic data collection
will be accessed by structural and spectral features. Thus, the user
will have to input (partial) structures and spectral data. Input
of chemical structures poses no fundamental problems, but a convenient and easy to use system requires large and sophisticated programs and/or expensive hardware. Spectral data should be input as
a digital image of the complete curve trace to avoid introduction
of artifacts and biases by the user. Consequently, analytical instruments delivering digital output are required, and a convenient
procedure to transfer the digitally acquired data from the instrument to the data bank system has to be available.

HOW MANY SPECTRA SHALL BE STORED?

To discuss the question of how many spectra should be included
in a reference data collection the storage requirements for various
spectra types have to be considered. In principle, one would like
to digitally store the data in such a way that all information contained in the analog spectrum is fully retained. However, this leads
to excessive storage space requirements. For example a modern routine
infrared spectrometer allows a resolution of the wave number scale
of better than 1 cm^{-1}. Thus, for the standard range about 4'000
intensity values have to be recorded. The digital resolution of the
intensity scale should be of the same order of magnitude as the re-
producibility of the instrument. If a resolution of about 1 in 250
is assumed, 8 bits are necessary. Total storage space thus amounts
to 32'000 bits per infrared spectrum. To store compound identification, chemical structure and technical parameters require another
8'000 bits. So the total storage space amounts to about 40'000 bits.
Therefore, an average disk cassette, as today commonly used with
popular minicomputers, will hold a few hundred infrared spectra only.

For bar type spectra as e.g. noise decoupled ^{13}C-NMR spectra or mass spectra, where the line shape conveys little or no useful information, storage space requirements are somewhat less demanding.

The reference data collection should as a base contain representative model compounds covering the full range of organic substances. It is very difficult to give a realistic estimate as to the number of compounds necessary. An order of magnitude may be specified, however. Textbooks treating organic chemistry on an advanced level are expected to give an overview of the full field. They generally have around 1'000 pages where on the average some 2 to 3 compounds are discussed. Thus, a realistic estimate for the size of the base collection is some thousand spectra. These will have to be enriched with spectra of compounds having special relevance to the users. This may result in a doubling of the size of the collection. We thus arrive at a spectra library containing between some 5'000 to 20'000 entries.

It is, therefore, obvious that even for a relatively small data collection a huge amount of storage space is needed if the full information content shall be retained. To relax the requirement for the full information is not an acceptable solution. If only the information currently known to be relevant is stored, all further development is blocked, and the data collection's fate is programmed for premature obsolescence. It is, however, appropriate to delete all data that is known to contain no useful information, and all tricks and gimmicks for data compression should be used.

IMPLEMENTATION

In the foregoing sections various aspects of computer readable compilations of spectroscopic reference data have been discussed. The identified requirements result in severe conflicts.

The data have to include a representation of the chemical structure of the reference compounds. The code should be compact, easy to manipulate, and simple to perceive on output. CAS registry numbers are very compact but completely unsatisfactory in all other respects. Connectivity tables are relatively easy to manipulate but are quite voluminous and difficult to interpret. Pictorial diagrams can provide optimal output but are otherwise unsatisfactory. In our experience the Wiswesser Line Notation (WLN) presents an acceptable compromise in that it is reasonably compact and directly readable. Furthermore, conversion into connectivity tables as well as into pictorial diagrams is in most cases not unduly difficult.

The spectral data have to be recorded in full. This requires adequate instrumentation on the spectrometer side. On the data system side appropriate input ports have to be provided and enough mass storage facilities have to be supplied. To output the data in a format that is optimally matched to the analyst's line of thinking, versatile and expensive computer peripherals are necessary. All this calls for a large and powerful computer system. However, the reference spectra collection and its associated programs will utilize only a small part of the available computing power, making a dedicated system highly inefficient. If the computer system has to be shared with many other users from widely different fields to justify the costs, chances are high that communication and various logistic problems will prevent a significant percentage of potential users from exploiting the data bank. This might be a temporary problem since serious attempts are made today to introduce the subject of digital computers and data processing into the chemistry curriculum. However, for the time being we have to live with the fact that there are psychological barriers to the widespread and general use of computers in chemistry. These barriers are definitely lower with dedicated mini systems, however.

We believe that large dedicated systems are not cost efficient for spectroscopic data banks. For local operation, investment and operation costs are excessively high. If operated in time share mode with a large common data base accessed by many different user groups, the data collection will in general contain mostly spectra of minor or no relevance to the individual user, with detrimental effects on quality and costs. For user groups with closely similar ranges of interest this solution might however be acceptable. The main disadvantages of large general purpose computing centres are communication problems and large turn-around times. As stated before, the former problem is probably a temporary one. However, at present we have to live with it and take it into consideration. In his daily work the analyst often needs almost instant answers to his spectroscopic problems. The result should become available when the sample is still at hand in the laboratory to allow for rerunning some measurements without going through all sample preparation steps. Not all large computer installations can provide such quick service. Furthermore, a direct link between a remote spectroscopic instrument and a large central computer also is not without problems. Regarding communications, system availability and turn-around time, small dedicated systems are optimal. However, the costs to equip a mini system with the necessary peripherals, and in particular with adequate mass storage capacity, are out of proportion. It is possible that the fulminant evolution in the field of personal computers will improve the feasibility of the "small but mine" approach.

At present the most appropriate solution seems to be a dedicate satellite connected to a large centralized system. We have, however, no experience with such a system yet. Our own implementation is realized on a large general purpose computer installation which, besides some rudimentary on-line capabilities, offers fast remote batch processing with turn-around times on the order of some 10 minutes. In this environment it is possible to arrive at an acceptable performance level.

ORGANIZATION OF THE OCETH-SYSTEM

The data compilations incorporated in the OCETH-System currently comprise about 6'500 mass spectra, 3'500 ^{13}C-NMR spectra, and 1'200 UV spectra, originating from various sources including our own laboratories. Initially, the number of compounds documented was almost twice as high, but elimination of duplicates and reference compounds of low utility have reduced the number to the present value. In addition, some 50 infrared spectra are also included to allow for developing and testing of the respective program segments.

The main objective of our implementation is to provide almost unlimited possibilities for combined processing of all data items without regard to the data type or to the spectroscopic method involved. This is realized by having a rigorously standardized format common to all spectroscopic methods presently represented and suitable for future expansion. For practical reasons the data related to different spectroscopic methods are kept on separate files. At present there are but few compounds documented with data from more than one spectroscopic method. Thus, overlap between the different spectra files is rather limited. Furthermore, as our collections are still growing fast, we have to give due consideration to the input side. New data tend to become available in batches of spectra from one method. Updating is thus more easy with specialized files. Finally, separate files for the different spectroscopic methods provide the specialized analyst with specialized data compilations for his special pet method.

This set of files, encompassing the full spectroscopic and supplemental information, are referred to as Library Files. As their detailed structure is to some extent influenced by the computer hardware and the operating system, the following description is limited to system-independent aspects.

Each spectrum documented corresponds to one data record in the library file. Each record consists of three segments. The first segment is of fixed length and holds all information related to the sample identity, e.g. identification number, chemical name, CAS registry number, structure code (WLN notation), empirical formula, and nominal molecular mass. Furthermore, it gives the key for interpreting the second segment. The format of the first segment is identical in all library files. The second segment includes the data relating to the spectrum registration. It is again of fixed length. Here, some entries have different meanings for different spectroscopic methods; the key being given in the first segment. In addition to instrumental data, the key for interpreting the third segment is given. The third segment is of variable length and holds the spectroscopic data in highly compressed form. The length of the third segment is specified in the second segment as well as the mode of compression used. The file is headed by a header record that identifies the file, gives its length, source, and history as well as other data necessary and/or convenient for processing the file. This data structure, where each part contains the information necessary to correctly process and interpret the following parts, makes it possible to write a unified set of programs that can handle the data from all files. To retrieve any data item from any file, the user just accesses the central processing program which will take care of the various codes and data compression schemes used with the different spectroscopic methods. The central processing program consists of many subprograms designed for various applications. These include routines to output full or partial data sets in standardized formats on various periferals, programs to generate images of the library in a format suitable for data exchange with other institutions, and for generating various index files and subfiles.

Even though the data in the library files are highly compressed, the files are still rather voluminous. The mass spectra file, with 6'500 data sets, has a length of roughly 5 megabytes, or 800 bytes per compound. The ^{13}C-NMR data require somewhat less storage space, namely about 500 bytes per compound. The length of the corresponding file with about 3'500 documented compounds is thus 1.75 megabytes. UV spectra require the same number of bytes per spectrum. IR spectra, however, require about 2'500 bytes per compound.

For most standard applications the library files are not accessed directly. Rather, a specialized index file comprising an appropriate subset of the full data is used. For example the chemical name, permuted WLN code, empirical formula, and nominal molecular mass are directly available via index files. As experience has shown, most chemists at our institutions still prefer to deal with a hard copy index rather than doing a computer search. Consequently, we supply the index files in printed form or on microfiches whenever this is economically and technically feasible. Furthermore, we also provide the user with truncated spectral data in hard copy form when

this seems appropriate. This somewhat conservative approach is
justified by the fact that computer terminals are not yet ubiquitous
and that a large proportion of all queries involve the search for
a fully specified entry, where a manual "telephone directory"
search is quite adequate.

More complex search problems are done with the computer. The
most common applications include the retrieval of reference compounds
exhibiting spectra similar to the spectrum of a sample of unknown
structure, substructure search, and multidimensional searches. For
the retrieval of spectra similar to a given model we use a special
search file where the spectral attributes are encoded in binary
form. The spectral attributes are selected so as to emphasize structural similarities rather than individual differences between reference spectra. The system accepts the spectrum of the sample as
input, compares it to all reference spectra in the file using a
self-optimizing search strategy. On output a list of the 20 compounds believed to be structurally most similar to the sample is
produced. In addition, the spectra of the retrieved reference compounds may be plotted either on hard copy with a digital plotter or
on a video screen. The plotting routines may, of course, also be
used directly for processing spectral data from other sources. Substructure search programs are still in the planning stage. However,
programs for the conversion of WLN codes into connectivity tables
and pictorial diagrams have been acquired recently and will be in-
tegrated into the system. Multidimensional search problems, where
entries meeting several conditions at the same time have to be retrieved, arise mainly in connection with the study of structure/
spectra correlations. For these applications we use inverted files.
The respective programs are currently under development.

The system has been well accepted by the chemists. At present,
it is most heavily used for retrieval of reference spectra of
specified structure and for plotting spectra. Substructure search
and retrieval of spectra similar to a given model are used less often.
This is most probably due to the fact that most of our chemists do
purely synthetic work where one is only very rarely confronted with
compounds of completely unknown structure.

The spectroscopic data bank also provides the base for various
research projects. For example, one project aims at a spectroscopy
oriented classification scheme for organic compounds. The classification schemes universally used today date from the beginning of our
century. They are based primarily on the reactivity of compounds and
functional groups, which makes them optimal for the discussion of
e.g. syntheses and reaction mechanisms. The modern analytical chemist,
however, sees the compounds he deals with, rather, from a spectros-

copic point of view and thus, thinks in other categories. In a large
compilation of spectra of compounds covering a wide range of compound classes (in the classical sense), one seeks natural clusters
of compounds with common spectral properties. These clusters then
correspond to spectroscopic compound classes.

The structural features common to the compounds in such a
cluster will be closely related to the spectral features that are
common to their spectra. So, new structure spectra correlations may
be discovered. Initial feasibility studies have shown promising results.

In another project improved methods for the estimation of
chemical shifts in ^{13}C-NMR spectra using additivity rules will be
developed. The structure handling programs are used to generate a
list of atoms or atom groups adjacent to every carbon atom. Using
least squares methods shift increment values are then assigned to
each type of neighboring atom so that the additivity rule optimally reproduces the measured chemical shift of the central carbon
atoms.

Spectra comparison methods and similarity measures to be used
with library search systems are also further investigated. The study
of these problems originally initiated the present work. Tests on
various algorithms and methods disclosed that the results were determined predominantly by the errors and idiosyncrasies of the data
base rather than by the methodology.

Consequently, elimination of errors and biases currently takes
up the major part of our work. The collection includes several
spectra compilations that are commercially available and believed
to be of fairly high quality. However, in a large collection of mass
spectra some 40 % of all data sets were either duplicates or contained obvious and uncorrectable errors and have been removed. In
a compilation of about 4'000 ^{13}C-NMR spectra 10 % of all shift values
were miscoded, 12 % of the literature references were erroneous, and
even 3 % of the molecular weights were wrong. It has to be said, however, that the data sets originating from our own laboratories had
error rates of the same order of magnitude.

Most of the problems encountered with computer readable spectra
compilations can be traced back to three main sources. First of all,
there is a strong tendency to overemphasize the number of spectra included in the compilations. Thus, data sets of doubtful value are included indiscriminately just to get an impressive growth rate. The
result is a data collection with many duplicate entries and with
spectra from large series of homologous compounds. Secondly, in many

data compilations "facts and fiction" are not clearly separated. For example in some ^{13}C-NMR data compilations no distinction is made between line multiplicities actually observed and inferred from line assignments. In order to save storage space many mass spectra compilations record selected peaks only, the selection being most often based on the relative intensity. Intensity, however, is not a good selection criterium. If a mass spectrum exhibits a strong peak at m/z 128, there is no simple way to decide if this fragment corresponds to a Naphtalene ion or to Hydrogen iodide if the weak isotope peak at m/z 129 has been deleted. Similar problems arise with other selection criteria. The only safe way is to include all observed data and observed data only. Thirdly, setting up computer-readable spectra compilations requires expertise in data processing as well as in applied spectroscopy. If data processing experience is missing, the project most probably never materializes. If know-how in applied spectroscopy is absent, the data may get corrupted, as shown in the following example. The ^{13}C-NMR data collection we use is the result of a world wide co-operation. Each contributor sends in his spectra to a central office where they will be formatted and (hopefully) checked. He subsequently gets back a copy of all spectra received. In this copy one record contains the shift values of all lines in descending order. Another record lists the respective line multiplicities together with the assignment in the same order. The assignment is given as a number referring to a connectivity table given in a third segment. In the copy we received, however, the multiplicity assignment pairs were sorted in ascending order of assignment numbers which, of course, results in complete loss of any relation between multiplicity, assignment and chemical shift values.

CONCLUSION

It is the authors' firm belief that spectroscopic data bank systems will eventually become extremely useful tools for the solution of analytical problems in routine as well as in research applications. The necessary hardware is already available today at reasonable costs or will be available in the near future. However, on the software side, much remains to be done. To build up systems that are adapted to the analyst's true needs requires personnel trained and experienced in data processing as well as in analytical chemistry. To be able to use a data bank system efficiently, the user needs some fundamental training in data processing and computers. Consequently, the respective subjects should be included in the chemistry curriculum. Even though some universities have already done so, in most cases the respective courses emphasize computational aspects rather than handling of non-numerical information.

Setting up a large computer-readable compilation of spectral data is a time-consuming and expensive task. It is thus very important that the product will meet not only the present requirements but also the future ones. Thus, the data sets have to include all information currently available even if part of it is presently not yet useful. Otherwise, all further development and progress is efficiently blocked. Criteria for very high quality should be applied and rigorously enforced to insure highly reliable data, and the contents of the data collection should be constantly monitored to avoid bias. Particularly, large sets of structurally and/or spectroscopically similar compounds as well as exotic compounds should be excluded.

The procurement of a large number of top quality spectra of model compounds suitable for inclusion in a spectral library at reasonable costs requires world-wide co-operation of analytical laboratories in industry and universities. A recent US government regulation requires identification of all impurities present in pharmaceutical products above a specified, rather low concentration level. Thus, analytical chemists will be confronted with the problem of identifying a large number of trace compounds, a job where spectroscopic data banks are of high utility. The above mentioned government regulation will provide a strong driving force for concerted action and so will have at least one positive aspect.

THE SOLUTION TO THE GENERAL PROBLEM OF

SPECTRAL ANALYSIS ILLUSTRATED WITH EXAMPLES FROM NMR

Gerhard Binsch

Institute of Organic Chemistry
University of Munich
Karlstrasse 23, 8000 Munich 2, Germany

ABSTRACT

A general iterative method for locating the global minimum of an error functional $\phi = (\underline{s}-\underline{\hat{s}})^T(\underline{s}-\underline{\hat{s}})$ in the presence of a multitude of local minima is presented. It relies on the introduction of a square matrix $\underline{W}$ in data space whose off-diagonal elements can be exploited for inducing continuously adjustable correlation between the residuals by way of $\phi' = (\underline{s}-\underline{\hat{s}})^T\underline{W}(\underline{s}-\underline{\hat{s}})$. The general structure of $\underline{W}$ is derived from symmetry, boundary and continuity requirements. In spectroscopic applications the vector $\underline{s}$ in data space consists of n digitized signal intensities s_i and the theoretical model is expressed as a function $\underline{\hat{s}} = f(\underline{p},\omega)$ of a discrete vector $\underline{p}$ in parameter space and a continuous frequency variable ω. The solution has been fully worked out for NMR spectra (high-resolution isotropic; anisotropic; exchange-broadened), including practical aspects such as automated data acquisition, data format conversion, pretruncation, baseline flattening, smoothing and posttruncation, and will be illustrated by a variety of examples. It is concluded that with the availability of a set of well-documented computer programs the rigorous analyses of complicated NMR spectra need in the future only rarely engage the attention of the spectroscopic specialist; in many if not most instances such tasks can now be entrusted to technicians to be performed as a routine service.

INTRODUCTION

At a Symposium devoted to the computer-chemistry interface it is fitting to remind ourselves of the noblest role the computer ought to play in an experimental science such as chemistry, which is to aid us in reducing the overwhelming and constantly swelling flood of raw data to the comparatively few pieces of information worthy of being preserved for posterity. There is not much that can be done in this respect with such classical chemical characteristics as melting points, boiling points and refractive indices beyond burying them unprocessed in the archives of Beilstein and handbooks. The situation is different for molecular spectroscopic data, not only owing to their great intrinsic detail, but also because upwards of 90% of that information is nowadays collected by chemists who lack the expertise as well as the inclination to subject it to analysis, with the result that much of it is deposited in the literature either in raw form or as loose qualitative descriptions whose permanent worth is questionable at best.

It was against this backdrop that in 1975 we began to become interested in the development of a spectral analysis procedure automated to such an extent that its practical usefulness would not be confined to the specialists, and we now venture to say that we have solved the problem for NMR spectra. The examples I shall show are exclusively taken from that area, but it should become apparent that the method is in fact quite general and in principle applicable to any kind of spectroscopy for which one can find a feasible procedure to express the exact or decently approximate theoretical model in the form of a single functional entity. How much of this potential can be exploited in practice remains to be seen, and I therefore welcome the opportunity of subjecting our ideas to the scrutiny of the experts in diverse fields gathered at this Symposium.

THE PROBLEM

Suppose that a research worker has recorded a high-resolution NMR spectrum of the general degree of complexity shown in Figure 1. Being the archetypal chemist alluded to in the Introduction, he might report this as δ 7-8 (m,5H), meaning that there is a 5-proton multiplet between 7 and 8 ppm downfield from TMS, hide that "information" in the experimental part of his paper and be

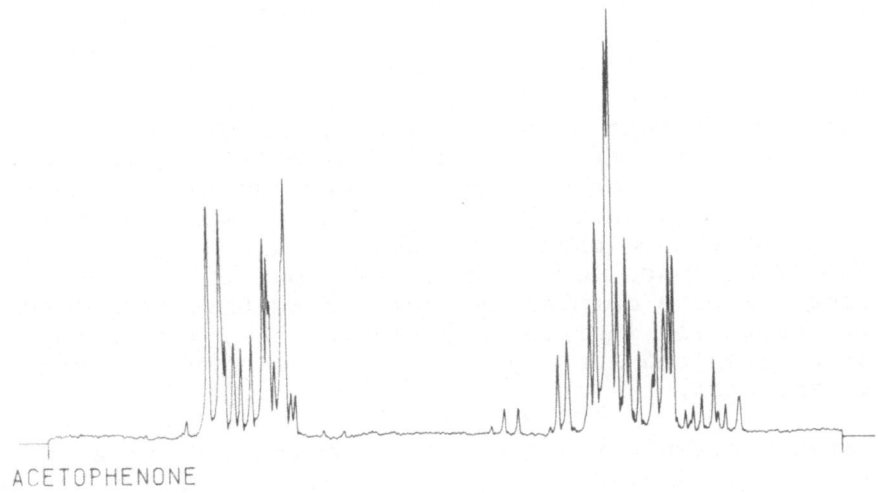

ACETOPHENONE

Figure 1.   100 MHz NMR spectrum of the aromatic protons of acetophenone.

done with it. On the other hand, if he happens to be an NMR spectroscopist intent on extracting the spectral parameters and if he has access to a standard iterative NMR computer program such as LAOCOON or one of its descendants, he would probably adopt the following three-stage procedure. First decide on the nature of the spin system and count the number of parameters entering the problem. In the present instance we are obviously dealing with an [AB]$_2$C system for the aromatic protons, characterized by 3 chemical shifts, 6 distinct coupling constants and the line width, of which the latter can be directly read off from the spectrum without calculation. Next calculate a few synthetic spectra from sets of guessed parameters. On the basis of a vast body of empirical knowledge about the proton NMR spectra of substituted benzenes one can be fairly confident to be able to estimate the 6 distinct coupling constants to within 1 or 2 Hertz. The downfield group of signals has an area corresponding to 2 of the 5 protons and elementary chemical reasoning establishes beyond doubt that it must be attributed to the ortho protons, whose chemical shift should be close to the center of gravity of that signal group. So the only major uncertainty involves the chemical shifts of the meta and para protons, and since these two parameters are also the most critical ones for the appearance of the spectrum, a good policy would be to

start with three trial values for each, resulting in a
total of 9 synthetic spectra. With some luck one might
find that one of these trial spectra resembles the experimental spectrum closely enough to permit the assignment of a sufficient number of lines, and with this information one can then proceed to the third, iterative
stage of the calculation. Given a spectrum such as that
of Figure 1, a moderately experienced NMR spectroscopist
should stand a good chance of successfully completing
the job in not much more than a week. But if one has an
NMR spectrum of a complexity typified by that of toluene
and if one really insists on analyzing that also, one
better be prepared for a year of work using the conventional approach.

When contemplating some of the past attempts at
automating the analysis process, in particular the heroic
effort undertaken by the late T.R.Lusebrink of the IBM
Research Laboratories at San Jose, it gradually became
clear to us that their failures were rooted in an unfor-
tunate formulation of the problem. In analogy to the
orthodox approach Lusebrink concentrated his attention
on the assignment step, but ended up in such a frightful
morass of ambiguities that nobody has since dared to
continue along these lines. We became convinced that to
make progress it was necessary not only to do away with
line assignments completely, in agreement with a conclusion reached by Diehl four years ago (1), but also
to dispose of another deeply entrenched and purportedly
indispensable notion of the spectroscopist, namely the
very concept of a line itself.

With this in mind, an experimental NMR spectrum is
regarded as a collection of digitized signal intensities
s_i represented (in a computer) by a vector $\underline{s}$ in data
space. By the same token, a theoretical NMR spectrum,
no matter how complicated, can always be expressed as a
function

$$\hat{s} = f(\underline{p}, \omega)$$

of a discrete parameter vector $\underline{p}$ and a continuous frequency variable ω, from which one can calculate a corresponding discretized estimator $\hat{\underline{s}}$ in data space. Then the
task consists in finding a $\underline{p}$ such that it minimizes the
scalar functional

$$\phi = (\underline{s} - \hat{\underline{s}})^T (\underline{s} - \hat{\underline{s}}) \ .$$

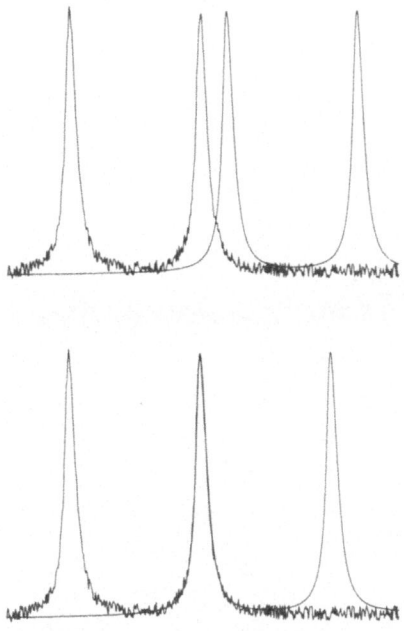

*Figure 2.   Behavior of standard least-squares method in
fitting a theoretical to an experimental doublet spectrum.
Top: starting situation. Bottom: situation after convergence.*

We have arrived at the standard formulation of a least-
squares problem and both objectionable concepts have indeed disappeared.

    The crucial problem one is up against can now be
identified: it resides in the properties of ϕ, the error
hypersurface in error space. For typical NMR spectra
consisting of sharp lines this functional is just about
the most pathological mathematical entity one can imagine.
To see why this must be so consider an experimental doublet spectrum and a trial spectrum computed with the
correct (and fixed) doublet spacing and linewidth, but
with the chemical shift displaced as shown in the top
part of Figure 2. After a few iteration cycles on the
chemical shift one arrives at the situation depicted
in the bottom part of Figure 2. Obviously, no further
progress can be made, since in proceeding to the global
minimum on ϕ one would first have to go uphill, and no
minimization algorithm can perform such a feat. In

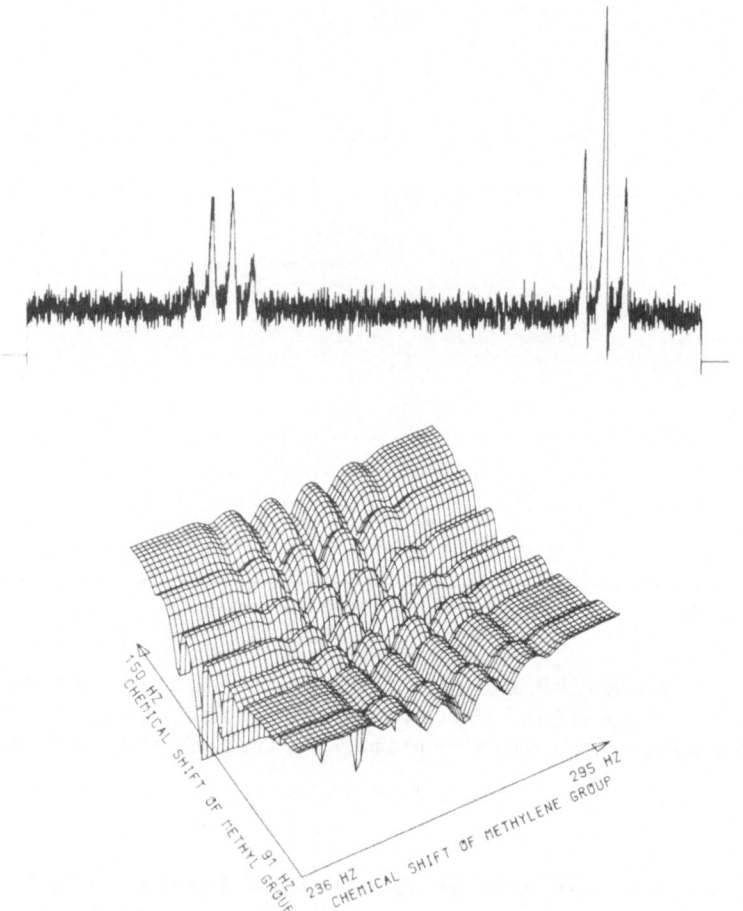

Figure 3.   Ethyl group ^1H NMR spectrum (90 to 340 Hz) and section of the corresponding error surface ϕ in 3-space.

Figure 3 I show the slightly more complicated case of the ethyl group spectrum of ethyl benzene, where again the coupling constant and the line width have been frozen at their known values. My coworker Dr. David S. Stephenson, with whom I had the great pleasure of colla- borating on this project for four years, calls the shown surface an eggbox. Naturally, it can only provide a faint impression of how complicated things may become in gene- ral. There may literally be billions of local minima on the ϕ hypersurface in multidimensional error space, out

of which one must locate just one, i.e. the global minimum corresponding to the correct solution; more precisely, one must locate at least one out of a set of global
minima which are mathematically equivalent.

THE SOLUTION IN PRINCIPLE

Let us rewrite the error functional as

$$\phi = (\underline{s}-\underline{\hat{s}})^T \underline{I}(\underline{s}-\underline{\hat{s}}) \ ,$$

where the unit matrix $\underline{I}$ in data space has been inserted
between the vector quantities on the right-hand side.
We are committed to $\underline{I}$ for an unbiased judgment of the
quality of the fit and hence also for the final stages
in the parameter refinement process, but there is no
such commitment for the early or intermediate phases
of the minimum search. Consequently we may write

$$\phi' = (\underline{s}-\underline{\hat{s}})^T \underline{W}(\underline{s}-\underline{\hat{s}}) \ ,$$

where $\underline{W}$ is a general square matrix in data space, whose
off-diagonal elements provide a mechanism for establi-
shing correlations between the residual at one point and
the residuals at all other points in the spectrum. By
choosing these elements judiciously we should be able
to alleviate the pathology of the error hypersurface,
and by making the degree of correlation an increasing
function of the iteration distance to the global minimum
we can then hope to evade the multitude of traps in the
course of the minimization process.

Rigorously speaking, the mathematical conditions
that $\underline{W}$ must satisfy in order to serve our purposes are
as follows. Define an ordered sequence of ℓ correlation
matrices $\underline{W}_k$

$$\underline{W}_1, \ \underline{W}_2, \ \dots \ , \ \underline{W}_k, \ \dots \ , \ \underline{W}_\ell = \underline{I}$$

with associated modified ϕ_k' hypersurfaces

$$\phi_1', \ \phi_2', \ \dots \ , \ \phi_k', \ \dots \ , \ \phi_\ell' = \phi \ .$$

Start with $k = \ell$ and the global minimum on $\phi_\ell' = \phi$. Then
each successive ϕ_{k-1}' must be monotonically convex down-
ward in a local region such that the corresponding minimum falls inside the convergence radius of the ϕ_k' minimum already identified, the region of local convexity
must increase with increasing correlation and it must

finally, for the last $\phi_{k-1}^! = \phi_!^!$, cover the entire specified domain of the parameter vector $\underline{p}$. In this way we
are assured of the existence of a unique pathway to the
global minimum and, furthermore, we know that this global minimum will actually be found in proceeding from
left to right in the sequences.

To state these requirements is one thing, to satisfy them in practice another. We can immediately forget
about any attempt to do that rigorously. Since the exact
conditions for $\underline{W}$ can only be specified in reverse, i.e.
in proceeding from right to left in the above sequences,
the best we could hope for would be a convergent iterative scheme whose complexity vastly exceeded that of the
problem we set out to solve to begin with. But even if
this obstacle could somehow be overcome, we would still
be faced with the necessity of recalculating the W_k
matrices in each particular case explicitly from the
experimental vector $\underline{s}$ and the specific form of the
theoretical model.

We conclude that to obtain a viable formalism large-
ly independent of the specifics of a particular case we
must be content with deriving some rather general though
still serviceable characteristics of the W_k matrices.
Fortunately, this is easily accomplished on the basis of
plausible symmetry, boundary and continuity arguments,
and the results are summarized in Table 1, where k stands
for zero or a positive integer. Of particular importance
is the so-called pulling criterion, whose function can
be visualized by considering a one-line spectrum. If the
computed line has no overlap with the experimental, the
contribution to ϕ' arising from the diagonal elements of
$\underline{W}$ is a constant. To establish a mechanism for pulling
the theoretical line all the way in to its experimental
counterpart, the error contributions arising from the
off-diagonal elements of $\underline{W}$ must be made distance dependent in such a way that $\phi^\top$ decreases monotonically along
the corresponding direction in parameter space, which in
turn requires that the W_{ij} be a monotonically decreasing
function of their absolute distance from the principal
diagonal. Since we cannot and indeed do not wish to make
any specific statements as to the magnitudes of the W_{ij},
we have no choice but to invoke a principle ascribed to
the scholastic philosopher William of Occam in treating
all lines, or more precisely all elements of $\underline{s}$, on the
same footing.

Table 1. General Properties of the Correlation Matrix $\underline{W}$

Property	Derived from
W_{ii} = const.	Absence of bias
W_{ii} = 1	Scale invariance
$W_{ij} = W_{ji}$	Frequency reversal invariance
$W_{ij} (i \neq j) < 1$	Nonnegative ϕ'
$W_{i,i+k} > W_{i,i+k+1}$	Pulling criterion
$W_{ij} (i \neq j) \geq 0$	Continuity
$W_{ij} = W_{i+k,j+k}$	Occam's razor

The last property of Table 1 leads to the general functional form

$$W_{ij} = f(|i-j|) ,$$

i.e. the correlation matrix must be a band matrix. The other properties can for instance be satisfied with the specific choices of an exponential

$$W_{ij} = \exp (-\alpha |i-j|)$$

or a Lorentzian

$$W_{ij} = [1+\beta (i-j)^2]^{-1}$$

decay, where the positive control parameters α and β can be used for monitoring the pulling force as well as the morbidity of ϕ'. The effect on ϕ' can be gleaned from Figure 4, which also illustrates the computational strategy: one starts with a high correlation, locates a minimum on the corresponding ϕ' and then reduces the correlation gradually, eventually reverting to the unit matrix $\underline{I}$. Our problem is thus solved in principle if we can in addition find a stable and efficient minimization

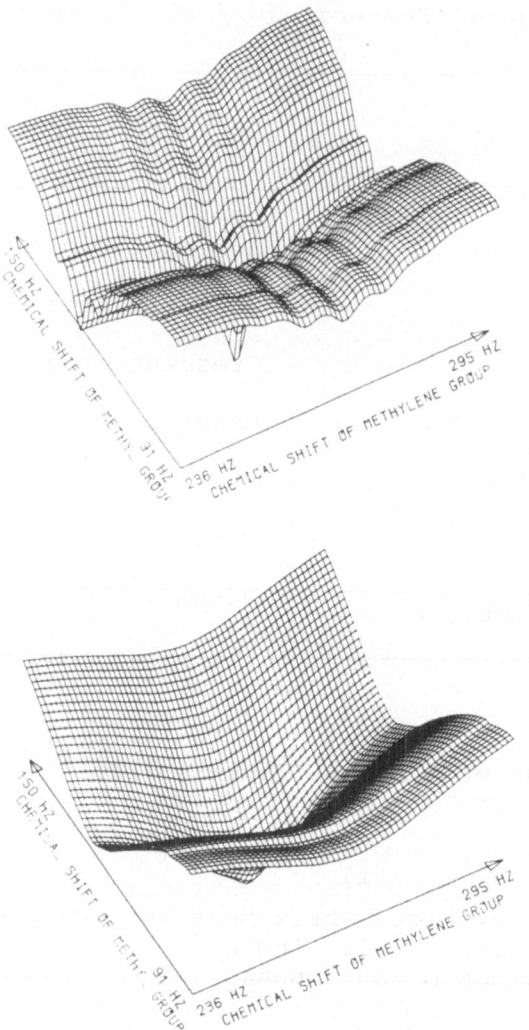

Figure 4. Modified error surfaces ϕ' for the example of Figure 3 with exponential correlation. Top: $\alpha = 0.1$. Bottom: $\alpha = 0.01$.

procedure applicable to modified as well as unmodified error hypersurfaces. About 65% of our four-year effort has been invested in the development of such an algorithm and we have reasons to believe that the result represents the optimum in speed for spectroscopic applications and is generally unsurpassed in its stability,

but in this lecture I do not wish to go into the technical details. Those interested are referred to a summary
account of this aspect of our work which has already
appeared in the literature (2).

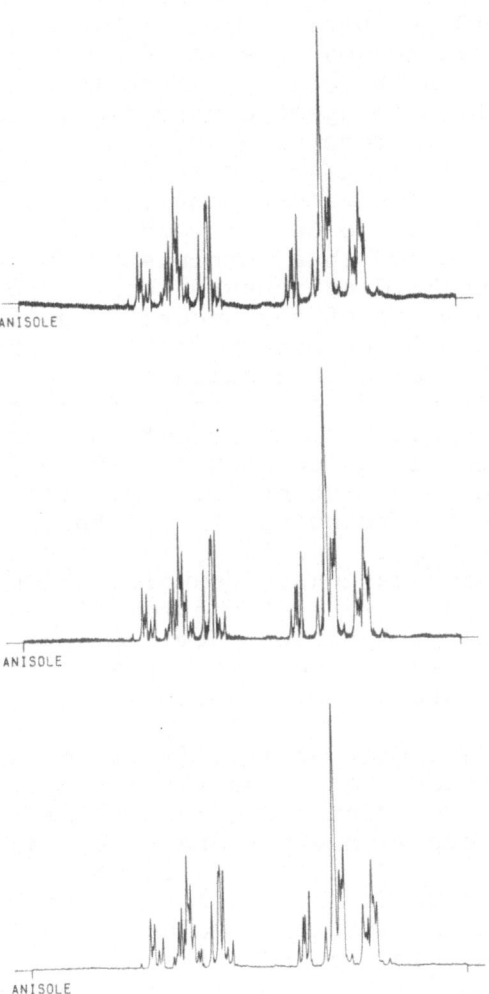

*Figure 5.   Top trace: experimental 100 MHz CW NMR spec-
trum of the aromatic protons of anisole. Middle trace:
experimental spectrum after baseline flattening. Bottom
trace: baseline-flattened experimental spectrum after
smoothing and posttruncation.*

THE SOLUTION IN PRACTICE

The described method pertains to idealized spectra, whereas actual NMR spectra are of course noisy, are distorted by instrumental and other artifacts, contain impurity signals, etc. In Figure 5 I give an example of how our computer program DAVINS (*Direct Analysis of Very Intricate NMR Spectra*) deals with such problems in practice. The top trace shows an experimental CW NMR spectrum consisting of 10200 16-bit intensity points automatically recorded and transferred to a standard 9-track magnetic tape by our NMR instrument system software. This tape, on which normally all spectra recorded during a work session on the spectrometer have been written, separated by end-of-file marks, then serves as an input medium for a CYBER 175. The data format conversion, from the original 16-bit words to the 60-bit words of a CYBER, is handled within DAVINS in a completely general way, easily adaptable to any kind of computer configuration. Unwanted features in the wings of the spectrum can then be cut off by pretruncation. An arbitrary section of baseline may be specified for the automatic calculation of an RMS noise figure. Application of an automatic flattening procedure, in which a least-squares variable-term Fourier series baseline function is calculated iteratively and subtracted out in a self-consistent fashion, yielded the middle trace of Figure 5. The bottom trace, the so-called preprocessed experimental spectrum, was obtained after smoothing and simultaneously reducing the number of points to those desired in the automatic analysis procedure (999 in the present instance). Unless stated otherwise, all experimental spectra displayed in the remainder of this lecture are of this preprocessed type.

In Figure 6 I show an example of an actual analysis by DAVINS. The starting values for the coupling constants were taken from the Munich Telephone Directory and the solution, complete with error analysis, was produced in about 2 minutes of CPU time in a single pass (called a "grand cycle") through the program, using exponential correlation with an initial α value of 10^{-6} and a correlation factor multiplier of 5. Visual comparison of the theoretical spectrum recalculated from the final parameter values with the preprocessed experimental spectrum confirms beyond doubt that the correct solution has indeed been found, which is further corroborated by the global R factor of 0.024.

This performance of DAVINS is typical for NMR spectra which exhibit complete or at least partial clustering of

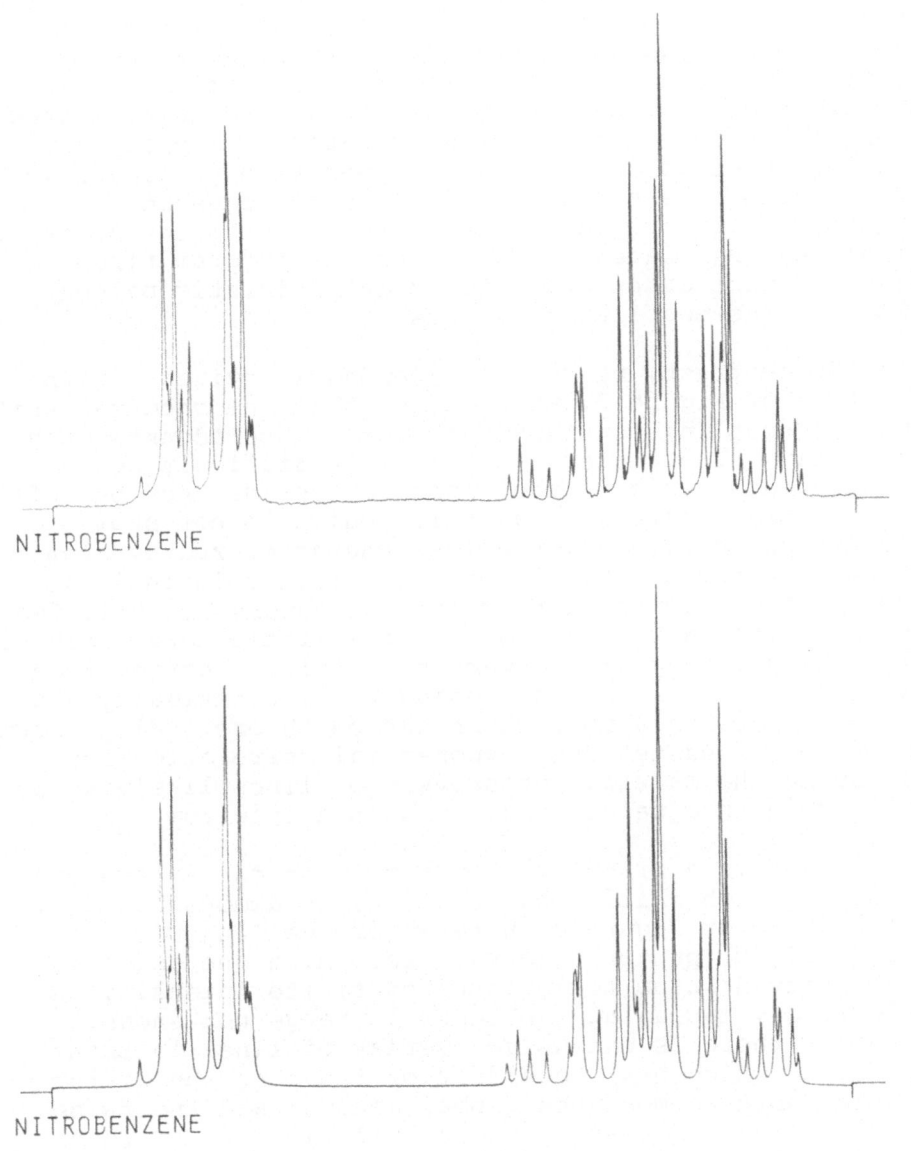

NITROBENZENE

NITROBENZENE

*Figure 6.   Top trace: preprocessed experimental 100 MHz
CW ^{1}H NMR spectrum of nitrobenzene. Bottom trace: computed spectrum.*

the fine-structure lines around the respective chemical
shifts. We refer to them as "normal" spectra; the experimental examples of Figures 1 and 5 also belong in

this category. In such cases whole groups of theoretical
lines need be pulled toward their experimental counter-
parts, and since the mean frequency separations between
experimental and theoretical groups are roughly aligned
along the chemical shift axes in parameter space, the
pulling force is cumulative and therefore particularly
strong along these directions. As a consequence, the
chemical shifts rapidly migrate to the vicinity of their
final resting places, after which the program finds it
easy, in most cases, to enter a deterministic pathway
also for the coupling constants.

In contrast, no such clustering is evident in the
spectrum of Figure 7, which means that the chemical shifts
and the coupling constants all need to be adjusted con-
currently. As expected, this is more difficult and a
single grand cycle of the standard type may not be sufficient. For spectra such as this, which do not show
clustering but are nevertheless characterized by plenty
of well-resolved fine structure, a blindfold analysis
from randomly selected starting parameters is still feasible if one is willing to invest a little more computer
time, by starting with a very high initial correlation
(α or β of 10^{-10}, say) and using the intrinsically slower
Lorentzian correlation, which can be theoretically shown
to be better suited than exponential correlation for
effecting the multiple crossovers of lines likely to be
inevitable in a calculation on such a spectrum.

There is a practical limit to what can be achieved
by a very high initial correlation, as demonstrated by
the example of Figure 8. Here we are dealing with a 7-
spin $[AB]_2CD_2$ proton spectrum, for which one expects
967 single-quantum transitions of finite intensity in
absorption, including the benzylic range not shown.
Clearly, there is extensive overlap of lines in this
spectrum, which has the effect of reducing the differences in depth between the global minimum and the false
local minima. Increasing the correlation to the point
where all the latter are guaranteed to be smoothed out
may thus become tantamount to throwing the child out
with the bath water. Also, with α or β $<10^{-10}$, rounding
errors within the computer start to become significant,
unless one wants to go to double or multiple precision.
If one is prepared to relinquish the attitude of deli-
berate ignorance, however, most of these difficulties
can easily be bypassed. In the present instance the
following palpable three-stage procedure led to a quick
success. In the first, a high-correlation exponential
grand cycle was performed for the aromatic proton

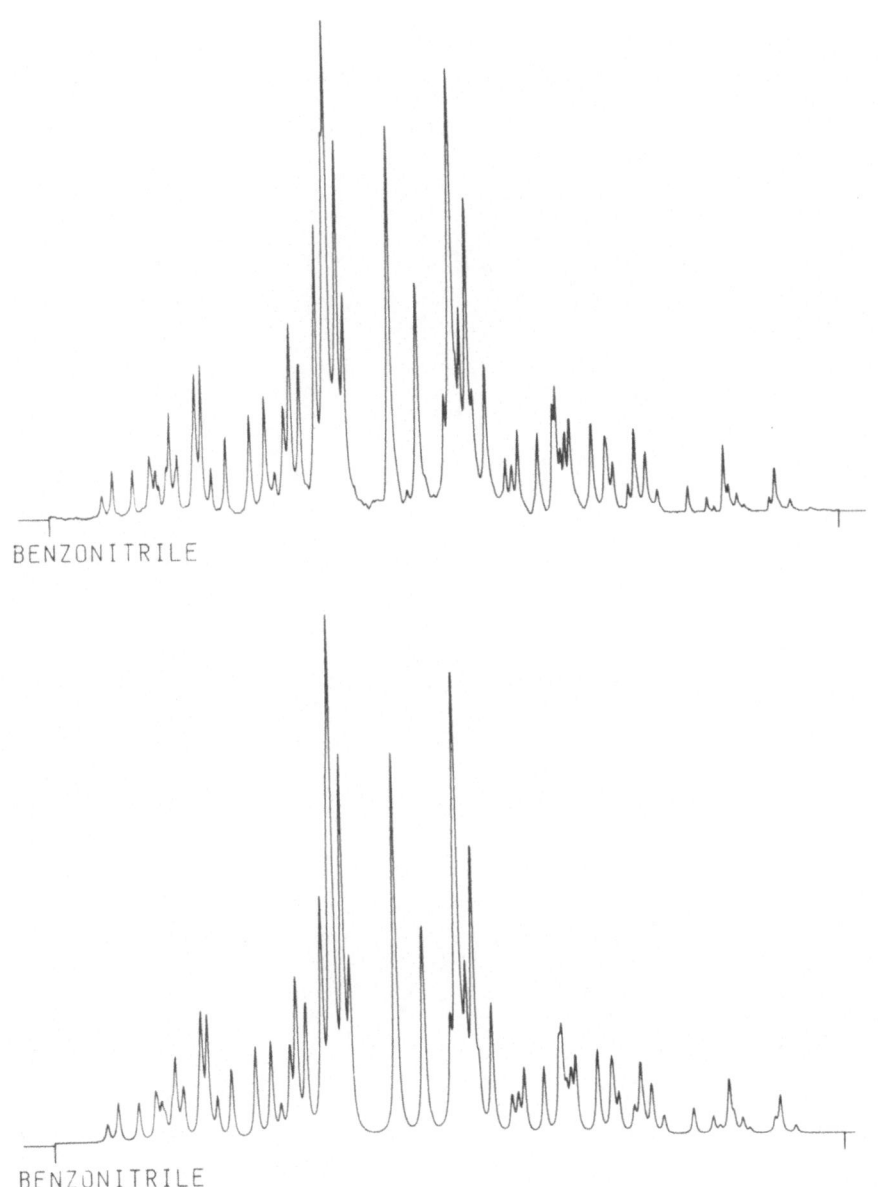

Figure 7. Top trace: preprocessed experimental 100 MHz FT-mode ^{1}H NMR spectrum of benzonitrile. Bottom trace: computed spectrum.

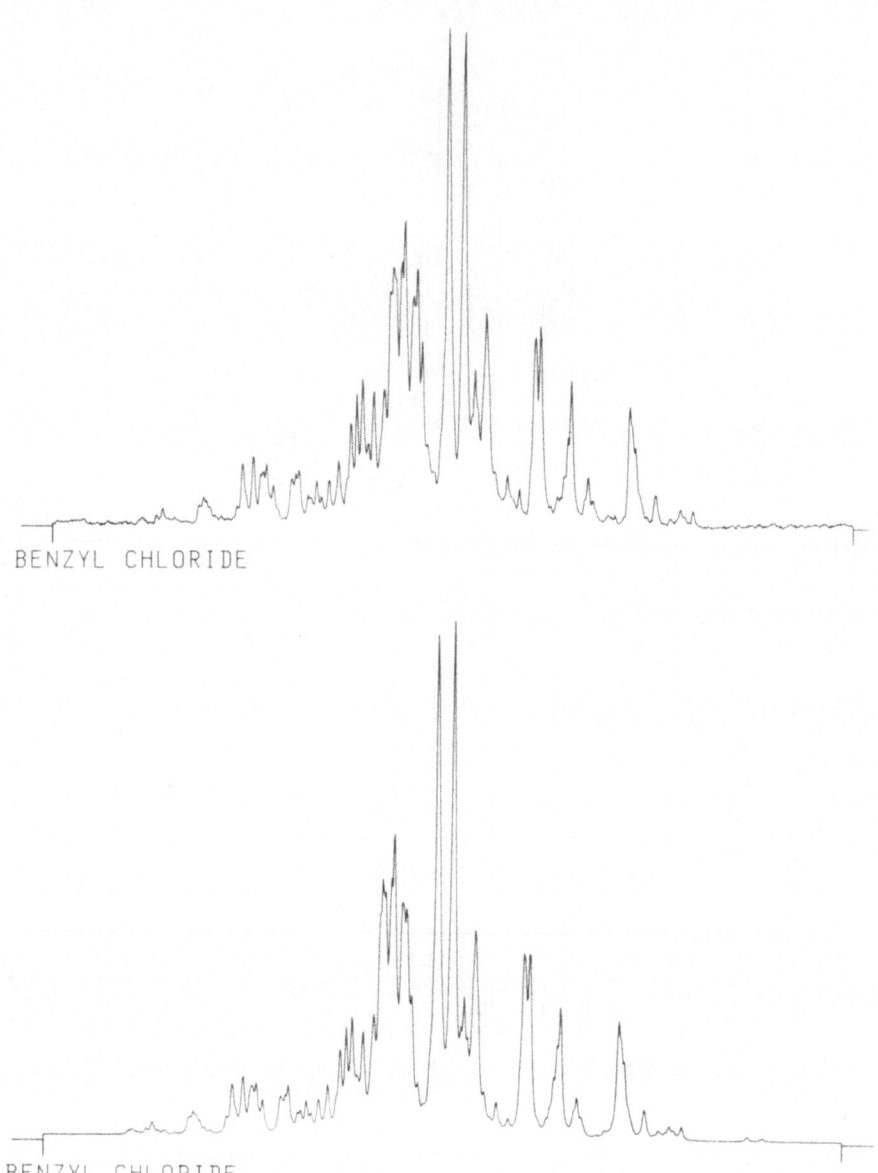

Figure 8.   Top trace: preprocessed experimental 100 MHz FT-mode NMR spectrum of the aromatic protons of benzyl chloride. Bottom trace: computed spectrum.

chemical shifts on the benzyl-methylene homonuclear decoupled spectrum, using fixed coupling constants set
equal to those extracted from benzonitrile. Next, a pass
through a small-correlation Lorentzian grand cycle on
all parameters of the decoupled spectrum produced an
excellent fit to the latter; a relatively small correlation was chosen in this calculation in order to prevent
the chemical shifts from wandering off again into un-
charted territory. In a final calculation on the unde-
coupled spectrum, again using Lorentzian correlation with
$\beta = 0.01$, only the newly introduced long-range coupling
parameters were specified as variable together with the
spectral origin. The latter trick has the effect of main-
taining the chemical shift differences, but allowing the
individual shifts to migrate in unison, which takes good
care of the small Bloch-Siegert shifts introduced by the
homonuclear decoupling field. The resulting agreement
between theory and experiment is superb, with an R factor
of 0.008, and may, if one so wishes, be regarded as one
of the most detailed experimental proofs of the time-
independent Schrödinger equation that has been reported
in the literature.

A last example of a high-resolution NMR spectrum in
isotropic solution is shown in Figure 9. It is again
part of a 967-line $[AB]_2CD_2$ spectrum, but this time the
degree of overlap is truly excessive. In a sense, the
interest in this problem is principally of an academic
nature. Under normal circumstances no sane spectroscopist
would waste a thought on it, but would send a sample to
Professor A.A.Bothner-By at Carnegie-Mellon University
with a friendly covering letter asking for a 600 MHz
spectrum. That the 100 MHz spectrum can in fact be analyzed, by a technique identical to that described for
benzyl chloride, simply serves to delineate to what ex-
tremes the application of DAVINS can be pushed in practice.

It hardly needs to be stressed that the method does
not *demand* sharp lines. The molecule 1,4-dinitrosopipera-
zine can exist as two stereoisomers. The room-temperature

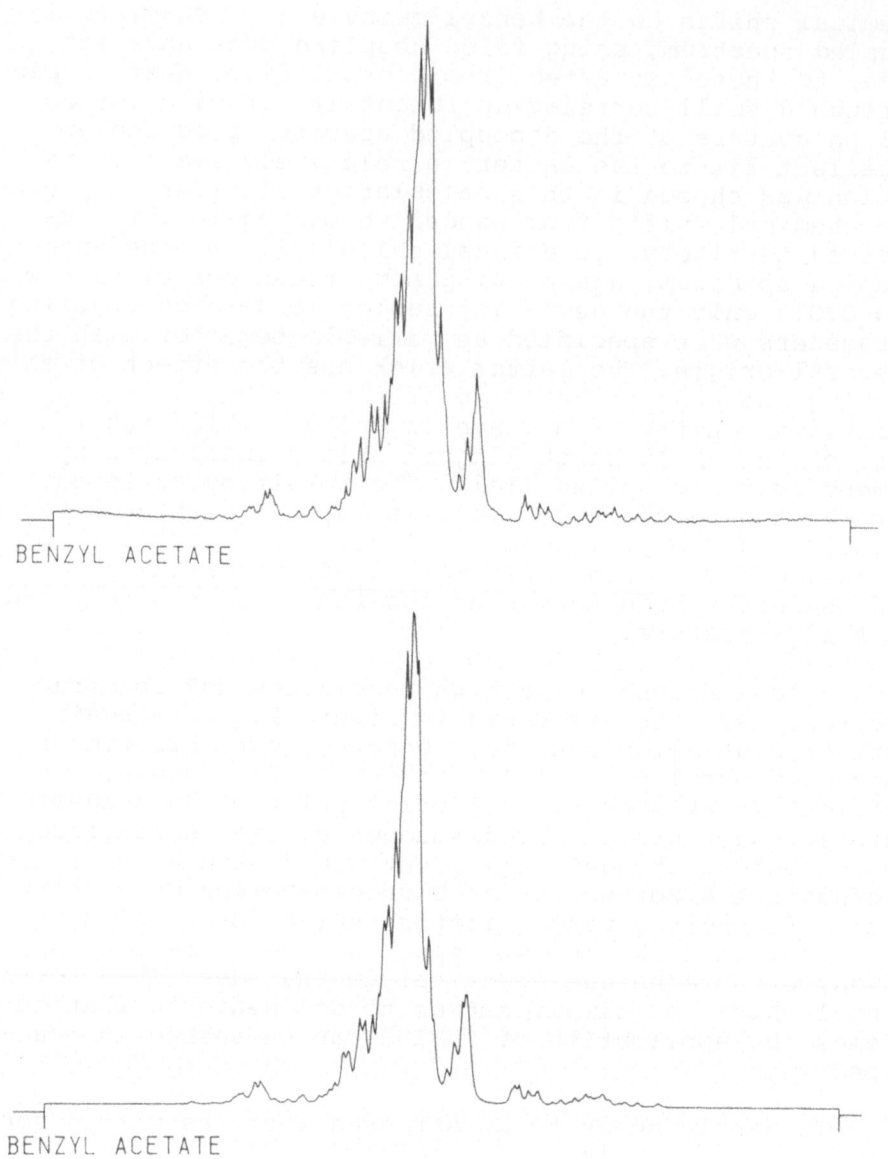

Figure 9. Top trace: preprocessed experimental 100 MHz FT-mode NMR spectrum of the aromatic protons of benzyl acetate. Bottom trace: computed spectrum.

spectrum consists of two singlets from the cis isomer superposed on an [AB]$_2$ from the trans isomer; at elevated

1,4-DINITROSOPIPERAZINE 130C

*Figure 10.   Superposed traces of the preprocessed ex-
perimental 100 MHz CW ^{1}H NMR spectrum of 1,4-dinitroso-
piperazine at 130°C and of the corresponding computed
spectrum.*

temperatures one detects the onset of internal rotations
giving rise to exchange-broadened spectra. The spectrum
recorded at 130°C is shown in preprocessed form in Figure 10 and the corresponding computed spectrum is
plotted on top of it. Apart from the theoretical model,
which inevitably has to be more complicated in the
present instance, the analysis was even easier than for
sharp-line spectra, since some of the local minima are
already smoothed out by the physics of the problem and
also because it is a simple matter to extract good
guesstimates of the chemical shifts and coupling constants from the static spectra. The dynamic analysis
(3), which also yielded the equilibrium constant, could
in fact be performed with the computer program DNMR5
(4) without using the correlation feature.

In recent weeks we have also developed and tested
a program version called DANSOM (*D*irect *A*nalysis of *N*MR
*S*pectra of *O*riented *M*olecules) for NMR spectra in liquid
crystal solvents, and an application is shown in Figure
11. Such spectra are typically spread over several kHz
and therefore require 4 to 10 times as many data points
in the iteration as normally needed for DAVINS. In addition, there are many more parameters owing to the
appearance of direct couplings between the spins. A
DANSOM grand cycle for the example of Figure 11 consumed
about half an hour of computer time.

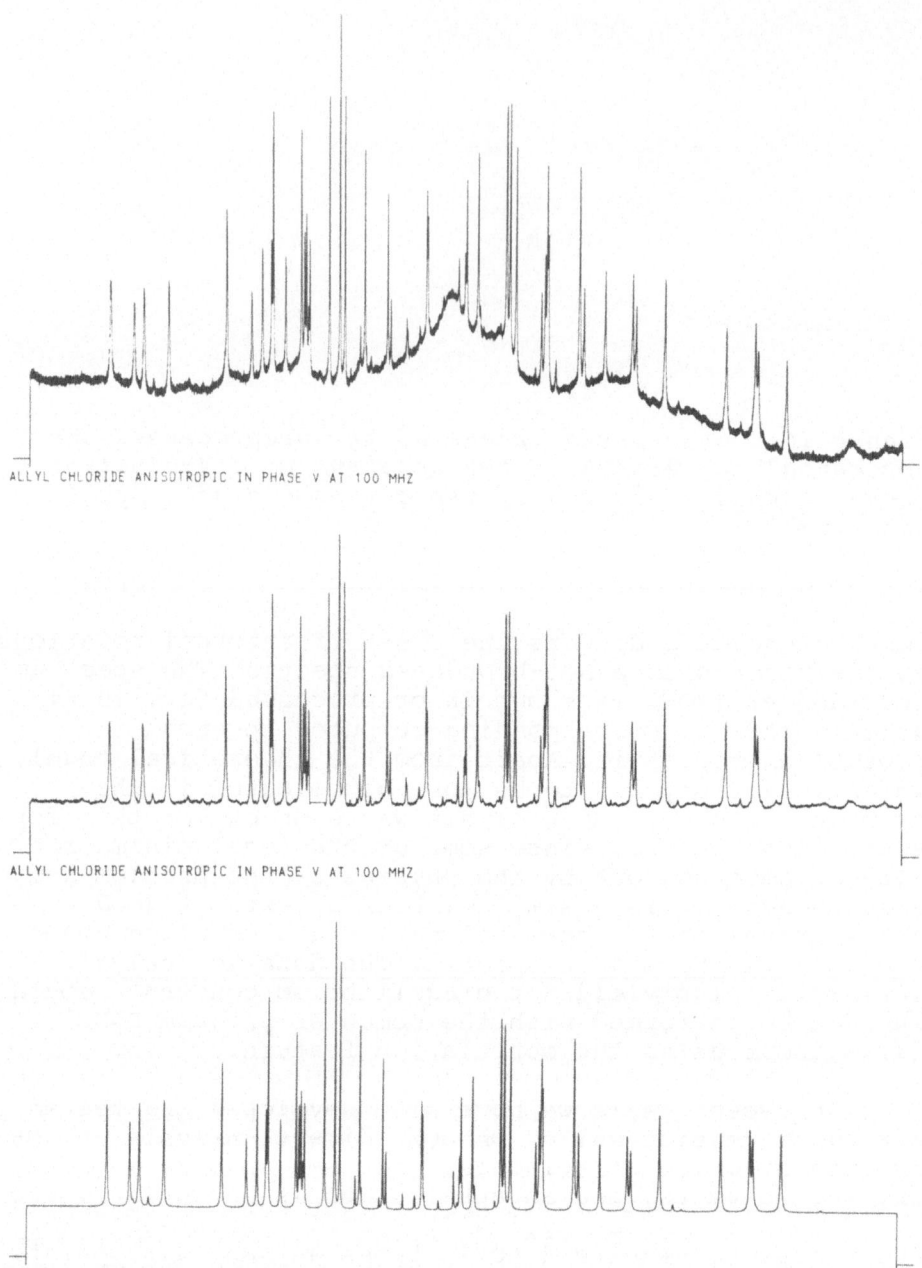

ALLYL CHLORIDE ANISOTROPIC IN PHASE V AT 100 MHZ

ALLYL CHLORIDE ANISOTROPIC IN PHASE V AT 100 MHZ

ALLYL CHLORIDE ANISOTROPIC IN PHASE V AT 100 MHZ

Figure 11. Top trace: experimental 100 MHz ^1H NMR spectrum of partially oriented allyl chloride. Middle trace: preprocessed experimental spectrum. Bottom trace: computed spectrum.

CONCLUSION

The computational strategy discussed and illustrated in this lecture has been implemented in the three well-documented computer programs DAVINS (5) for ordinary high-resolution NMR spectra in isotropic media, DANSOM (6) for high-resolution anisotropic NMR spectra and DNMR5 (4) for exchange-broadened NMR spectra. Two papers in press (7) contain further details about the formalism of DAVINS and numerous applications, and our liquid crystal work is presently being written up for publication (8).

We estimate that about 70% of the NMR spectra for which complete analyses have been reported in the past 25 years belong to the class of "normal" spectra as previously defined, and that an additional 10 - 20% could be converted to it by the now ubiquitous high-field spectrometers. Spectra of this type can in the future be analyzed routinely by technicians possessing a modicum of practical experience but devoid of theoretical background. In the few instances where the expertise and judgment of the spectroscopist is still indispensable, the task is greatly simplified by virtue of the fact that it no longer requires the computation of trial spectra and the laborious and frustrating assignment of lines.

As already mentioned, it should be self-evident that the general technique is not limited to NMR. Besides ESR, to which the extension is utterly trivial, any kind of spectroscopy is a candidate in principle. Nor are the conceivable applications exhausted with spectroscopy. At present we have no plans to branch out into any of these directions, but I would be anxious to learn how the prospects are realistically appraised by the diverse specialists attending this Symposium.

REFERENCES

1. P.Diehl, S.Sykora and J.Vogt, *J.Magn.Reson.* **19**, 67 (1975); P.Diehl and J.Vogt, *Org.Magn.Reson.* **8**, 638 (1976).
2. D.S.Stephenson and G.Binsch, *J.Magn.Reson.* **32**, 145 (1978).
3. D.Höfner, D.S.Stephenson and G.Binsch, *J.Magn.Reson.* **32**, 131 (1978).
4. D.S.Stephenson and G.Binsch, *DNMR5: Iterative Dynamic Nuclear Magnetic Resonance Program for Unsaturated*

Exchange-Broadened Bandshapes, Program 365, Quantum
Chemistry Program Exchange, Indiana University (1978).
5. D.S.Stephenson and G.Binsch, *DAVINS: Direct Analysis
    of Very Intricate NMR Spectra*, Program 378, Quantum
    Chemistry Program Exchange, Indiana University (1979).
6. D.S.Stephenson and G.Binsch, *DANSOM: Direct Analysis
    of NMR Spectra of Oriented Molecules*, Document avail-
    able from the authors.
7. D.S.Stephenson and G.Binsch, *J.Magn.Reson.*, in press.
8. D.S.Stephenson and G.Binsch, to be submitted to *Org.
    Magn.Reson.*

DETERMINATION OF THE STRUCTURES OF ORGANIC MOLECULES BY COMPUTER EVALUATION AND SIMULATION OF INFRARED AND RAMAN SPECTRA

Bernhard Schrader, Daniel Bougeard, and Werner Niggemann

Dept. of Theoretical and Physical Chemistry
University of Essen, D 4300 Essen 1, W.-Germany

A non-linear molecule with n atoms is able to perform $3n-6$ different normal vibrations. Depending on the molecular symmetry these vibrations are visible as bands in the infrared and the Raman spectrum, each with a typical intensity. This means: the frequency and intensity values of both vibrational spectra supply about $3(3n-6)$ different data for each definite kind of molecules.

All these data are sensitive to changes in the molecular structure: every change of a bond length, a bond angle or an atomic mass is reflected by changes of frequencies and intensities. In reality the spectra show more than $3n-6$ bands (due to combinations or overtones) or less bands (due to symmetry-forbidden vibrations, or when bands are very weak).

Nevertheless, the amount of information, given by the spectral data, is considerably larger than the information, which characterizes the molecule and its structure (i.e. n atomic numbers and $3n-6$ Cartesian coordinates). Each sort of molecules is, therefore, uniquely characterized by its vibrational spectra; consequently these spectra are applied effectively as a 'fingerprint' for the identification of molecules.

In addition, one may hope to find a direct way to transform the spectral information into the structural parameters. As it will be shown later in this paper this is - for mathematical reasons - not possible. But there is an alternative: For molecules with supposed structures frequencies and intensities of the vibrational spectra can be calculated. These model structures are modified and refined until the best fit is found between the real and the simulated spectra. This procedure is described in the last part of this paper.

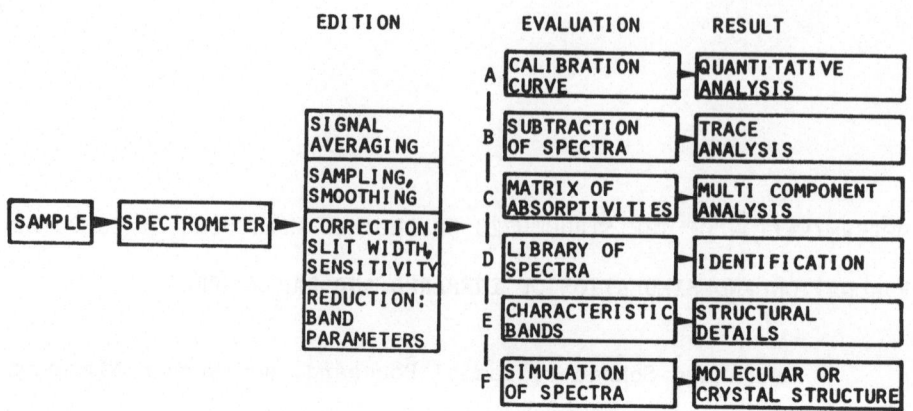

Fig. 1. Computer applications in vibrational spectroscopy.

During the last years technical development induced considerable impact of computer-techniques on vibrational spectroscopy.
There is a surprising increase of the application of infrared
spectrometers, which are directly coupled to mini computers.
Most of these systems have the ability to perform the different
editing procedures, shown in figure 1. The supplied programs
allow further several ways of quantitative analyses (A, B, C,
in figure 1). Especially the signal averaging and spectra sub-
traction techniques (B) have expanded the applications of IR-
spectroscopy considerably. These procedures will not be discussed
in this paper. From the other possible applications the 'library
search' for identification is applied quite seldom till now (D).
Only very few laboratories have developed and use the automatic
evaluation of characteristic bands (E). The last way of evaluation is the most powerful, but also the most expensive one.
It is only practicable with the help of large off-line computers
(F). Therefore it is really an 'academic' non routine procedure.
The procedures D, E, and F are only practicable with the help
of computers. We are sure, that in future they will become an
essential tool in analytical spectroscopy. The procedures D,
E, and F are described and discussed in the following paragraphs.

AUTOMATIC EVALUATION OF VIBRATIONAL SPECTRA

Prior to the evaluation process the recording conditions
are to be selected to record within a given time a spectrum with
maximum signal/noise-ratio and minimal systematic errors (1,2).
If the signal/noise-ratio is not sufficient it depends on the
nature of the predominant noise ('white' or '1/f' noise) whether
a larger time constant or multiple scanning with subsequent signal averaging is the best way to invest time for a higher signal/

noise-ratio (3). With the help of polynomial smoothing (4,5) and band fitting procedures band parameters can be calculated with optimal accuracy (6). The last mentioned procedure is part of a large collection of FORTRAN programs, developed by R.N.Jones and his coworkers for the use with off-line computers. Our automatic evaluation programs are derived for desk-top calculators, directly coupled to spectrometers. They will be described in detail elsewhere (7,8). In the following paragraphs only a short description is given.

The reduction of the spectra may be expanded to different levels, depending on the subsequent evaluation procedure. All procedures begin with an automatical correction of the base line which takes into account the usual stray light in the short wave length part of the IR-spectrum, the absorption of the cell material in the long wave length region or the fluorescence background in the Raman spectra (figure 2). At the same time the IR spectra are standardized, in order to give the strongest band 95 % absorption.

The spectral data are then fitted to a cubic polynome. Its coefficients are calculated by the method of Savitzky and Golay (4). Zero first derivatives in the interval $\pm$ 1 data point define the relative extreme values; maxima and minima are defined by the negative or positive second derivatives. A descrimination level is adjusted to select 'bands' from 'noise'. The half width of the bands is measured in the spectra whenever it is possible, otherwise it is determined from the extrema of a band, assuming a Lorentz contour. These data concentrate the significant spectral information. They need only a fraction of the storage, compared with the original spectrum. Nevertheless, the spectra may be redrawn from these data, giving a realistic picture of the original spectra. In our interpretation program these spectral data are used directly. For the identification program the spectra are reduced further to give two code numbers, defining frequency and intensity (cf.fig. 2). Finally, the last code number represents the most reduced form of a spectrum, it is representative for the ASTM spectra collection which gives by a '1' or a '0' the information of bands present or absent in spectral channels with a width of 0.1 μm. More than 100.000 IR-spectra in this simple form are the base of identification programs which work with surprising success (9).

Automatic Identification

Our identification process is developed on the base of a paper of Rann (10). In this paper a spectrum is encoded as a number of ten digits. Figure 3 shows that each digit stands for a spectral region of 1.000 cm^{-1} for the spectrum between 4.000

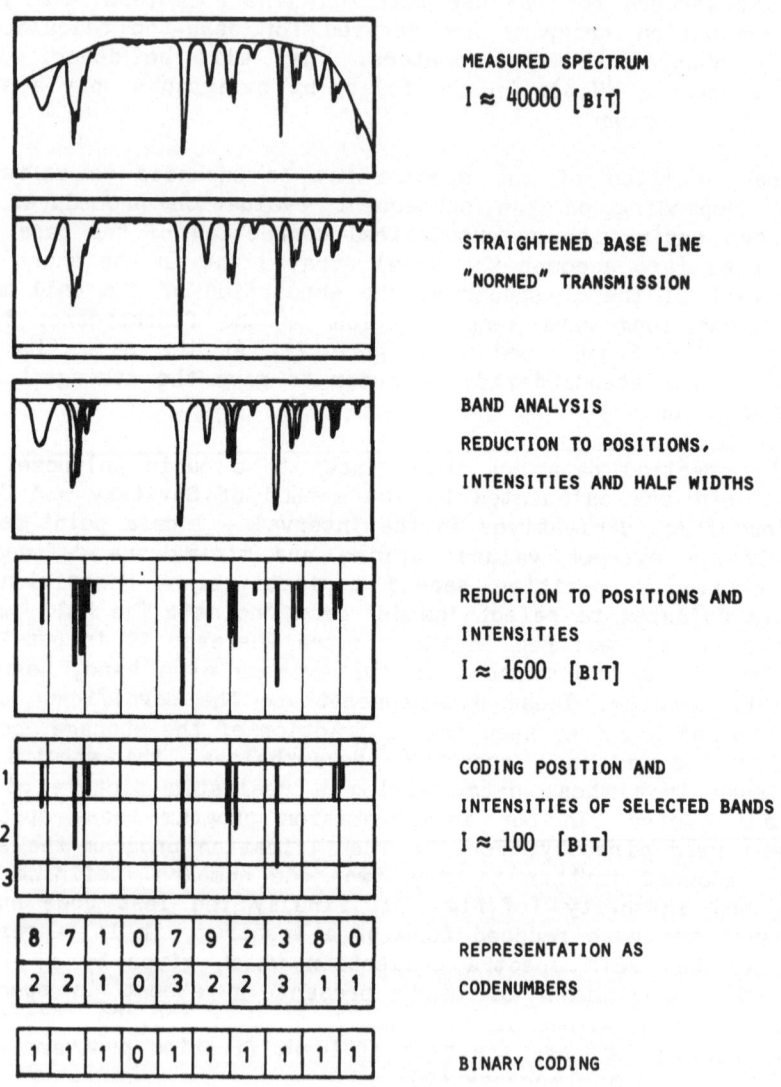

MEASURED SPECTRUM
$I \approx 40000$ [BIT]

STRAIGHTENED BASE LINE
"NORMED" TRANSMISSION

BAND ANALYSIS
REDUCTION TO POSITIONS,
INTENSITIES AND HALF WIDTHS

REDUCTION TO POSITIONS AND
INTENSITIES
$I \approx 1600$ [BIT]

CODING POSITION AND
INTENSITIES OF SELECTED BANDS
$I \approx 100$ [BIT]

REPRESENTATION AS
CODENUMBERS

BINARY CODING

Fig. 2. Correction and reduction of vibrational spectra.

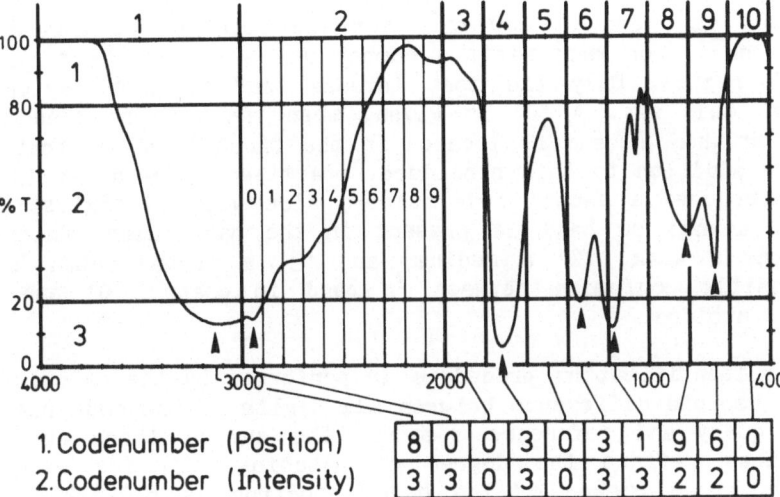

Fig. 3. Definition of the two codenumbers.

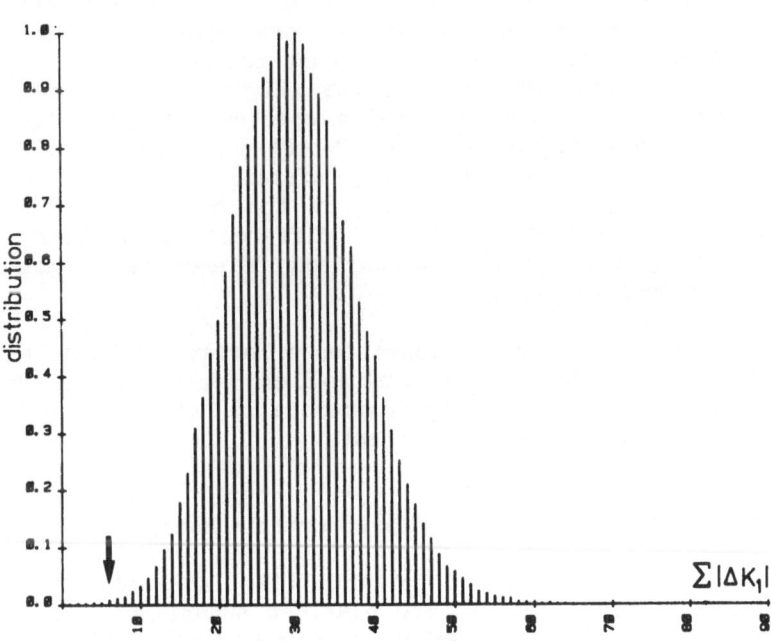

Fig. 4. Distribution of the absolute difference of the first codenumber for all possible comparisons of a collection of 360 infrared spectra.

and 2.000 cm^{-1} and of 200 cm^{-1} for the spectrum between 2.000 and 400 cm^{-1}. For each digit a number 0 - 9 represents a tenth of each region. Only the most intense band in each region is coded in this way. A '0' is given, when there is no band in a region or when a band is located in the first tenth of that region. In addition to this procedure, practised by Rann, we (7,11) have introduced a second code number, also with 10 digits. Here the '0' means: no band is present in the particular region, a '1' means a weak, '2' a medium, and '3' a strong band. Thus, the possible confusion between 'a band in channel 0' and 'no band' is avoided.

The identification procedure is performed by the calculation of the absolute difference between all digits of the code numbers of the query and reference spectra. The sum of these absolute differences defines the degree of matching. If, for example, the first code number of the query spectrum is 6004612225 and the code number of the reference is 6004614333, the absolute difference is 0000002112 and its total is 6. Figure 4 shows the result of a comparison of the first code number for all possible combinations in a collection of 360 coded spectra of drugs and other organic compounds. The maximum of the distribution is near a total difference of 30. Only 1 %o of the spectra have a total of < 6. Therefore, a total of < 6 means matching of both spectra. The second code number allows now a more detailed comparison and gives an error sum taking into account accidental fluctuation of band positions and band intensities (Figure 5). In Figure 6 a typical result of a comparison of a spectrum with the library is shown. The query sample has been 2-pentanole which was automatically reduced and encoded. The reference with the lowest error sum is, in fact, 2-pentanole. By the original method of Rann sometimes references to quite different substances are found, wheras by the detailed comparison references to homologues are found preferably.

This program has been developed for a Hewlett-Packard Calculator 9825 cooperating with a Perkin-Elmer Spectrometer Model 580. Similar configurations are commercially available or can be set up with less expensive components.

Automatic Interpretation of Characteristic Bands

Since the pioneer work of Coblentz (12) infrared bands are known which are characteristic for distinct substituents or skeletons. Most IR work in the past has been the evaluation of characteristic bands, which are documented in several well-known books (13,14). When IR-spectra are run automatically, e.g. for process control or by an instrument, coupled to a chromatograph, automatic evaluation procedures are very useful (15,16). Our

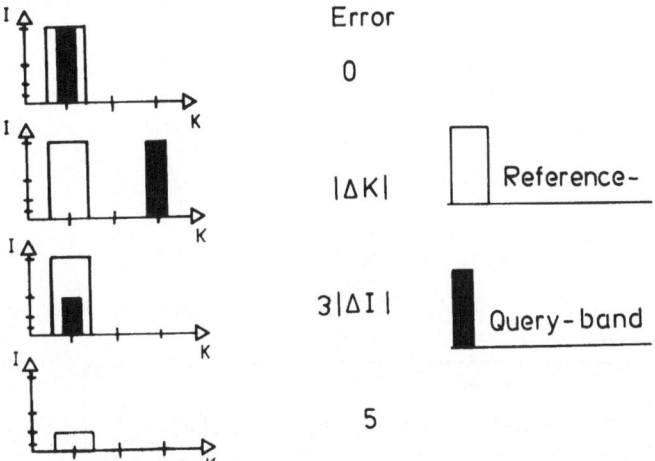

Fig. 5. Comparison of frequencies and intensities: Definition of the 'error'.

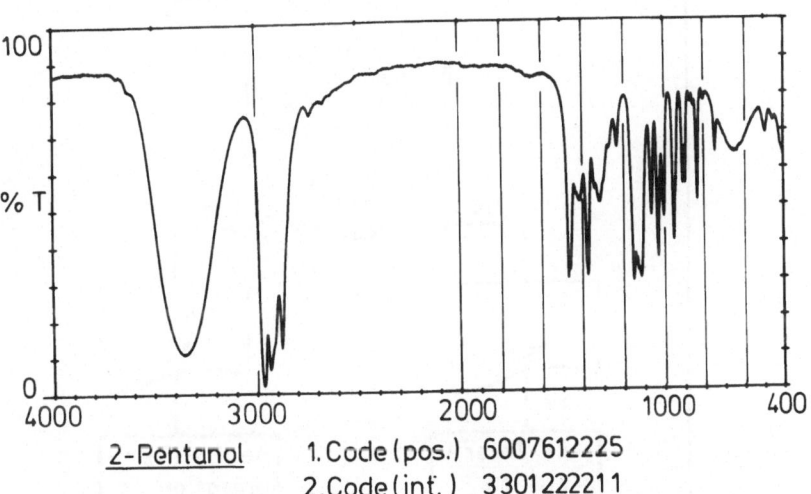

2-Pentanol   1. Code (pos.) 6007612225
              2. Code (int.) 3301222211

COMPARISON AFTER RANN/LEUPOLD		COMPARISION OF POSITION AND INTENSITY	
REFERENCE	DIFFERENCE	REFERENCE	ERROR
2-PENTANOL	0	2-PENTANOL	7
2,2,4,-TRIMETHYLPENTANE	4	4-METHYLHEPTANE	25
2-METHYLBUTANE	5	2-PROPANOL	26
2,2,3,3-TETRAMETHYLBUTANE	5	2-BUTANOL	26
METHYLCYCLOPENTANE	5	2,5-DIMETHYLHEXANE	27
3-PENTANONE	6	1-BUTANOL	30
2-OCTANOL	6	3,4-DIMETHYLHEXANE	30
CYCLOHEPTANONE	6	CYCLOHEPTANONE	30

Fig. 6. Encoding and result of comparison of the first and of both codenumbers respectively.

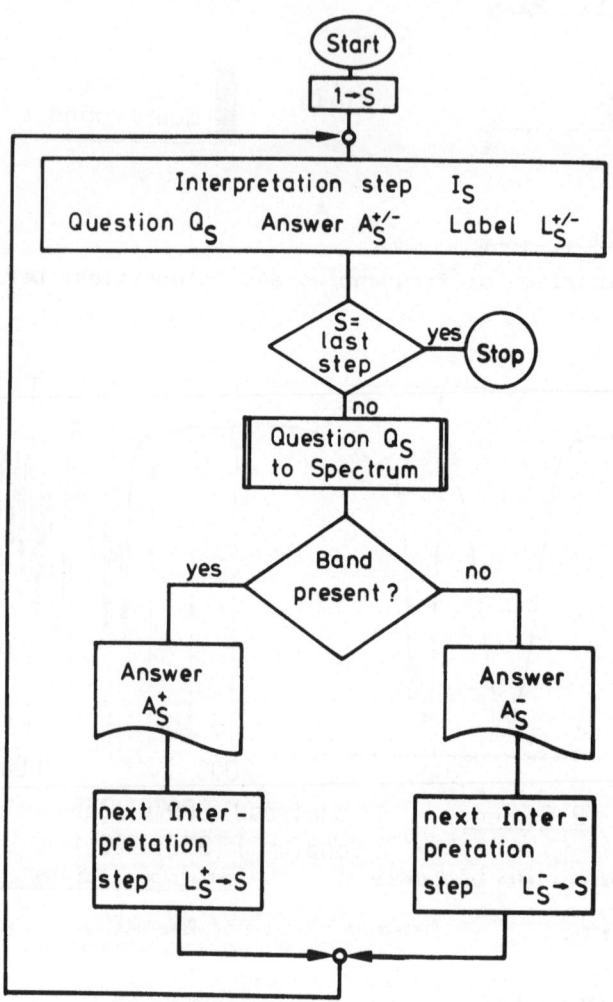

Fig. 7. Interpretation of vibrational spectra: algorithm.

program is developed for a simultaneous interpretation of infrared and Raman spectra. Such programs give the most reliable answers since they make use of the complementary information from
both spectra. Similar programs may be developed for the interpretation of infrared or Raman spectra alone.

   The algorithm of our program is given in figure 7. It looks
quite simple. Its 'intellectual' content is collected in the
'list of interpretation steps', the 'list of the questions' and
the 'list of the answers'. The evaluation is performed by questions regarding bands present or absent in a definite region
with distinct relative intensity and half width. Answers give
information about groups present or absent. Infrared and Raman
spectra of a typical example are shown in figure 8, with the
series of characteristic bands, which are evaluated. A typical
output is represented by figure 9. A non-present feature is usually excluded with a reliability of somewhat more than 90 %.
A present feature is found with a reliability between 50 and
100 %, depending on the nature of this feature.

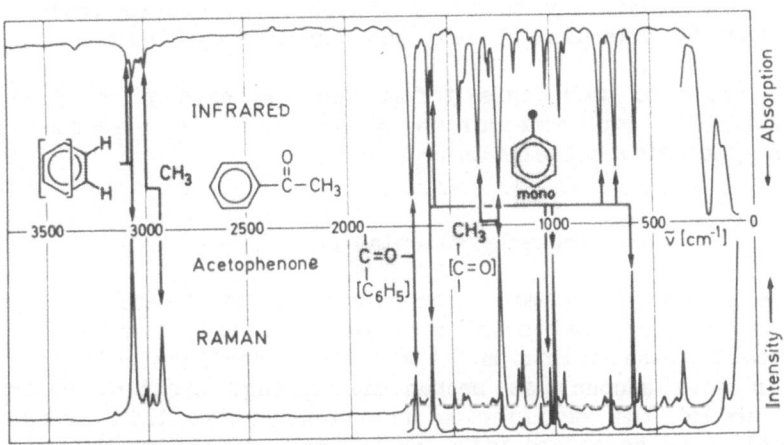

Fig. 8. Typical groups of bands in the infrared spectrum (above)
and the Raman spectrum (below) which are evaluated.

INTERPRETATION

+ SAT. HYDROCARBON
+ METHYL
+ UNSAT. HYDROCARBON
- ISOL. DOUBLE BOND
+ BENZENE
+ MONO SUBST.
+ KETONE

Fig. 9. Typical output of an interpretation.

SIMULATION OF INFRARED AND RAMAN SPECTRA

In opposition to the previous procedures which can be considered as an empirical treatment of the measured spectral data
the following method consists in the theoretical simulation of
the spectra on the base of the microscopic molecular properties.
On the one hand we expect from this approach a better theoretical background for the well-known experimental assignment
criteria such as the frequency shift upon isotopic substitution,
and, further, the concept of characteristic frequencies or different physical effects like temperature or pressure dependence.
On the other hand, we want to get a better knowledge of the molecular properties: the vibrational frequencies depend on the intramolecular potential through its second derivatives, called
force constants, and the intensities are correlated to the dipole
moment μ and the polarizability α, and to their variations during
the vibrations. The total energy of a molecule can be calculated
from the energy of the individual molecular orbitals, while the
charge distribution can be obtained from the wave functions upon
a gross population analysis. Thus, with a spectrum simulation
we have the possiblity to check indirectly different models re-
presenting the molecules at rest and during the vibration.

In order to gain this information we have to perform the
calculation in two consecutive steps, (i) the frequency and,
(ii) the intensity calculation.

Frequency Calculation

The theory of the molecular vibrations of molecules is based
on a simple mechanical model assuming that the atoms can be represented by mass points and that the intramolecular forces can
be taken into account as mechanical springs creating restoring
forces around the equilibrium position. Under this assumption
the problem can be solved with the Lagrange equation

$$\frac{d}{dt} \frac{\partial L}{\partial \dot{q}_i} - \frac{\partial L}{\partial q_i} = 0 \qquad \text{with} \qquad L = T-V$$

where T and V are the kinetic and potential energies respectively, while q designes a coordinate set, $\dot{q}$ meaning dq/dt.

Mathematical tool. Such a model has been first proposed
by Wilson (17) and is known as GF matrix method. The kinetic
energy T can be written

$$T = \frac{1}{2} \sum_{i}^{n} m_i v_i^2$$

n is the number of atoms. Using the Cartesian coordinates x we may write T in the matrix form

$$2T = \overset{\sim}{\dot{\mathbf{X}}} \mathbf{M} \dot{\mathbf{X}}$$

with

$$\dot{\mathbf{X}} = (\frac{dx_i}{dt})$$

In order to simplify this expression, one introduces the mass-weighted Cartesian displacements

$$x_{mi} = \sqrt{m_i}\, x_i \qquad\qquad 2T = \overset{\sim}{\dot{\mathbf{X}}}_M \dot{\mathbf{X}}_M$$

The potential energy is expanded as a Taylor series. The zero order term is chosen as the origin and thus set to zero; the first order term is zero at the equilibrium if the basis set is not redundant; terms of order equal or greater three are neglected, thus leading to the harmonic approximation which retains only second order terms. Finally we get

$$2V = \overset{\sim}{\mathbf{X}}_M \mathbf{M}^{-1/2} \overset{\sim}{\mathbf{B}} \mathbf{F} \mathbf{B} \mathbf{M}^{-1/2} \mathbf{X}_M$$

where
B is the matrix which transforms the Cartesian coordinates into the internal coordinates according to

$$q_i = \sum_j b_{ij}\, x_j$$

$\mathbf{M}^{-1/2}$ is a matrix whose components are of the form $1/\sqrt{m_i}$
F is the force constant matrix with the elements

$$f_{ij} = (\frac{\partial^2 V}{\partial q_i\, \partial q_j})_0$$

q_i are the so-called internal coordinates: bond stretching, angle bending or bond torsions; for complete definition, see (17, 18). The F matrix is a symmetrical one. Its diagonal elements represent "classical" force constants, according to Hooke's law. The off-diagonal elements, however, represent the interaction of two different internal coordinates in the potential energy.

After applying and solving the Lagrange equations of motion for these expressions of T and V it turns out that the solution has the form

$$\det (\mathbf{M}^{-1/2} \overset{\sim}{\mathbf{B}} \mathbf{F} \mathbf{B} \mathbf{M}^{-1/2} - \mathbf{E}\, \lambda) = 0 \qquad\qquad (1)$$

where the variable λ is connected to the vibration frequency by the relation

$$\tilde{\nu} \ (cm^{-1}) \ = \ \frac{\sqrt{\lambda}}{2\pi c}$$

If we define

$$\mathbf{G} \ = \ \mathbf{B} \ \mathbf{M}^{-1} \ \tilde{\mathbf{B}}$$

and transform equation (1), it follows

$$\det \ (\mathbf{G}\,\mathbf{F} \ - \ \mathbf{E}\,\lambda) \ = \ 0 \qquad\qquad\qquad (2)$$

which is the original form proposed by Wilson. Under this form the G matrix represents the kinetic energy, the F matrix the potential energy. From eq. 1 or 2 we can see that the following data must be known in order to perform a calculation:

- the force constants necessary for the F matrix and
- the structure parameters for the computation of the M and B matrices.

The structure will be supposed to be experimentally known or defined with standard values or by a quantum mechanical calculation; the problem of the force field will be discussed later.

From the form of eq. 1 and 2 we recognize an eigenvalue problem which can be solved by diagonalization of the GF matrix; the eigenvalues λ_i leads to the frequencies $\tilde{\nu}_i$, while the eigenvectors yield the corresponding normal coordinates Q_i, i.e. the form of the vibration. By working with eq. 2 the vibration is described in terms of internal coordinates, while eq. 1 yields the Cartesian displacements of all atoms around the equilibrium position.

Several computer programs working with one of these equations have been developed and are currently available (19-23). The problem can be treated in two different directions: (i) Assuming a set of given force constants the frequencies are calculated (direct eigenvalue problem); (ii) Starting from experimental data the force constants are determined (inverse eigenvalue problem). Most of these programs allow the fitting of an initial set of force constants on the experimental frequencies by a least square method (24,25). Special problems of the calculation (definition and sign of the internal coordinates, redundancy, difference between linear and curvilinear coordinates) (18,26-28) cannot be discussed in this frame. Further, in order to simplify the interpretation in terms of symmetry and to reduce the volume of the numerical problem use can be made of the symmetry of the molecule by introducing the so-called 'symmetry coordinates'. In this way the problem is solved in submatrices which can be diagonalized separately (17). The mathematical technique being known we can now turn to the force field.

Molecular Force Field. The force field can be defined in any basis of coordinates, but in order to find a proper physical interpretation, the force constants are usually determined in internal coordinates. Thus, each force constant corresponds to the spring for a bond stretch or angle bend in the molecule or to their interactions.

For a general asymmetrical molecule with n atoms, we need $3n-6$ internal coordinates so that the symmetrical F matrix has $(3n-5)(3n-6)/2 = (9n^2-33n+30)/2$ components. The same molecule has only $3n-6$ degrees of freedom, so that only a maximum of $3n-6$ frequencies can be observed. In other words: we have a problem with more parameters (force constants) than measured data (frequencies) and it can only be exactly solved for very special cases. For example, for some highly symmetrical molecules the same force constant is assigned to different internal coordinates, thus leading to a reduction of the number of adjustable parameters. Another means is the evaluation of supplementary data: isotopic substitution, Coriolis coupling constants and rotational distortion. For large molecules such a way is not possible and one has to do some approximations. A recent example is given by the molecule NSCl (37) for which all IR and Raman data and Coriolis coupling constants of ^{32}S, ^{34}S, ^{14}N, ^{15}N, ^{35}Cl and ^{37}Cl derivates are known. They determine all force constants except the v_{NS}/v_{SCl} interaction which could be 0.04 or 1.18 mdyn/A; only personal preference based on physical intuition can decide which value will be retained.

Choice of the force field. Historically, at the very beginning of the normal coordinate analysis a great subject of discussion was the choice between UBFF (Urey-Bradley Force Field) (29) and the GVFF (General Valence Force Field) and all modified or intermediate solutions between both (30,31). The UBFF does not contain explicitly any quadratic cross term; they are represented by force constants related to stretches of nonbonded distances. At the present time this force field seems to disappear and to be replaced by the GVFF in which all second derivatives of the potential energy are considered.

'Classical' determination of the force field. The next problem is to determine the structure of the force field, i.e. which force constants have to be considered, which can be set to zero and by which arguments. In order to get significant results one has to find a choice strategy independent from individual feeling. A usual method is to try to transfer some force fields determined for smaller molecules to larger ones under the assumption that such parts are not strongly perturbed by the inclusion in another molecule. As an example a calculation of hexamethyl-

benzene (32) can be done by combining the GVFF of benzene obtained from chlorinated benzene derivates (33) or from polycon-
densed aromatics (34) with the GVFF for a methyl group in paraf-
fins (35). For the force constants which cannot be fixed in this
way some other criteria are employed such as the fact that no
interaction is considered between internal coordinates having
no common atom or that a given interaction force constant can
be neglected, if it does not influence strongly the calculated
values for all frequencies.

This type of method has been applied by different authors
and leads to a number of reliable and transferable force fields
(22,31,36). Anyway all these force fields suffer from the approximations made during the development. At this point and in order
to support the results on small molecules a quantum mechanical
approach seems to be reasonable in order to decide with physical
arguments about the parameters of the force field.

Quantum mechanical determination of the force field. There
are three ways to calculate the force constants as second derivatives of the potential energy (38):

- two consecutive analytical derivations
- two consecutive numerical derivations
- one analytical, followed by one numerical derivation

Pulay has shown that the first way is rather unpracticable and
that the second one could lead to large errors due to the double
numerical derivation, if care is not taken to obtain a precise
potential hypersurface. He prefers the third way which he proposes to call 'force method', for the force is the first derivative of the potential. Anyway, the last two methods have been
applied in the past ten years with different quantum chemical
calculations: ab initio (39-42), CNDO/2 (43-45), MINDO (46,47),
and PCILO (48) particularly. A survey of these papers enables
some conclusions:

If the orbital basis used is large enough, rather accurate force
constants can be obtained, thus leading to satisfactory vibration
frequencies (39). Blom could show for example that for obtaining
accurate force constants the Gaussian 70 STO 4-31G version should
be prefered to the STO-3G version (41). With such a basis all
magnitudes and signs are determined satisfactory (see columns
1-3 and 7 of table 1 for an example).

Semiempirical methods (CNDO, INDO, MINDO) lead to force constants with systematical errors which can be corrected by introducing scale factors, thus getting quite reliable values (column 5). For the most significative interaction force constants

they give also after such a correction the right order of magnitude and the right sign.

Particularly for the interaction force constants such calculations lead to new information. For example some values which where assumed to be 0 (table 1) are shown to have a non negligible value ($v_{CC}-v_s$ and $v_s-\delta_s{}'$) thus showing that the approximations mentioned above (no common atom or small participation to the frequency determination) have to be used very carefully.

Finally, it should be mentioned that ab initio methods are rather computer intensive and expensive and cannot be used for larger molecules. For these cases semi empirical methods seems to lead to reliable values. Blom (41) proposes INDO; Pulay (43) prefers CNDO/2 to MINDO/2 because of the determined values of some interactions. It is possible that the new parametrization of MINDO/3 leads to much better results.

Table 1. Comparison of some of the force constants obtained with different methods for the potential field of ethane. v, δ, and ρ stay for stretching, bending, and rocking respectively while s and d mean symmetric and degenerate. The prime makes the difference between both methyl groups.

	4-31G[a]	3G[a]	ab initio[b]	INDO[a]	CNDO[b]	MINDO[b]	exp.[c]
v_{CC}	4.861	6.662	5.073	12.328	4.67	5.10	4.450
v_s	5.543	7.095	5.678	13.657	5.27	5.91	4.900
v_d	5.374	7.302	5.543	13.080	5.05	5.47	4.764
δ_s	0.711	0.832	0.756	0.803	0.811	0.522	0.606
δ_d	0.719	0.889	0.688	0.825	0.769	0.368	0.560
ρ	0.823	0.941	0.85	0.816	0.809	0.529	0.682
v_{CC}/v_s	0.140	-0.008	0.140	0.546	0.212	0.444	0[d]
v_{CC}/δ_s	0.394	0.343	0.390	0.443	0.283	0.357	0.346
v_s/δ_s	-0.161	-0.113	-0.123	-0.239	-0.136	-0.091	-0.050
δ_s/δ_s	0.031	0.030	0.033	0.039	0.043	0.039	0.033
δ_d/δ_d	-0.005	-0.007	-0.005	-0.008	-0.007	-0.008	-0.005
ρ/ρ	0.168	0.201	0.193	0.153	0.142	0.022	0.139
$v_s/\delta_s{}'$	0.022	-0.007	0.024	0.033	0.021	0.057	0[d]

[a] ref. (41);   [b] ref. (43);   [c] ref. (49);   [d] assumed

Calculation of the Intensities

After having determined the parameters of the first dimension of the spectra, the frequencies, we can use the obtained normal coordinates in order to calculate the intensities. This information is important because it can be very helpful to decide between two possible force fields (example of the NSCl molecule), between two possible assignments of the bands for larger molecules or finally in order to get a better insight into the behaviour of the molecules during a vibration. We have here again two possibilities: the classical and the quantum mechanical one.

General formulas for the IR- and Raman intensity. The intensity of an infrared absorption band is defined by:

$$A = \frac{1}{cd} \int \ln \left(\frac{I_o}{I}\right) d\nu,$$

where c means the number of absorbing molecules pro cm^3 and d the thickness of the sample. I_o and I is the incident and the emerging light flux, respectively. A can be expressed as a function of molecule parameters:

$$A = \frac{8\pi^3 N_o \tilde{\nu}}{3h} \quad g \quad b^2 \left(\frac{\partial\mu}{\partial Q}\right)_o^2, \tag{3}$$

where N_o is the Avogadro number, g the degeneracy and $\partial\mu/\partial Q$ the first derivative of the dipole moment with respect to the normal coordinate Q.

Similarly the Raman scattering coefficient used as measure of the intensity of a Raman band, is given by

$$S = \frac{2^4 \pi^4}{1-\exp(-hc\tilde{\nu}/kT)} \quad (\tilde{\nu}_o - \tilde{\nu})^4 g \, b^2 (\alpha'^2 + \frac{7}{45}\gamma'^2) \tag{4}$$

for measurement with linear polarized light; for other experimental arrangements different factors have to be used with α'^2 and γ'^2 (50). In this equation $\tilde{\nu}_o$ and $\tilde{\nu}$ are the frequencies (in cm^{-1}) of the exciting line and of the vibrating molecule respectively,

$$b = \left(\frac{h}{8\pi^2\nu}\right)^{1/2}$$

is the zero point amplitude, T the temperature and h and k are the Planck and Bolzmann constants. α' and γ' are the two quantities of the α' tensor which remain invariant under rotation. They are called mean value and anisotropy respectively and are defined as:

$$\alpha' = 1/3 \left(\frac{\partial\alpha_{xx}}{\partial Q}\right)_o + \left(\frac{\partial\alpha_{yy}}{\partial Q}\right)_o + \left(\frac{\partial\alpha_{zz}}{\partial Q}\right)_o$$

$$\gamma'^2 = 1/2 \left\{ \left[\left(\frac{\partial \alpha_{xx}}{\partial Q}\right)_o - \left(\frac{\partial \alpha_{yy}}{\partial Q}\right)_o \right]^2 + \left[\left(\frac{\partial \alpha_{yy}}{\partial Q}\right)_o - \left(\frac{\partial \alpha_{zz}}{\partial Q}\right)_o \right]^2 \right.$$

$$+ \left[\left(\frac{\partial \alpha_{zz}}{\partial Q}\right)_o - \left(\frac{\partial \alpha_{xx}}{\partial Q}\right)_o \right]^2 + 6 \left[\left(\frac{\partial \alpha_{xy}}{\partial Q}\right)_o^2 + \left(\frac{\partial \alpha_{yz}}{\partial Q}\right)_o^2 \right.$$

$$\left. \left. + \left(\frac{\partial \alpha_{xz}}{\partial Q}\right)_o^2 \right] \right\}$$

These expressions for the Raman intensities are only valid for measurements far from all resonances between the exciting line and the electronic or vibrational levels of the molecule. For the IR as well as for the Raman spectrum the intensity depends on the variation of the dipole moment vector and the polarizability tensor during the vibration represented by the normal coordinate Q. Thus, in order to determine the intensities we have to determine the variation of μ and α by a vibration.

Classical models, electro optical theory and polar tensors. Starting from the measured intensities one has two problems with the determination of $\partial\mu/\partial Q$ and $\partial\alpha/\partial Q$. The first one lies in the determination of the sign since the intensities are related to the square of those values. The second difficulty arises from the fact that one tries to obtain some intensity parameters which are transferable from one molecule to another one (by analogy with the force constants) and that for this reason one has to define, how a molecular property can be divided into contributions from bonds or atoms. These methods have been developed for the IR spectra and are sometimes extended for the Raman effect. The first parameter which can be used is the variation of the molecular dipole moment with some internal or symmetry coordinates (51,52), but these parameters cannot be transfered to other molecules properly.

The second way so far tried was proposed by Wolkenstein (53), further improved by different authors (see for example ref. 54 and references therein) and is still recently applied by Zerbi and his group (55-57). This method, the electro optical theory, makes use of the variation of the μ and α of each bond with the different internal coordinates. The total dipole moment μ and polarizability tensor α are written as sums over all the n bonds of the molecule of bond dipole moments m and polarizabilities a:

$$\vec{\mu} = \sum_i^n m_i \vec{e}_i \qquad\qquad \alpha = \sum_i^n a_i$$

where $\vec{e}_i$ means a unit vector in the direction of the ith bond. Because of the tensorial nature of the polarizability the equations for the Raman effect are somewhat complicated, so that only the IR intensities will be introduced as an example in the following (for Raman see (56)).

The derivation with respect to Q leads to:

$$(\frac{\partial \vec{\mu}}{\partial Q})_0 = \sum_i^n (\frac{\partial m_i}{\partial Q} \vec{e}_i + m_i \frac{\partial \vec{e}_i}{\partial Q})$$

Using the eigenvectors L obtained from the normal coordinate analysis we get:

$$\frac{\partial m_i}{\partial Q} = \sum_1^K \frac{\partial m_i}{\partial R_1} \frac{\partial R_1}{\partial Q} = \sum_1^K \frac{\partial m_i}{\partial R_1} L_1$$

and similarly:

$$\frac{\partial \vec{e}_i}{\partial Q} = \sum_1^K \frac{\partial \vec{e}_i}{\partial R_1} L_1$$

where the index 1 identifies one of the K internal coordinates R (bond stretches, in- and out-of-plane angle bending). Now we have:

$$(\frac{\partial \vec{\mu}}{\partial Q})_0 = \sum_1^K \{ \sum_i^n (\frac{\partial m_i}{\partial R_1} \vec{e}_i + m_i \frac{\partial \vec{e}_i}{\partial R_1}) L_1 \}$$

This expression can be divided in two parts:

- one part depending on the equilibrium geometry, usually called the valence part where the parameters are of the type $\partial m_i/\partial R_1$

- another part depending on the distorded molecule with the parameters m_i.

The parameters $\partial m_i/\partial R_1$ and m_i, $\partial \alpha_i/\partial R_1$ and α_i for the Raman effect all together are called the electro optical parameters, EOPs, and can be determined from spectral intensities with methods analogous to those used for the force constants.

Zerbi and his group (55-57) applied this approach to the IR and Raman spectra of n-alkanes and have shown that the derived parameters can be used to predict the spectra of cyclohexane or polyethylene. Although the validity of the transfer concept in other chemical environment has not yet been very well studied this technique appears still very promising.

The third method has been proposed simultaneously with the preceding one in the hope of getting less difficulties for the transfer of the parameters between different molecules. This 'polar tensor method' has been developed by Morcillo et al (58) and was later rediscovered by Person (59); so far it has been only applied to the IR spectrum. In that theory the parameters used are the changes of the dipole moment with the Cartesian displacement of each atom of the molecule. If μ_x, μ_y, and μ_z are the components of the total dipole moment and x^{th}, y, and z the Cartesian coordinates of the α atom, the atomic polar tensors are defined as P_α:

$$P_\alpha = \begin{pmatrix} \partial\mu_x/\partial x_\alpha & \partial\mu_x/\partial y_\alpha & \partial\mu_x/\partial z_\alpha \\ \partial\mu_y/\partial x_\alpha & \partial\mu_y/\partial y_\alpha & \partial\mu_y/\partial z_\alpha \\ \partial\mu_z/\partial x_\alpha & \partial\mu_z/\partial y_\alpha & \partial\mu_z/\partial z_\alpha \end{pmatrix}$$

The tensor for the whole molecule is obtained by juxtaposition of the atomic tensors P_α:

$$P_x = (P_1 \quad P_2 \quad \cdots \quad P_\alpha \quad \cdots \quad P_n)$$

Knowing $\partial\mu/\partial Q$ from the experiment, it is possible to obtain the P_α in a manner analogous to the derivation of the electro optical parameters. One of the possible applications of these tensors is their use in order to get further information about the atomic population densities (60).

Anyway, for both methods the main problem remains the determination of the sign to enter the calculation. Usually a quantum mechanical determination of the sign is performed. Thus we find the transition to the next class of intensity calculations. In fact, if we have to introduce such a theoretical method for one step of a calculation, the tentation is great to make all the calculation quantum mechanically.

Quantum mechanical models. As we have already seen the calculation of the intensities reduces to the determination of μ and α for different distorted structures of the molecule in order to determine $\partial\mu/\partial Q$ and $\partial\alpha/\partial Q$ which can be introduced in equation 3 and 4 in order to get the intensities. The dipole moment is given by all standard programs available as a result of the analysis of the charge distribution. The polarizability is less well known and the calculation methods which are still in the development will be discussed later.

Infrared spectrum. Already at the very beginning of the development of computational theoretical chemistry Segal (61) has tried to apply the new CNDO/2 method to the determination of the IR intensity of simple molecules. By determining μ for the equilibrium and a structure distorted according to the normal coordinates, he had the possibility to determine $\partial\mu/\partial Q$; it was in fairly agreement with the measured data. Later the same method has been applied by different authors with different quantum mechanical approximations for a number of molecules. In Table 2 some results have been collected for comparison. Two conclusions can be drawn:

- the overall agreement for all methods is satisfactory,
- the observed discrepancies cannot be correlated simply to the level of approximation used in the calculation.

A comparative study with MINDO/3 (62) leads to a poorer agreement than with CNDO/2. Anyway, this type of calculation appears to be rather powerful. The next development should be a systematic study of the discrepancies and of the possible corrections. For the choice of the quantum mechanical model other arguments have to be considered such as the fact that ab initio calculations for larger molecules are limited by the capacity of the computer or by the time necessary for the calculation. As an example, CNDO/2 was used for the simulation of the spectrum of hexamethylbenzene leading to a quite satisfactory agreement in fig. 1 (32). This method can be decisive in the determination of an assignment when some bands overlap (region 300-400 cm^{-1}).

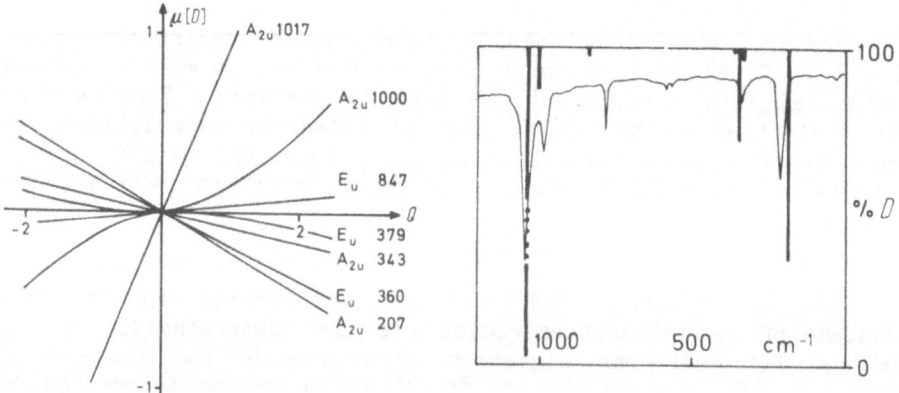

Fig. 10. Calculated and experimental IR spectrum of hexamethylbenzene

Table 2: Observed and calculated IR intensities of some molecules

Frequency	Experimental	Intensity $A(10^{16} cm^2 sec^{-1} mol^{-1})$	
		Calculated	
		semi-empirical	ab initio
H_2O			
ν_s	$0.66^{a)}$	$0.67^{b)}$	$1.08^{c)}$
δ_s	$16.08^{a)}$	$9.03^{b)}$	$30.02^{c)}$
ν_a	$13.38^{a)}$	$4.18^{b)}$	$14.61^{c)}$
C_2H_4 $^{d)}$			
949 Γ_{CH_2}	$23.93^{d)}$	$12.68^{e)}$	$39.30^{g)}$
3105 ν_{CH_2}	7.47	20.32	11.70
826 ρ_{CH_2}	0.16	1.03	0.19
3021 ν_{CH_2}	4.05	7.87	7.05
1444 δ_{CH_2}	2.93	0.11	2.52
Cyclopropane $^{f)}$			
3102 ν_{CH_2}	$9.06^{f)}$	$24.79^{e)}$	-
854 ρ_{CH_2}	0.15	0.16	-
3024 ν_{CH_2}	11.56	14.89	-
1438 δ_{CH_2}	0.54	0.06	-
1028 Γ_{CH_2}	6.10	6.75	-
869 Γ_{ring}	9.34	10.43	-

a) S.A. Clough, Y. Beers, G.P. Klein, L.S. Rothman, J. Chem. Phys. 59, 2284 (1973).
b) CNDO; D. Bougeard, S. Brüggenthies, B. Schrader, J. Mol. Struct. (in press).
c) I.G. John, G.B. Bacskay, N.S. Hush, Chem. Phys. 38, 319 (1979).
d) R.C. Golike, I.M. Mills, W.B. Person, B. Crawford, J. Chem. Phys. 25, 1266 (1956).
e) CNDO; M. Spiekermann, D. Bougeard, B. Schrader, J. Mol. Struct. (in press).
f) I.W. Levin, R.A.R. Pearce, J. Chem. Phys. 69, 2196 (1978).
g) Gaussian 70 4-31G; C.E. Blom, C. Altona, Mol. Phys. 34, 177 (1977).

Raman spectrum. The dipole moment is well determined by standard methods, this is not the case for the polarizability tensor which is defined by the following equation:

$$\vec{\mu}_{ind} = \alpha \ \vec{E}$$

where the polarizability α express the proportionality between the electric field E acting on the molecule and the dipole moment μ_{ind} it induces. Therefore in order to determine α we have to perturb the molecule with an electric field.

Calculation of α . We have two different possibilities to introduce this pertubation. The first one consists in doing a second order perturbation calculation and leads to expressions sometimes called 'sum over states' equations where the polarizability is given as:

$$\alpha_{ij} = 2 \sum_{k}^{occ} \sum_{1}^{virt} \frac{< 1| \mu_i |k > < k | \mu_j |1 >}{E_1 - E_k}$$

The sums run over all occupied k and virtual 1 levels; μ_i and μ_j are the dipole moments and E the energy of the levels. [1] This calculation method has the great advantage to use integrals which have already been calculated for the dipole moment. Alternatively the perturbation is introduced directly into the Hamiltonian and leads to some correction terms in the Fock matrix after the application of the variation theory. This method, including a finite field in the Hamiltonian, is sometimes defined as 'Finite Perturbation Theory' (FPT). The theory has been developed by Davies (63) for the CNDO approximation level. He has shown that the corrections to the Fock matrix are

$$h_{rr} = - e \ E_\sigma \ \sigma_A \quad \text{for the diagonal terms,}$$

$$h_{rt} = - e \ E_\sigma \ < s_A | \sigma | p_{\sigma A} > \quad \text{for the off-diagonal terms,}$$

where $\sigma = x, y, z$, E is the field and r and t atom orbitals of atom A. The SCF-problem being solved for the undisturbed and disturbed molecule, the polarizability is obtained as

$$\alpha = (\mu_{dist} - \mu_{undist}) / E$$

Sometimes, when strong fields are used, it is necessary to take account of the nonlinearity of the response in expanding the polarizability into a series:

$$\mu_{ind} = \alpha E + \beta E^2 + \gamma E^3 + \dots$$

the polarizability being then the first order term, β and γ are the hyperpolarizabilities. The fields observed with usual argon ion lasers have a magnitude of about 5.10^4 V/m. In order to observe the effects of the hyperpolarizability giant puls lasers have to be used which create a field of about 10^9 V/m (64). Several authors have used this method for the calculation of α but only few have gone ahead to the calculation of Raman intensities. Shinoda (67) proposed a third very sophisticated method by computing the intensity from the original vibronic equations of Albrecht (68) thus taking into account also the resonance effects. He compares his results with those of the FPT method. The differences between SOS and FPT methods have been theoretically derived and discussed by Ditchfield (69).

Raman intensities. We can apply the same method as for the dipole moment to the polarizabilities in order to obtain the first derivative with respect to the normal coordinate. We calculate the equilibrium geometry with field and without field and repeat the same operation for the geometry distorted according to the normal coordinate. Bleckmann has published the first application of this method using the FPT formalism in a CNDO-program for the spectra of cyclopropenone (70). In the last years other groups have applied it to different groups of molecules (71,72,73). This field is actually in rather fast expansion but some conclusions and unsolved questions can already be drawn:

- satisfactory results have been obtained with all methods concerning the relative intensities.

- for the IR as well as for the Raman intensities the reliability of the different quantum mechanical methods is not well established. Apparently it turns out that the CNDO or MINDO semi empirical methods yield results which are as good as those obtained by ab initio STO-3G calculations.

More sophisticated ab initio methods like Gaussian STO 4-31G seems to lead to quite better results (71,72). A way to improve the polarizability as well as the Raman intensity is the inclusion of polarization functions to the usually used valence or minimal basis sets. By using the standard basis sets the electrons are constrained to remain in the basis orbitals (1s for H, 2s + 2p for Li to F, etc.) and thus cannot response to the field like in a real molecule. In order to enable a greater 'mobility' of the electrons 2p orbitals for H or 3s, 3p or 3d orbitals for Li to F can be added (71,73). The improvement for the polarizability is very significative (74) and leads to better intensities for the Raman spectra.

Finally it should be mentioned that the factors determining the intensities are not yet all well understood. Shinoda (67)

has shown the influence of the force field on his calculation
of ethylene; such a sensitivity has been confirmed by Spiekermann et al. (73) for the spectra of cyclopropane. On this basis
one can expect that such methods are very useful as a help for
dealing ambiguous assignments or overlapping bands.

A further factor is the determination of the geometry to
be used in the calculation. Research is still in development
in this field based mostly on a proposition from Schwendeman
(75) who found that the experimental geometry should be used
instead of the quantum mechanically optimized one but this should
be tested for the different methods because some bands are very
sensitive to variations in the structural parameters (61,76).

Conclusive example and final remarks. A conclusive example
is the conformational problem of α,β - unsaturated aldehydes and
ketones which is resumed in fig. 11. Apart from a frequency
shift which can be reproduced by the normal coordinate analysis strong intensity effects are observed. As indicated in the
box the ratio of the intensities of the $\nu C=0$ and $\nu C=C$ bands
are low (0.6-3.5) for the s-cis and larger (5.2) for the s-trans
form, while in the Raman spectrum the $\nu C=0$ band of the s-trans
conformer is stronger than the same band for the s-cis compound.
A model calculation with methylvinylketone using the CNDO-FPT
method (73) shows a very good agreement by turning the dihedral
angle from 0 to 180° , thus allowing the conclusive statement
that a combination of frequency and intensity calculation is
a very powerful tool to support and develop assignment criteria
and possibly allow to estimate some geometrical parameters, which
are difficult to account for.

The development of the simulation of infrared and Raman spectra
by combined frequency and intensity calculations is still at
a very early stage. Since this tool may be of great practical
help for structural studies in all fields of chemistry some effort is justified to find a compromise between the necessary
computer time and expense and the attainable level of absolute
or relative accuracy. In addition, such calculations may also
help in the development of suitable computer programs and the
optimization of their parameters for the frequency as well as
especially the quantum mechanical calculation of the intensities.

All the calculations mentioned so far have been derived for an
isolated molecule. Similar methods have been developed or extended to treat condensed phases but due to their complexity
they cannot be treated extensively here. The interested reader
should look at the papers by Califano, Luty or Sanquer to get
an insight in these particular problems particularly for the
solid state (77-81).

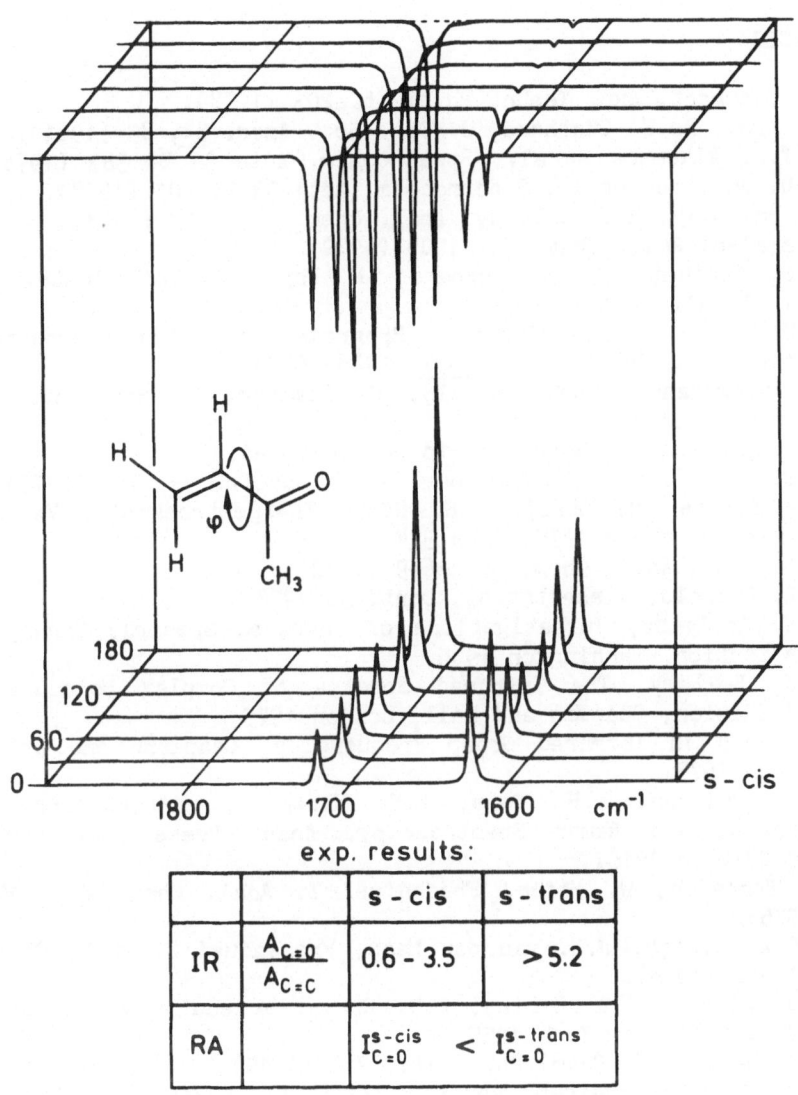

		s - cis	s - trans
IR	$\dfrac{A_{C=O}}{A_{C=C}}$	0.6 - 3.5	> 5.2
RA		$I_{C=O}^{s-cis}$	$< \quad I_{C=O}^{s-trans}$

Fig. 11. Variation of the intensities of the $\nu_{C=C}$ and $\nu_{C=O}$ bands of methylvinylketone with the conformation.

Acknowledgement. Financial support by the Deutsche Forschungsgemeinschaft, the Fonds der Chemischen Industrie and the Minis-
ter für Wissenschaft und Forschung, Nordrhein-Westfalen is gratefully acknowledged.

REFERENCES

(1) M. Schubert, Exp. Techn. Physik 6, 203 (1958).
(2) K. Frei, Hs.H. Günthard, J. Opt. Soc. Amer. 51, 83 (1961).
(3) C.T.J. Alkemade et al., Spectrochim. Acta 33 B, 383 (1978),
    G.D. Butilier et al, Spectrochim. Acta 33 B, 401 (1978).
(4) A. Savitzky, M.J.E. Golay, Anal. Chem. 36, 1627 (1964);
    see also: Anal. Chem. 44, 1906 (1972).
(5) J.P. Porchet, Hs.H. Günthard, J. Phys. E. Sci. Instr. 3,
    261 (1970).
(6) R.N. Jones et al., Computer Programs for Infrared Spectro-
    photometry, NRCC Bull., 11-17 (1968-1977).
(7) W. Niggemannn, W.R. Leupold, C. Domingo, B. Schrader, to
    be published.
(8) W. Niggemann, B. Schrader, to be published.
(9) G.T. Rasmussen, T.L. Isenhour, Appl. Spectr. 33, 371 (1979),
    see further H. Günzler, H. Böck, IR-Spektroskopie, Verlag
    Chemie, Weinheim 1975, page 324.
(10) C.S. Rann, Anal. Chem. 44, 1669 (1972).
(11) W.R. Leupold, Dissertation, Dortmund 1978.
(12) W.W. Coblentz, Investigations of Infrared Spectra, Carnegie
     Institution, Washington 1905.
(13) L.J. Bellamy, The Infrared Spectra of Complex Molecules,
     3rd Edition, Chapman and Hall, London 1975;
     Advances in Infrared Group Frequencies, Chapmann and Hall,
     London 1975.
(14) N.B. Colthup, L.H. Daly, S.E. Wiberley, Introduction to
     Infrared and Raman Spectroscopy, Acad. Press, New York,
     2nd Edition 1975.
(15) B. Schrader, W. Meier, Fresenius' Z. Anal. Chem. 275, 177
     (1975).
(16) C.G.A. v.Eijk, J.H. van der Maas, Fresenius' Z. Anal. Chem.
     291 308 (1978).
(17) E.B. Wilson, J.C. Decius, P.C. Cross, Molecular Vibrations,
     McGraw Hill, New York 1955.
(18) J.C. Decius, J. Chem. Phys. 17, 1315 (1949).
(19) J.H. Schachtschneider, Reports 231/64 and 53/65 Shell Devel-
     opment Co., Emeryville California (1964) (1965).
(20) T. Shimanouchi, Computer Programs for Normal Coordinate
     Treatment of Polyatomic Molecules, Tokyo (1968).
(21) R.N. Jones, Computer Programs for Infrared Spectrophotometry
     NRCC Bull. 15 (1976).

(22) T. Shimanouchi, H. Matsuura, Y. Ogawa and I. Harada, J. Phys. Chem. Ref. Data 7 (4), 1323 (1978).
(23) D.F. McIntosh, M.R. Peterson, QCPE 10, 342 (1977).
(24) D.E. Mann, T. Shimanouchi, J.H. Meal, L. Faro, J. Chem. Phys. 27, 43 (1957).
(25) J. Overend, J.R. Scherer, Spectrochim. Acta 19, 1567 (1963).
(26) B. Crawford, J. Overend, J. Mol. Spectrosc. 12, 307 (1964).
(27) M. Gussoni, G. Zerbi, Chem. Phys. Lett. 2, 145 (1968).
(28) W.M.A. Smit, F.A. Roos, Mol. Phys. 36, 1017 (1978).
(29) H.C. Urey, C.A. Bradley, Phys. Rev. 38, 1969 (1931).
(30) T. Shimanouchi, Pure Appl. Chem. 12, 131 (1963).
(31) S. Califano, Pure Appl. Chem. 18, 353 (1969).
(32) D. Bougeard, P. Bleckmann, B. Schrader, Ber. Bunsenges. Phys. Chem. 77, 1059 (1973).
(33) J.R. Scherer, Spectrochim. Acta 17, 719 (1961); ibid. 20, 345 (1964); ibid. 23 A, 1489 (1967).
(34) N. Neto, M. Scrocco, Spectrochim. Acta 22, 1981 (1966).
(35) J.H. Schachtschneider, R.G. Snyder, Spectrochim. Acta 19, 117 (1963).
(36) H.J. Becher, Fortschritte der Chemischen Forschung 10, 156 (1968).
(37) A. Müller, N. Mohan, S.J. Cyvin, J. Mol. Spectrosc. 59, 161 (1976).
(38) P. Pulay, Mol. Phys. 17, 197 (1969).
(39) P. Pulay, W. Meyer, J. Mol. Spectrosc. 40, 59 (1971).
(40) P. Pulay, W. Meyer, Mol. Phys. 27, 473 (1974).
(41) C.E. Blom, P.J. Slingerland, C. Altona, Mol. Phys. 31, 1359 (1976).
(42) C.E. Blom, C. Altona, Mol. Phys. 31, 1377 (1976).
(43) K. Kozmutzza, P. Pulay, Theor. Chim. Acta 37, 67 (1975).
(44) F. Török, A. Hegedüs, K. Kosa, P. Pulay, J. Mol. Struct. 32, 93 (1976).
(45) G. Fogarasi, P. Pulay, J. Mol. Struct. 39, 275 (1977).
(46) J.W. McIver, A. Komornicki, Chem. Phys. Lett. 10, 303 (1971).
(47) M.J.S. Dewar, G.P. Ford, J. Amer. Chem. Soc. 99, 1685 (1977).
(48) C. Coulombeau, A. Rassat, J. Chim. Phys. 74, 220 (1977).
(49) J.L. Duncan, Spectrochim. Acta 20, 1197 (1964).
(50) J. Brandmüller, H. Moser, Einführung in die Raman-Spektroskopie, Dr. D. Steinkopf Verlag, Darmstadt (1962).
(51) R.G. Snyder, J. Chem. Phys. 42, 1744 (1965).
(52) W.B. Person, D. Steele, Mol. Spectrosc. 29, 357 (1974).
(53) M.W. Wolkenstein, Dokl. Akad. Nauk. SSSR 30, 791 (1941).
(54) L.M. Sverdlov, M.A. Kovner, E.P. Krainov, Vibrational Spectra of Polyatomic Molecules, J. Wiley & Sons, New York (1974).
(55) M. Gussoni, S. Abbate, J. Chem. Phys. 65, 3439 (1976).
(56) M. Gussoni, S. Abbate, G. Zerbi, J. Raman Spectrosc. 6, 289 (1977).
(57) S. Abbate, M. Gussoni, G. Masetti, G. Zerbi, J. Chem. Phys. 67, 1519 (1977).

(58) J. Morcillo, L.J. Zamorano, J.M.V. Heredia, Spectrochim. Acta 22, 1969 (1966).
(59) W.B. Person, J.H. Newton, J. Mol. Struct. 46, 105 (1978).
(60) W.T. King, G.B. Mast, J. Phys. Chem. 80, 2521 (1976).
(61) G.A. Segal, M.L. Klein, J. Chem. Phys. 47, 4236 (1967).
(62) P.K.K. Pandey, P. Chandra, P.L. Prasad, S. Singh, Chem. Phys. Lett. 49, 353 (1977).
(63) D.W. Davies, Mol. Phys. 17, 473 (1969).
(64) D.A. Long, Raman Spectroscopy, McGraw Hill, London (1977).
(65) N.S. Hush, M.L. Williams, Chem. Phys. Lett. 5, 507 (1970).
(66) H. Meyer, A. Schweig, Theor. Chim. Acta 29, 375 (1973).
(67) H. Shinoda, Bull. Chem. Soc. Japan 49, 1267 (1976).
(68) J. Tang, A.C. Albrecht 'Raman Spectroscopy' Vol. 2, H.A. Szymanski ed. Plenum Press, New York (1976).
(69) R. Ditchfield, N.S. Ostlund, J.N. Murrell, M.A. Turpin, Mol. Phys. 18, 433 (1970).
(70) P. Bleckmann, Z. Naturforsch. 29a, 1485 (1974).
(71) I.G. John, G.B. Baskay, N.S. Hush, Chem. Phys. 38, 319 (1979).
(72) C.E. Blom, C. Altona, Mol. Phys. 34, 177 (1977).
(73) M. Spiekermann, D. Bougeard, B. Schrader, J. Mol. Struct. (in press).
(74) E. Norby-Svendsen, T. Stroyer-Hansen, Int. J. Quant. Chem. XIII, 235 (1978).
(75) R.H. Schwendeman, J. Chem. Phys. 44, 2115 (1966).
(76) D. Bougeard, S. Brüggenthies, B. Schrader, J. Mol. Struct. (in press).
(77) G. Taddei, H. Bonadeo, M.P. Marzocchi, S. Califano, J. Chem. Phys. 58, 966 (1973).
(78) R. Righini, N. Neto, S. Califano, S.H. Walmsley, Chem. Phys. 33, 345 (1978).
(79) V. Schettino, S. Califano, J. Chim. Phys. 76, 197 (1979).
(80) T. Luty, A. Mierzejewski, R.N. Munn, Chem. Phys. 29, 353 (1978).
(81) M. Sanquer, O. Contreras, Mol. Cryst. Liq. Cryst. 39, 7 (1977).

PHENOMENA IN PHOTOELECTRON SPECTROSCOPY

AND THEIR THEORETICAL CALCULATION

W. von Niessen
Institut für Physikalische Chemie
Technische Universität Braunschweig
D-33 Braunschweig, W.-Germany

L.S. Cederbaum, W. Domcke, and J. Schirmer
Fakultät für Physik, Universität Freiburg
D-78 Freiburg, W-Germany

Abstract

The method of the one-particle Green's function for the calculation of ionization energies and electron affinities as well as
of the vibrational structure and vibronic coupling effects in
photoelectron spectra is introduced. Selective examples are given
for the calculation of ionization energies in cases where Koopmans' approximation fails badly. The vibrational structure is
discussed for C_2N_2 and vibronic effects are calculated in the
photoelectron spectrum of HCN. In the inner valence region the
familiar molecular orbital model of ionization breaks down
completely. The intensity becomes distributed over numerous lines
and a main line ceases to exist. Several spectra are interpreted
in this way and it is demonstrated how the photoelectron spectra
in the inner valence region look like. In the case of para-nitroaniline it is seen that the splitting of a line can occur also for
core orbitals. In this case the shake-up energy is found to be
negative.

Introduction

Excitation of molecules by sufficiently energetic radiation or
electrons leaves the molecules in the ground state and excited
states of the ion. The transition energies to the different
ionic states are the ionization energies. The electronic excitation can be accompanied by the excitation of vibrations and

rotations. The vibrational excitation is expected to be strong
in case a strongly bonding or antibonding electron (bonding
or antibonding with respect to a given normal coordinate not
with respect to two neighbouring nuclei) is ejected and weak if
a nonbonding electron is ejected. The vibrational structure can
be resolved or - due to its complexity and the limited experimental resolution - appear only as a broadening of the bands.
In case the vibrational structure is resolved, this adds significantly to the information one can deduce from a photoelectron
spectrum (PES). The rotational structure is, except for the case
of the hydrogen molecule, not resolvable at present. In addition
to the energies of the electronic transitions and the vibrational
structure, a PES contains in principle much more information: e.g.
the ionization cross section, relative cross sections or branching
ratios, angular dependencies, and line widths. These quantities are
in principle measurable, but in the normal case the corresponding
quantities are properties of the apparatus, in particular of the
analyzer, and show sometimes little relation to the physical
quantities. It is certainly a major aim of present day photoelectron
spectroscopy to deduce this information. In this and in other
respects photoelectron spectroscopy is a young field of research,
which is only in the beginning of its development.

Given a PES, one would like to assign it; i.e. one would like to
know the symmetry identification of the ionic states, their energies, and the interpretation of the vibrational structure with
respect to which vibrational modes are excited, and what are their
frequencies. There are purely experimental means to assign a PES,
but frequently these methods fail to completely assign a given
spectrum and theoretical calculations are used. The close link
necessary between photoelectron spectroscopy and theoretical
calculations has proved profitable for experimentalists as well
as theoreticians.

Theoretical Calculation of Ionization Energies

The simplest method to calculate ionization energies and by this
calculation assign the PES is based on Koopmans' theorem [1].
Koopmans' theorem itself is a stability theorem for ionic wave
functions and is of little concern to us here; but a consequence
of this theorem is an approximation for the i-th ionization
energy. We shall call this approximation Koopmans' approximation.
Let us consider a closed shell system. In the Hartree-Fock (HF)
approximation the wave function consists of a Slater determinant

$$\Psi_0 = det \left| \varphi_1 \overline{\varphi_1} \cdots \varphi_i \overline{\varphi_i} \cdots \varphi_n \overline{\varphi_n} \right| \qquad (1)$$

built from molecular spin orbitals $|\varphi_i\rangle$ ($|\varphi_i\rangle$ without the bar denotes that the orbital is associated with a spin function with $m_s = +\frac{1}{2}$, and with the bar that the orbital is associated with a spin function with $m_s = -\frac{1}{2}$) which are solutions of the HF equations

$$F \, |\varphi_i\rangle = \mathcal{E}_i \, |\varphi_i\rangle \tag{2}$$

with $\mathcal{E}_i$ the orbital energy and F the HF operator

$$F = h + \sum_i 2 J_i - K_i \quad , \tag{3}$$

where h is the one-electron operator, $J_i = \langle \varphi_i(2)|1/\tau_{12}|\varphi_i(2)\rangle$ the Coulomb and $K_i = \langle \varphi_i(2)|1/\tau_{12}|\varphi_i(1)\rangle$ the ex-change operator. The HF operator contains only an average inter-action between the electrons resulting from the summation and integration over the coordinates of the other electrons. Because of this approximate interaction, one has neglected an energy con-tribution, the socalled correlation energy. If one approximates the ionic wave function by

$$\Psi_i = det \, |\varphi_1 \bar{\varphi}_1 \cdots \varphi_i \cdots \varphi_n \bar{\varphi}_n| \, , \tag{4}$$

where the electron has been taken out of the spin orbital $|\bar{\varphi}_i\rangle$ leaving all other orbitals unchanged, one obtains for the energy difference

$$E_i(\Psi_i) - E_o(\Psi_o) = I_i = -\mathcal{E}_i \quad . \tag{5}$$

The i-th ionization energy is thus approximately given by the negative of the i-th orbital energy. This is Koopmans' approxi-mation which has proved to be very useful. With the ansatz eq (4) for the ionic wave function, in which the molecular orbitals of the neutral groundstate are used for the ionic state one has neglected the socalled reorganization energy of the electrons. Ejection of an electron will always lead to a charge rearrange-ment, and the ionic wave function should be constructed from orbitals appropriate for the ion. Thus Koopmans' approximation in-volves the neglect of the correlation energy both in the ion and the neutral ground state and the neglect of the reorganization

energy in the ion. In the outer valence region of many molecules
this approximation is quite acceptable, as these two neglected
effects tend to cancel to a certain degree, but there can be no
guarantee that the approximation is reliable. There are quite a
number of molecules, and there are whole classes of molecules,
where Koopmans' approximation fails badly in supplying the correct
ordering of ionic states. Ionic states can be quite close together in energy and thus one needs more accurate means of
calculating the ionization energies which take into account the
effects of electron correlation and reorganization. This can e.g.
be done by separate configuration interaction (CI) calculations
for the ionic states and the ground state, but there is also a
direct way to calculate the ionization energies, the method of
Green's functions [2,3].

Before discussing this method let us briefly mention another consequence of Koopmans' theorem which one could call Koopmans'
hypothesis. One can take an electron out of each orbital $|\varphi_i\rangle$.
A PES thus should contain as many lines as there are orbitals.
The reason for this is that the transition operator is a one-
particle operator. In addition to these one-electron transitions
two-electron transitions can also be observed which are ionization
combined with simultaneous excitation. These processes lead to
the socalled satellite lines. They usually have small intensities
and borrow this intensity from the main transitions via many-
body effects. This one-to-one correspondence between orbitals and
lines in a PES is well documented in the outer valence region of
the molecules, but we are going to see that it may completely
fail to describe reality in ionization from inner valence orbitals,
whereas in the core region it does again apply in general.

The one-particle Green's function is defined in time, state space
as the expectation value with respect to the exact ground state
wave function of a time-ordered product of annihilation and
creation operators for electrons in one-particle states

$$G_{k\ell}(t,t') = -i\langle \Psi_0^N | T\{a_k(t)a_\ell^+(t')\} | \Psi_0^N \rangle . \qquad (6)$$

$a_k(t), a_\ell^+(t')$ are operators in the Heisenberg representation

$$a_k^{(+)}(t) = e^{iHt} a_k^{(+)} e^{-iHt} \qquad (7)$$

with H the full Hamiltonian of the system. They annihilate
(create) electrons in one-particle states $|k\rangle$ according to
the relations

$$a_k^+ |0\rangle = |k\rangle \qquad\qquad a_k^+ |k\rangle = 0 \tag{8}$$

$$a_k |k\rangle = |0\rangle \qquad\qquad a_k |0\rangle = 0$$

where $|0\rangle$ is the one-particle state containing no electron with
quantum number k. These equations also take account of the Pauli
exclusion principle. The operators fulfill the anticommutation
relations $[a_k, a_\ell^+]_+ = \delta_{k\ell}$ with all other anticommutators
vanishing. T is Wick's time ordering operator which orders the
operators so that time increases from right to left. A permutation of the operators from the original ordering by the
action of T is accompanied by a change of sign. With the help of
the Fourier transformation one can go over from time, state space
to energy, state space

$$G_{k\ell}(\omega) = \int_{-\infty}^{\infty} G_{k\ell}(t,t')\, e^{i\omega(t-t')}\, d(t-t'). \tag{9}$$

By inserting the decomposition of unity and performing the inte-
gration one arrives at the spectral representation of the Green's
function

$$G_{k\ell}(\omega) = \lim_{\eta \to +0} \left\{ \sum_n \frac{\langle \Psi_0^N | a_k | \Psi_n^{N+1}\rangle\langle \Psi_n^{N+1} | a_\ell^+ | \Psi_0^N\rangle}{\omega + A_n + i\eta} \right.$$

$$\left. + \sum_m \frac{\langle \Psi_0^N | a_\ell^+ | \Psi_m^{N-1}\rangle\langle \Psi_m^{N-1} | a_k | \Psi_0^N\rangle}{\omega + I_m - i\eta} \right\} \tag{10}$$

with $A_n = E^N - E_n^{N+1}$ the vertical electron affinity
and $I_m = E_0^{N-1} - E_0^N$ the vertical ionization energy. By calculating the poles of the Green's function one thus obtains directly the ionization energies and the electron affinities. In
the spectral representation form of G this is certainly quite
difficult because of the infinite summation. Instead one obtains the desired quantities from the Dyson equation which is
formally equivalent to the spectral representation but amenable
to numerical calculations

$$G(\omega) = G^0(\omega) + G^0(\omega) \Sigma(\omega) G(\omega). \tag{11}$$

The Dyson equation connects the Green's function with the HF
Green's function $G^0_{k\ell} = \delta_{k\ell}/(\omega - \mathcal{E}_k)$ and the
quantity $\Sigma(\omega)$ which is called the self-energy potential.
$\Sigma(\omega)$ is the exact potential seen by an electron due to the
interaction with its surroundings. The Green's functions have
inverses

$$G^{-1} = (G^0)^{-1} - \Sigma(\omega) . \tag{12}$$

Thus instead of calculating the poles of G we calculate the
zeros of G^{-1} . In a diagonal approximation this takes the form

$$G^{-1}_{mm} = 0 = \omega - \mathcal{E}_m - \Sigma_{mm}(\omega) \tag{13}$$

The energies ω fulfilling this equation are the ionization energies and electron affinities. They can be calculated by obtaining
the intersection points of the straight line $y = \omega - \mathcal{E}_m$ with
$\Sigma_{mm}(\omega)$. This and the structure of Σ is represented
schematically in Fig. 1.

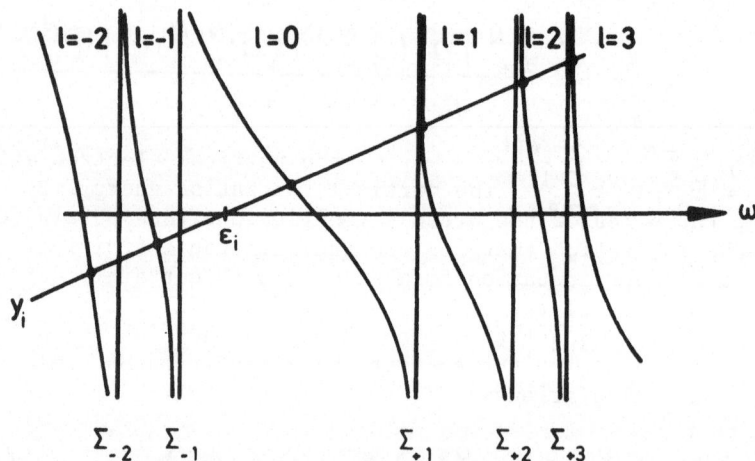

Fig.1: A schematic plot of Σ_{mm} as a function of ω and of the
       solution of the Dyson equation

Σ itself has poles and is a monotonically decreasing function of ω between the poles. For closed shell systems Σ has always a large interval free of poles. In this interval the outer valence ionization energies are found, and for their calculation high accuracy is required. In this region far away from the poles a perturbation expansion of Σ in the electron – electron inter-action is justified. We include all terms up to and including the third order terms. Higher order terms are taken into account by a renormalization procedure

$$\Sigma = \Sigma^{(1)} + \Sigma^{(2)} + \Sigma^{(3)} + \Sigma^{(R)} \quad . \qquad (14)$$

$\Sigma^{(1)}$ is zero if one starts from HF solutions. The diagrams of this perturbation expansion are given in Figs. 2 and 3.

Fig.2: The time-ordered self-energy diagrams of second order

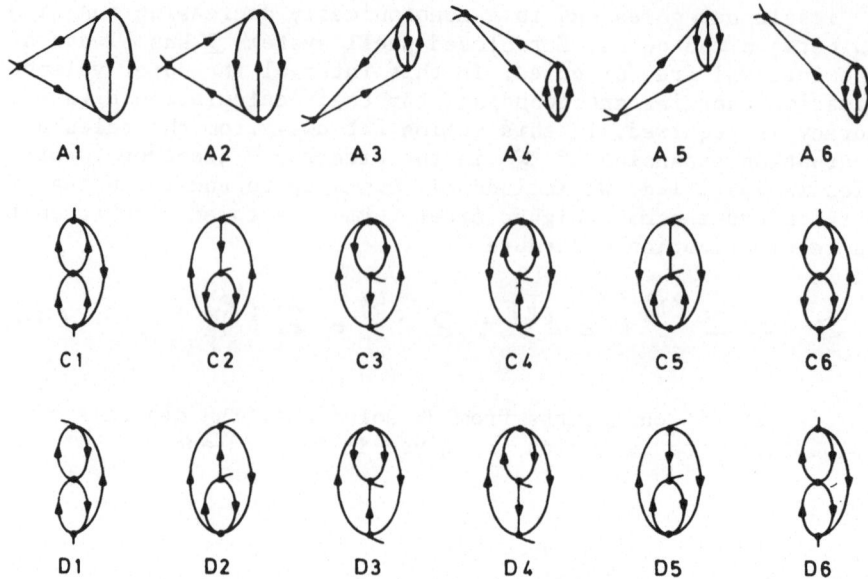

Fig.3: The notation of the time-ordered self-energy diagrams of
       third order

Such a finite perturbation expansion is not possible in the region
of the poles of Σ. In the pole region an approximation to Σ
must be used which correctly accounts for the pole structure and
the correct analytical behaviour of Σ (Σ has only simple poles,
but any finite perturbation approximation beyond the second order
introduces poles of higher order destroying in this way the correct
analytical properties of Σ). Such an approximation has been developped: the two-particle-hole RPA (random phase approximation
[3]) and the two-particle-hole Tamm-Dancoff approximation (TDA)
[3,4] which are used to calculate the ionization energies in the
inner valence region. As seen from Fig.1 there are many inter-
sections of the straight line y $= \omega - \varepsilon$ with $\Sigma(\omega)$. Thus one
obtains many ionization energies (in principle infinitely many)
for ionization out of a given orbital. Which of the energies
corresponds closest to Koopmans' approximation cannot be decided
by the energy in general but only by the relative intensities,
the pole strengths. These relative intensities are given by

$$p^{(s)} = \left(1 - \frac{\partial}{\partial\omega}\left(\varepsilon + \Sigma(\omega)\right)\Big|_{I_s}\right)^{-1}. \qquad (15)$$

If the pole strength of one solution is close to unity this will
be the main line. The other intersections will have small intensities and correspond to the mentioned satellite lines. But it
may also happen that there is no solution with a pole strength
close to unity. These solutions occupy us in the last part of
this article.

Vibrational Structure in Photoelectron Spectra

The vibrational structure is a prominent feature of molecular PES.
It is a very important source of information, and thus should be
included in a theoretical discussion of PES. Vibrational excitation reflects the bonding properties of the electrons. In polyatomic molecules the vibrational structure can become very complex,
and an analysis of which vibrations couple to the electronic motion
may present difficulties. This is a question which can be answered
by theoretical calculations. For the calculations of the vibrational structure one needs to know the potential surfaces of the ion
and of the neutral ground state. For polyatomic molecules this is
in general too expensive, and considerations have in general been
restricted to the harmonic approximation. The Green's function
method has been extended to include vibrational effects $[3,5]$.
In the derivation use has been made of the Born-Oppenheimer,
Franck-Condon, and harmonic approximations (for methods which go
beyond these approximations see the section on vibronic coupling),
but it should be mentioned from the outset that the harmonic
approximation used here is not identical to the harmonic approximation generally used. The philosophy behind the approach is, that
the neutral ground state of the molecules is well characterized,
but little or no information is available on the ionic states. We
thus use information on the ground state, in particular the force
field and the harmonic frequencies, and calculate the properties of
the ion.

In the one-particle approximation the Hamiltonian is given by

$$H = V_0 + \sum_s \omega_s \left(b_s^+ b_s + \frac{1}{2} \right) + \sum_i \varepsilon_i(Q) \left(q_i^+(Q) q_i(Q) - n_i \right), \quad (16)$$

where V_0 is a constant, ω_s are the ground state frequencies, b_s^+ and
b_s are boson creation and destruction operators, Q_s the normal
coordinates and $n_i = \begin{cases} 1 & \text{for i occupied} \\ 0 & \text{for i unoccupied.} \end{cases}$

The orbital energies and electronic operators depend on the normal
coordinates. The neglect of this dependence in the electronic

operators corresponds to the Born-Oppenheimer approximation which amounts to setting the commutators $[b_s^{(+)}, a_i^{(+)}] = 0$. $\mathcal{E}_i$ (Q) is expanded in a Taylor series where we include only the first-order term (for a more complete derivation see Refs. 3 and 5, for a simplified derivation Ref. 6)

$$\mathcal{E}_i(Q) = \mathcal{E}_i(0) + \sum_s \left(\frac{\partial \mathcal{E}_i}{\partial Q_s}\right)_0 Q_s + \cdots \qquad , \qquad (17)$$

where "0" denotes the equilibrium geometry of the neutral ground state. Introducing the first order coupling constants

$$K_s^{(0)}(i) = -\frac{1}{\sqrt{2}} \left(\frac{\partial \mathcal{E}_i}{\partial Q_s}\right)_0 \qquad (18)$$

we obtain for the Hamiltonian

$$H = V_{NN}(0) + \sum_s \omega_s \left(b_s^\dagger b_s + \frac{1}{2}\right) + \sum_i \mathcal{E}_i(0) a_i^\dagger a_i$$

$$- \sum_i \sum_s K_s^{(0)}(i) (a_i^\dagger a_i - n_i)(b_s + b_s^\dagger) , \qquad (19)$$

where $V_{NN}(0) = V_0 - \sum_i \mathcal{E}_i \mathcal{N}_i$. Many-body effects are included by replacing $\mathcal{E}_i$ by E_i and K_s^0 (i) by $k_s(i) = -\frac{1}{\sqrt{2}} \left(\partial E_i / \partial Q_s\right)_0$, where E_l is the exact pole of the Green's function. The spectrum is given by the transition probability per unit time and unit energy. For ionization out of orbital $|i\rangle$ and for one vibrational mode this is given by

$$P_i(\omega) = \sum_{n=0}^{\infty} e^{-a} \frac{a^n}{n!} \delta(\omega - E_i - a\omega + n\omega) \qquad (20)$$

with $a = (K/\omega)^2$. The first part is the Franck-Condon factor and the δ-function gives the position of the lines. $E_i + a\omega$ is except for the sign the adiabatic ionization energy (0-0 transition).

The Hamiltonian has been obtained by expanding all Q-dependent terms with the exception of the electronic operators about the equilibrium geometry of the electronic ground state. The calculation of the spectral function implies that the ground state and the ionic state potential surfaces are expanded about the ground state equilibrium geometry. It must be emphasized at this point that the traditional approach to calculate Franck-Condon

factors within the harmonic approximation is to expand both potential surfaces up to second order about their respective minima.
That this expansion is inappropriate is easily understood from an
inspection of Fig.4 which shows schematically the initial and
final state potential surfaces. The initial

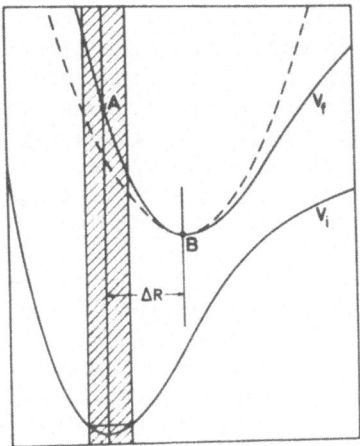

Fig.4: A schematic one-dimensional drawing of the initial (i) and
       final (f) potential surfaces. The Franck–Condon region is
       indicated by the shaded area. The harmonic expansion of V_f
       about its equilibrium geometry is represented by a broken
       line.

state vibrational wave function can be assumed to be well described
within the harmonic approximation. The minimum of the upper potential surface, however, may lie considerably outside the Franck–
Condon region as indicated by the shaded area in Fig.4. The overall shape of the spectrum depends only on the behaviour of the final
state potential surface within the Franck–Condon region. It is advantageous therefore to expand the final state potential surface
about a point within the Franck–Condon region, i.e. the ground
state equilibrium configuration, point A, in Fig.4. An expansion
about the final state equilibrium configuration (point B) will
give a poor description of the final state energy surface within
the Franck–Condon region if the coupling is strong. In contrast
to the intensities the vibrational energy levels in the final
state depend on the potential surface as a whole. Therefore, its
harmonic expansion about the initial state equilibrium geometry
does not necessarily lead to accurate line spacings in the calculated spectrum. It is our opinion, however, that for inter-
pretative purposes the accurate calculation of intensities is more

valuable than the accurate calculation of line spacings as the
intensities are necessary to assign the spectrum. The calculated
intensities are, as should be noted, independent of the line
spacings. In drawing the spectra the ionic frequencies may be
taken as those of the ground state or from the PES.

The success of this method can clearly be demonstrated in the
calculation of the vibrational structure in the first band in
the PES of NH_3 [7] . The calculation of the vibrational structure
of some Rydberg transitions and of the first band in the PES of
NH_3 has presented a problem. These transitions show an extended
progression in the bending mode ν_2. The intensity maximum occurs
at n = 6 in the UV absorption spectrum and at n = 7 in the PES.
The strong excitation of the bending mode is in qualitative
agreement with the transition from a pyramidal ground state to a
planar final state. Calculations in the traditional way have been
performed within the harmonic approximation and employing anharmonic potentials for the bending vibration. In the harmonic
approximation the intensity maximum was found to occur for n = 4
in the excitation spectrum in rather poor agreement with experiment. The use of anharmonic potentials did surprisingly not improve the agreement with experiment. If the vibrational structure
in the PES is calculated according to the procedure outlined above
very good agreement with experiment is obtained as can be seen
from Fig.5. The intensity maximum is calculated to occur at
the n = 7 line as is indeed observed in the high resolution
spectrum of Rabalais et al [8] . The calculation presented here
is an absolute one including the calculation of the position of
the band on the energy scale. The potential in the case of NH_3^+
is rather anharmonic. But as the Franck-Condon zone is narrow a
second order Taylor expansion about the center of the zone gives
a fairly accurate description of the final state potential surface within the zone. In the traditional approach this part of
the potential surface is poorly described, and a high order Taylor
expansion would have been necessary.

Applications

The numerical procedure for the calculation of ionization energies
and electron affinities with the Green's function method involves
several steps

a) evaluation of one- and two-electron integrals over the basis
   functions; Cartesian Gaussian function are used
b) solution of the SCF equations
c) transformation of the integrals from the basis of Gaussian
   functions to the basis of molecular orbitals
d) evaluation of the diagrams from the list of integrals and solution of the Dyson equation. This is an iterative process as
   a Brillouin-Wigner type perturbation theory has been used.

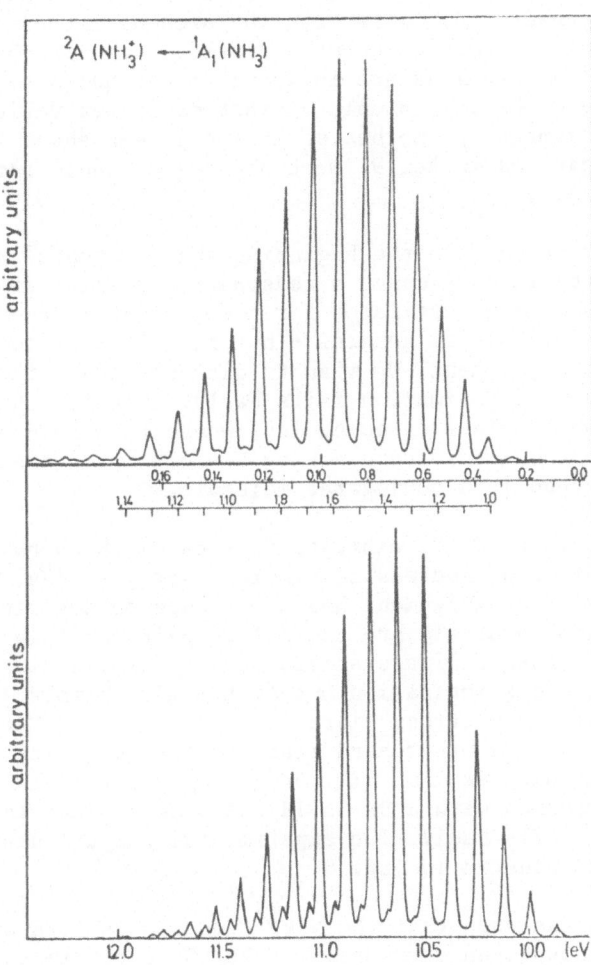

Fig.5: The first band in the PES of NH_3 as recorded by Rabalais
et al. [8] (upper part) and the calculated spectrum (lower
part)

The basis sets which are used in the calculations should at least be of double-zeta quality. For larger molecules this is the only practical choice. If one wants to achieve high accuracy in the calculated ionization energies one should at least add one polarization function per center and preferably use extended basis sets. The accuracy achievable with the present method is 0.2 eV in the ionization energies even in critical cases such as N_2 , if polarization functions are included in the basis set. The results prove to be very stable against basis set variations and further enlargement of the basis. This has been shown in extensive investigations in Ref.9. We thus have a stable and accurate method available.

It has been mentioned in the beginning that Koopmans' approximation supplies in many cases a reasonable ordering of ionic states and reasonable estimates for the ionization energies although the values are in general - but not always - too large compared with experiment. We wish to discuss here a few cases where Koopmans' approximation fails badly. Actually these cases belong to a whole class of molecules, where this method is useless. There are molecules containing N atoms in an aromatic ring or the $C \equiv N$ group in a conjugated molecule.

The HeI spectrum of $C_2 N_2$ exhibits four bands which have been assigned in order of increasing binding energy as $1\pi g$, $3\sigma g$, $2\sigma_u$ and $1\pi_u$ [10] . The assignment has been based on the vibrational structure. The σ-orbitals are the N-lone pair orbitals and ionization out of these orbitals should lead to little vibrational excitation, whereas ionization out of the π-orbitals (CN and CC bonding) should show strong vibrational excitation. Koopmans' approximation on the other hand leads to the following sequence of ionization energies $1\pi g$, $1\pi_u$, $3\sigma g$, $2\sigma_u$. (The calculations have been performed with a 9s 5p 1d basis set). Thus this approximation fails badly. The experimental and calculated spectra are reproduced in Fig. 6.

In addition it is seen that the vibrational structure of the $1\pi_u$ band is not given correctly in the SCF approximation. The spectrum calculated with the Green's function method is given in Fig.6 as well [11] . A double interchange of the two σ-ionization energies with the $1\pi_u$ ionization energy occurs in the many-body calculation and the computed spectrum is in quantitative agreement with the experimental one. The calculated ionization energies of $C_2 N_2$ are given in Table 1. It is seen that the maximum error is 0.16 eV which is very satisfactory.

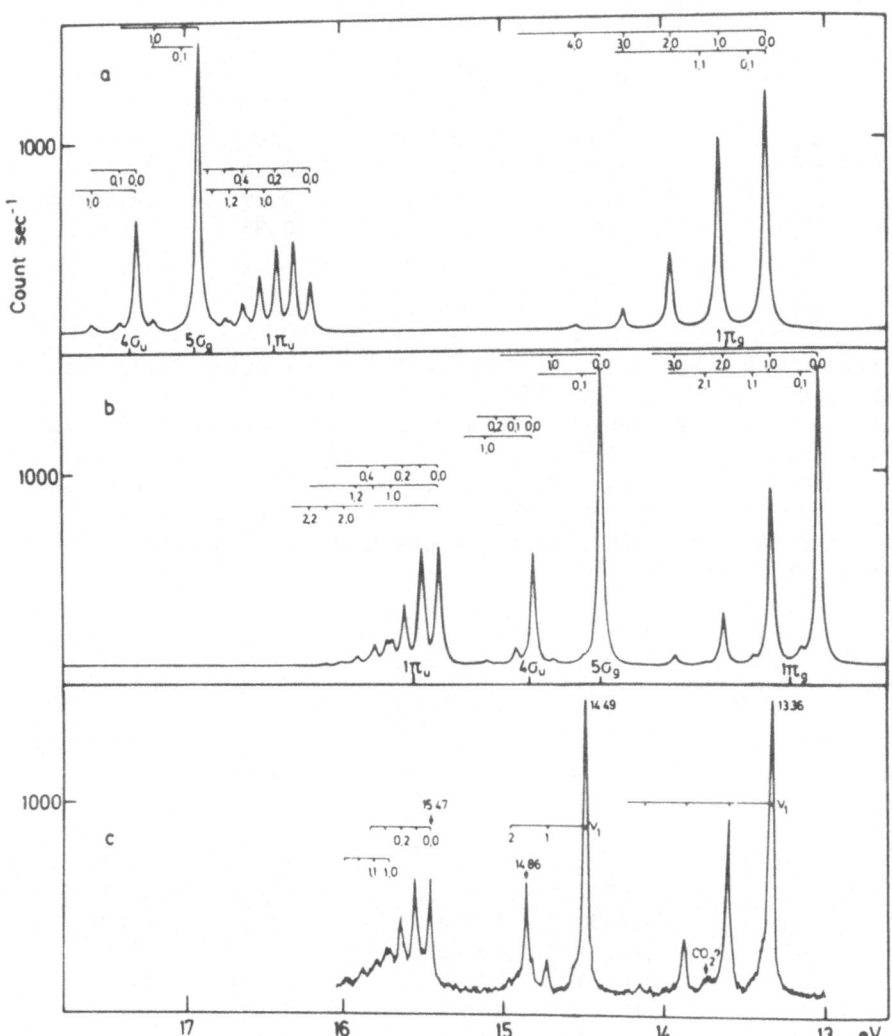

Fig.6: A comparison of the calculated and experimental PES of
$C_2 N_2$ a) spectrum calculated on the HF level, b) spectrum
calculated by including many-body effects c) experimental spectrum.

Table 1

Ionization energies of $C_2 N_2$ calculated with a 9s 5p 1d basis set. The P are the pole strengths. All energies in eV.

Symmetry	$-\varepsilon_i$	$I_i^{theor.}$	P_i	$I_i^{exp.}$
$1\pi_g$	13.60	13.20	0.91	13.36
$3\sigma_g$	16.93	14.40	0.90	14.49
$2\sigma_u$	17.35	14.80	0.89	14.86
$1\pi_u$	16.42	15.56	0.88	15.6

In the same way as for $C_2 N_2$ Koopmans' approximation fails for dicyanoacetylene, $C_4 N_2$. In this case the calculations have been performed only with a double-zeta basis set. The agreement with experiment is thus not as good as for $C_2 N_2$ [12] . Again the correct assignment is $2\pi_u$, $4\sigma_g$, $3\sigma_u$, $1\pi_g$, whereas Koopmans' approximation gives $2\pi_u$, $1\pi_g$, $4\sigma_g$, $3\sigma_u$ and a double interchange is necessary to obtain the correct ordering of ionic states as can be seen from Table 2.

Table 2

Ionization energies of $N\equiv C - C \equiv C - C \equiv N$ calculated with a 9s 5p basis set. The P are the pole strengths; all energies in eV.

Symmetry	$-\varepsilon_i$	$I_i^{theor.}$	P_i	$I_i^{exp.}$ [12]
$2\pi_u$	12.30	11.88	0.91	11.84
$4\sigma_g$	16.51	13.65	0.87	13.91
$3\sigma_u$	16.55	13.68	0.87	14.00
$1\pi_g$	15.28	14.41	0.89	14.16
$1\pi_u$	16.73	15.39	0.89	15.00
$3\sigma_g$	24.10	20.79	0.84	20.70
$2\sigma_u$	26.65	22.79	0.78	23.0

For a fairly large number of conjugated nitriles it has been found that Koopmans' approximation fails consistently in predicting the correct ordering of ionic states. But for this class of molecules it fails also in another respect and this may be even worse. It has been found that in an $X - C \equiv N$

molecule the ionization energy for ejection of an electron from
the N-lone pair orbital increases in the sequence X = CH_2, H,
CF_2 and F. This sequence is found from experiment and the many-
body calculations but Koopmans' approximation produces a different
sequence. In molecules of the type X - $C \equiv C - C \equiv N$ one would have
expected based on the known qualitative rules of organic chemistry and of substituent effects in photoelectron spectroscopy the
same trend in the n (N) ionization energies, although with reduced shift sizes. However, this is not the case. The n (N)
ionization energy in $F - C \equiv C - C \equiv N$ is found at lower ionization
energy than that of $CF_3 - C \equiv C - C \equiv N$, namely at the same position
as in $H - C \equiv C - C \equiv N$. The n (N) ionization energy thus increases
in the sequence X = CH_2, $H \approx F$, CF_3. This is reproduced by the
many-body calculation (the molecule $CF_3 - C \equiv C - C \equiv N$ has not been
calculated), but Koopmans' approximation again fails. This ordering also contradicts the rules of the perfluoro-effect. Thus, it
is dangerous to use Koopmans' approximation to predict such trends.
Even in such instances the inclusion of many-body effects is
necessary.

There is a simple model which permits to rationalize when Koopmans' approximation fails [13,14] . If a molecule possesses low-
lying virtual orbitals of non-diffuse character it is an
indication that considerable non-uniform many-body corrections
can be expected and thus the ordering of ionic states obtained
from Koopmans' approximation may not be the correct one. More
precisely, for a linear or planar molecule the existence of a
low-lying π (σ)-type unoccupied orbital of non-diffuse
character and of an outer valence π (σ)-type occupied orbital
leads to large many-body corrections for outer valence ionization
energies of σ (π)-type orbitals. The change of ordering
occurs in the RCN-type molecules so easily because the n (N) and
π-type ionization energies are close together in energy.

This same argument applies to the azabenzenes. In this class of
molecules the use of Koopmans' approximation to assign the PES
is a real desaster. The results have no relation to reality,
whereas the Green's function method proves to be a very powerful calculational tool [15] . We consider s-tetrazine. The He I
and He II spectra from Ref 16 are given in Fig.7 together with
the calculated results. In Koopmans' approximation the ordering
of ionic states would be (with increasing ionization energy)
1,2,3,4,5,6, where the detailed symmetry labels can be deduced
from Fig.7. The correct ordering of states is, however, 1,5,2,4,
6,3, i.e. there is a complete reordering necessary. Beyond the
sixth ionization energy the one-particle picture of ionization
breaks down and Koopmans' approximation is intrinsically inapplicable. This point will be discussed later. The results for
the first six ionization energies are given in Table 3.

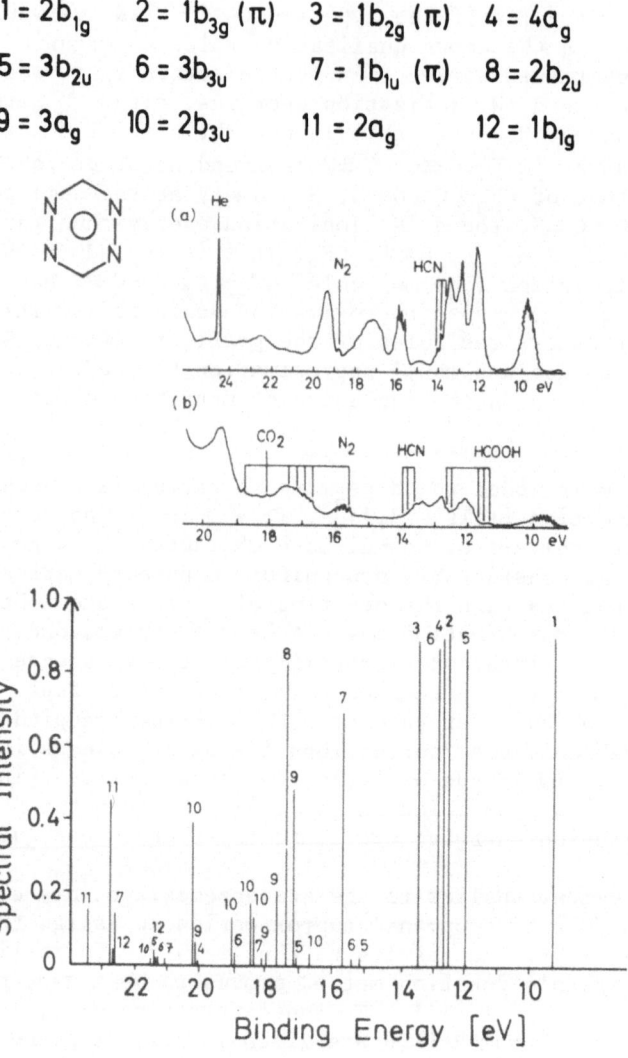

$1 = 2b_{1g}$ $2 = 1b_{3g} (\pi)$ $3 = 1b_{2g} (\pi)$ $4 = 4a_g$

$5 = 3b_{2u}$ $6 = 3b_{3u}$ $7 = 1b_{1u} (\pi)$ $8 = 2b_{2u}$

$9 = 3a_g$ $10 = 2b_{3u}$ $11 = 2a_g$ $12 = 1b_{1g}$

Fig.7: Ionization spectrum of s-tetrazine up to about 24 eV. The numbering of the orbitals is according to the HF sequence

Table 3

Ionization Energies of s-tetrazine

Symmetry	$-\mathcal{E}_i$	I_i^{theor}	P_i	I_i^{exp}[16]	I_i^{exp}[17]
2 b_{1g}	11.48	9.24	0.90	9.7	9.7
3 b_{2u}	14.64	11.97	0.87	11.9	12.1
1 $b_{3g}(\pi)$	12.69	12.52	0.91	12.1	12.1
4 a	14.57	12.64	0.90	12.8	12.8
3 b_{3u}	15.63	12.89	0.87	13.3	13.4
1 $b_{2g}(\pi)$	13.95	13.47	0.89	-	≈ 13.5

After so much disaster with Koopmans' approximation we would like
to supply one example of a fairly large molecule – norbornadiene –
where this approximation works well [18] . Norbornadiene is a key
compound for the study of through-space and through-bond interactions. The assignment of the first two ionic states which arise
from ejection of an electron from the symmetric and antisymmetric linear combination of the two ethylenic π bond orbitals
$\pi_\pm = \pi_a \pm \pi_b$ would give information on whether through-
space or through-bond interaction dominates. The He I and He II
PES were recorded by Bischof et al [19] and Bieri et al [20] .
The first two ionic states have been assigned by Heilbronner
and Martin [21] . They came to the conclusion that the $b_2(\pi)$
orbital (which is essentially the π_--combination) lies above the
a_1 (π) orbital (the π_+-combination) . Consequently it is
through-space interaction which dominates between the orbitals
π_a and π_b in norbornadiene. In Fig.8 we give the He I
spectrum and the results of the theoretical calculations.

Koopmans' approximation supplies the correct ordering of states
here and the spacing of the ionization energies is in fairly
good accord with the many-body calculation and the experimental
values. Both reproduce the ordering $I(5b_2(\pi)) < I(7a_1(\pi))$.
As is rather typical for such relatively large molecules with
closely spaced and sometimes not even resolved bands a detailed
assignment of every feature in the PES is not possible in all
cases. For details see Ref. 18.

Vibronic Coupling

Until now we have dealt with "normal" effects in photoelectron
spectroscopy, i.e. where there is a one-to-one correspondence
between orbitals and lines in the spectrum and where the

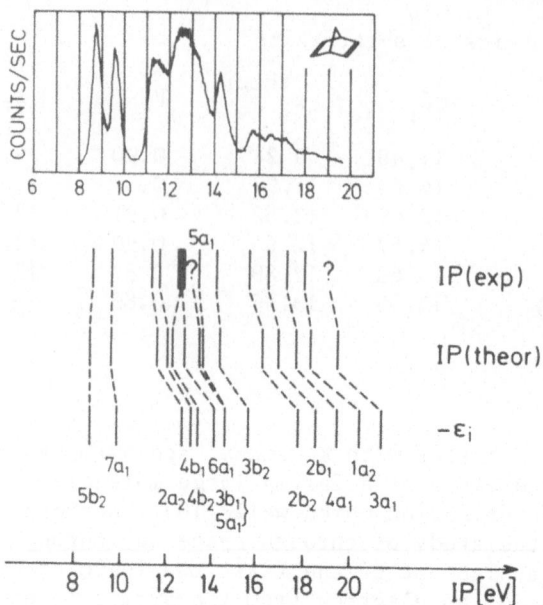

Fig.8: He I PES of norbornadiene and calculated ionization energies

vibrational structure can be calculated in the Born-Oppenheimer
and Franck-Condon approximations. Now we would like to venture into
more complicated fields.

The adiabatic approximation relies on the fact that the energy
difference between electronic states is large compared to the
spacing of the vibrational energies. A breakdown of this
approximation may therefore occur when two electronic states become degenerate or nearly degenerate. The Jahn-Teller effect
represents just one example. But vibronic coupling effects are
more general. We have dealt with these effects in the PES of butatriene [22] , where vibronic coupling is responsible for the
appearance of an additional band in the PES, and in ethylene,
where the strong excitation of the torsional mode in the first
band arizes from vibronic coupling [23] . Here we would like to
deal with the HCN molecule [24] . In the treatment of the vibrational motion as presented above we have neglected the Q-
dependence of the electronic operators. This is not permitted if
the energy difference between electronic states is not large

compared to the vibrational spacings. We start from the Hamiltonian equ. (19), where it is assumed that the electronic operators still depend on the normal coordinates. We transform the a_i (Q) to a Q-independent basis:

$$a_i (Q) = \sum_j \langle \varphi_i (Q) | \varphi_j (0) \rangle \, a_j (0) \, . \qquad (21)$$

We insert this expansion in equ. 19, expand $|\varphi_i (Q)\rangle$ in a Taylor series and collect all terms linear in Q to obtain

$$H = \sum_s \omega_s \left(b_s^+ b_s + \tfrac{1}{2} \right) + \sum_i \varepsilon_i (0) \, a_i^+ a_i$$
$$- \sum_{i,s} K_s^0 (i) \left[a_i^+ a_i - n_i \right] (b_s + b_s^+)$$
$$+ \sum_{i<j} \sum_s \lambda_s^0 (i,j) \left[a_i a_j^+ + a_j a_i^+ \right] (b_s + b_s^+) \, . \qquad (22)$$

For the vibronic coupling constants λ_s^0 (i,j) the following expression is obtained

$$\lambda_s^0 (i,j) = \frac{1}{\sqrt{2}} \lim_{Q \to 0} \left[\varepsilon_i (Q) - \varepsilon_j (Q) \right] \langle \varphi_j (Q) | \tfrac{\partial}{\partial Q_s} | \varphi_i (Q) \rangle$$
$$(23)$$

From the form of the Hamiltonian on can deduce the following selection rules for nonvanishing adiabatic coupling constants Ks(i):

$\Gamma_i \times \Gamma_i \times \Gamma_s \supset \Gamma_A$ and for the nonvanishing non-adiabatic ones (λ_s): $\Gamma_i \times \Gamma_j \times \Gamma_s \supset \Gamma_A$, where Γ_i is the representation of orbital $|\varphi_i \rangle$, Γ_s the representation of Q_s and Γ_A the totally symmetric representation. Vibronic coupling thus takes place between different electronic states by vibrations of appropriate symmetry (not necessarily totally symmetric ones).

For the spectrum one obtains the expression

$$P(\omega) = \int dt \, e^{i \omega t} \, \underline{\tau}^+ \langle 0 | \, e^{i \mathcal{H} t} \, | 0 \rangle \, \underline{\tau} \qquad , \qquad (24)$$

where $\underline{\tau}$ is the matrix of the electric dipole operator between the initial and final state.

$$|0\rangle = \begin{pmatrix} |0\rangle & 0 & 0 \\ 0 & |0\rangle & 0 \\ 0 & 0 & |0\rangle \end{pmatrix}$$

$|0\rangle$ is the vibrational groundstate of the neutral molecule. The dimension of $|0\rangle$ is three as in HCN we are dealing with three electronic degrees of freedom, the degenerate Π states and the nearly degenerate Σ state.

$\mathcal{H}$ is a matrix Hamiltonian free of electronic operators. For HCN it is given by

$$\mathcal{H} = \mathcal{H}_1 + \mathcal{H}_2$$

$$\mathcal{H}_1 = \omega(b_+^\dagger b_+ + b_-^\dagger b_-)\underline{I} - \begin{pmatrix} E_\pi & \lambda(b_- + b_+^\dagger) & 0 \\ \lambda(b_+ + b_-^\dagger) & E_\sigma & \lambda(b_- + b_+^\dagger) \\ 0 & \lambda(b_+ + b_-^\dagger) & E_{\bar\pi} \end{pmatrix}$$

$$\mathcal{H}_2 = \left(\sum \omega_{ig} b_{ig}^\dagger b_{ig}\right)\underline{I} + \begin{pmatrix} \sum K_i(\pi)(b_{ig} + b_{ig}^\dagger) & 0 & 0 \\ 0 & \sum K_i(\sigma)(b_{ig} + b_{ig}^\dagger) & 0 \\ 0 & 0 & \sum K_i(\bar\pi)(b_{ig} + b_{ig}^\dagger) \end{pmatrix}$$

$$(25)$$

The $b_{ig}^{(\dagger)}$ are the boson creation and annihilation operators for the totally symmetric modes and $b_+^{(\dagger)}$, $b_-^{(\dagger)}$ are the bosonic operators for the bending mode in an angular momentum adapted basis. $\underline{I}$ is a unit matrix. This Hamiltonian describes the vibrational motion in the ionic states as well as the vibronic coupling between these states. It should be noted that unless $K_i(\pi) = K_i(\sigma)$, i = 1,N the two parts $\mathcal{H}_1$ and $\mathcal{H}_2$ do not commute and, therefore the totally symmetric and bending motions cannot be treated separately.

The interpretation of the PES of HCN has presented problems. The vibrational structure in the first two bands ($^2\Pi$ and $^2\Sigma$) is so complex that it proved impossible to assign the lines to progressions in the familiar way [25,26] . In particular the excitation of the bending vibration has tentatively been attributed to vibronic interaction. The experimental spectra with the highest resolution are given in the upper part of Fig.9 for HCN and Fig.10 for DCN. (The HCN spectrum has been deconvoluted [26]). Using a basis set of 11s 7p 1d functions we obtain the following values for the energies and coupling constants $-E_\pi$ = 13.50 eV, $-E_\sigma$ = 13.92 eV, $K_1(\pi)$ = 0.004, $K_1(\sigma)$ = -0.080, $K_3(\pi)$ = -0.235, $K_3(\sigma)$ = 0.057, λ = 0.073. The excitation of γ_1 (C–H stretching) is negligible as $|K_1| \ll \omega_1$ and has been omitted. The excitation of CN stretching is, however, important and must be taken into account. Since $K_3(\pi) \neq K_3(\sigma)$ the bending and stretching modes cannot be decoupled. The spectrum calculated with these data is not in good agreement with experiment as the spectrum depends very sensitively on the energy

difference $E_\pi - E_\sigma$ and on λ as has also been found in the
case of butatriene [22] . Both quantities would have to be
calculated with an accuracy of 1 %. Although E_π and E_σ have
separately an accuracy of 2 % their difference has an error of
100 %. A best fit to the experimental spectrum gives a separation of 0.2 eV compared to the separation of 0.4 eV which has been
calculated. An accuracy of 1 % in λ cannot be reached at present
either. We have thus fitted E_σ , E_π and λ to best reproduce
the experimental spectrum; K3 (π) and K3 (σ) were left unchanged.
These best fit parameters are for HCN: $-E_\pi$ = 13.8 eV, $-E_\sigma$ = 14.0 eV,
λ = 0.090. The spectra calculated in this way are given in the
middle part of Fig.9 for HCN and of Fig.10 for DCN. The spectrum
for DCN has not been independently fitted, but the parameters were
obtained by using the isotope rules. The agreement of the calculated spectra with the experimental ones can be considered as very
satisfactory in view of the complexity of the spectra and the
simplicity of the Hamiltonian used. The equally good agreement obtained for the spectra of HCN and DCN which have not been fitted
independently shows that the agreement is not fortuitous. The
calculated vibrational structure agrees well up to about 14 eV with
experiment: above $\approx$14.0 eV the vibrational structure cannot be
reproduced better by simply readjusting the coupling constants
since the calculated line spacings differ from the experimental
ones. An analysis of the low energy portion of the experimental
spectrum shows that the CN stretching frequency ω_3 has decreased
upon ionization. This frequency change upon ionization is not
accounted for by the Hamiltonian. It can be included, however,
within the present framework in a simple way.
We allow the frequency to vary and keep the ratios, $K\alpha / \omega g$, which
determine the Franck-Condon factors, fixed. This procedure is
applied only to the ν_3 mode. All other quantities remain fixed.
The spectra calculated in this way are given in the lower parts
of Fig.9 and Fig.10, respectively. There is again an improvement
in the agreement with experiment. It should be mentioned that
the Renner-Teller effect cannot explain the observed structure,
(the relevant coupling constant has been calculated and has been
found to be small) although it may contribute to some details
of the structure. It is in fact the vibronic coupling between
the $^2\Pi$ and Σ electronic states via the bending and the CN
stretching mode which is responsible for the complex structure.

To demonstrate that the bending and stretching modes cannot be
decoupled we give in Fig.11 two spectra. The upper spectrum has
been calculated by only including the CN stretching mode and the
lower one by including only the bending mode. It is seen that the
spectrum resulting from the full vibronic solution is not simply
the convolution of a) and b). As conclusion we would like to note
that vibronic coupling effects can strongly modify the vibrational
structure of a given band and lead to the appearance of new bands.

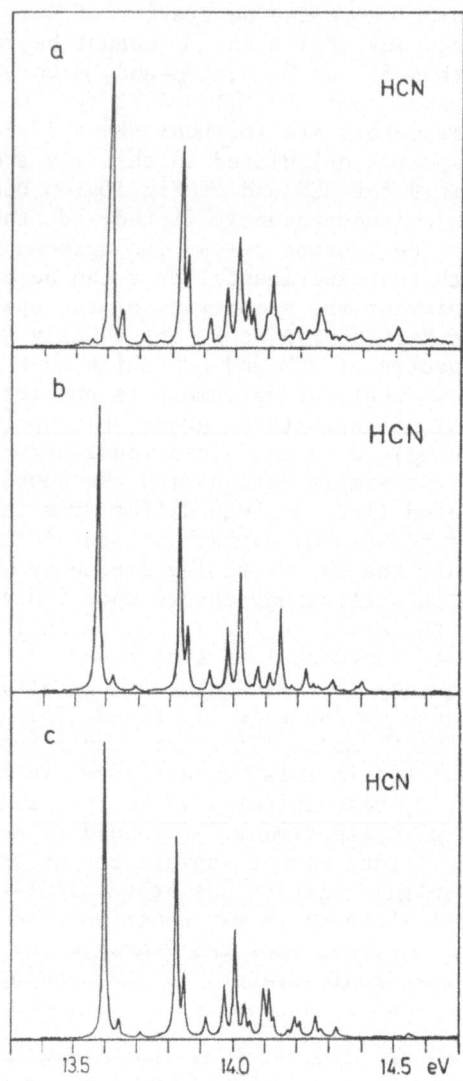

Fig.9: Experimental photoelectron spectrum of HCN (upper part) and calculated spectra (for details see text)

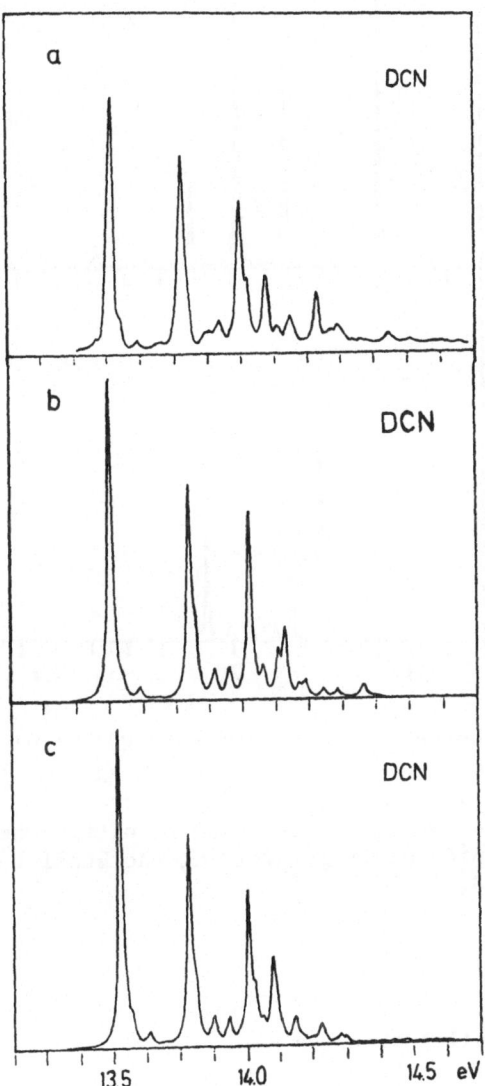

Fig.10: Experimental photoelectron spectrum of DCN (upper part)
and calculated spectra (for details see text)

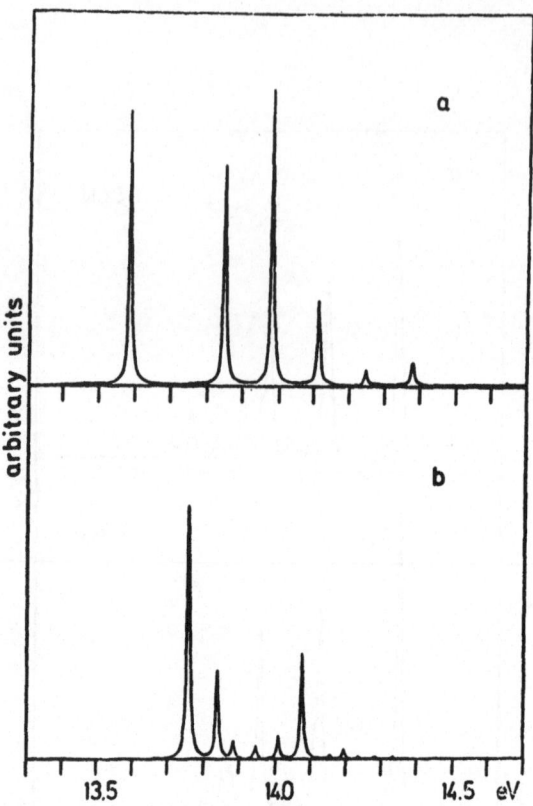

Fig.11: a) The photoelectron spectrum of HCN resulting when only
           the coupling to the CN stretching mode is included
           (i.e. λ = 0)
        b) The photoelectron spectrum resulting when only the
           vibronic coupling involving the bending mode is in-
           cluded.

Inner Valence Shell Ionization

There exist even stronger effects than vibronic coupling which
lead to a deviation from the familiar one-to-one correspondence
between orbitals and bands in the PES and to a complete breakdown of the molecular orbital model of ionization; these are
purely electronic effects. In the first part we have presented
the Green's function method as it can be applied to the calculation of outer valence ionization energies; i.e. to ionization

energies which are far from the poles of the self-energy. These
ionization energies have pole strengths close to unity. A consequence is that there is one line in the PES for each orbital.
Satellite lines will accompany these main lines but they have
only small relative intensities. The corresponding electronic
states are of a complicated nature. These solutions lie in the
pole region of the self-energy as do the inner valence ionization
energies. In this energy range a method is required which correctly accounts for the pole structure of the self-energy. We are
going to see that in the inner valence region the one-particle
picture of ionization may break down. The familiar concepts useful in the outer valence region and in the core region are without validity here. We do not wish to present the method as this
would lead into too lengthy details. The method has been presented in Refs. 3,4,6. The method gives qualitatively correct spectra over the entire energy range but its accuracy is not comparable to that of the outer valence Green's function method. We
only would like to mention that this method is not a finite order
perturbation method which necessarily breaks down in the region
of the poles, but an infinite summation implied in a system of
linear equations. The method is called two-particle-hole Tamm-
Dancoff Green's function approximation (TDA). We only would like
to give a physical interpretation. We denote a configuration where
an electron has been ejected out of orbital p by p^{-1} and configurations where one electron has been ejected out of orbital k
and another one simultaneously excited from orbital 1 to an unoccupied orbital j by $k^{-1} 1^{-1} j$. If the energy of configuration
p^{-1} , E (p^{-1}), is far away from the energy of other configurations
then one line will be found in the PES. If, however, E (p^{-1}) is
close to E $(k^{-1} 1^{-1} j)$ for some configuration $k^{-1} 1^{-1} j$ then there
will be configuration mixing and a redistribution of intensity.
Thus several lines of roughly equal intensities may appear in the
PES for ionization out of the orbital p. For molecules consisting of first row atoms E $(k^{-1} 1^{-1} j)$ is larger than the energy of
the neutral ground state by 15 to 25 eV. Thus these effects will
be found at higher energies affecting mainly the 2s lines of the
first row atoms. An important factor which enters E $(k^{-1} 1^{-1} j)$ is
the excitation energy of the molecule. If a molecule has very
low lying excitations these many-body effects will appear at
lower energy. The spectrum will depend on two other quantities;
the first one is the **interaction matrix element between the con-**
figurations p^{-1} and $k^{-1} 1^{-1} j$. This is approximately given by
V_{pjkl} . Such a matrix element is expected to be large only if
the virtual orbital j is localized in space as the occupied orbitals are. Therefore sufficiently large interaction between the
relevant configurations can be expected especially for those molecules that possess low lying valence type excited states. If V_{pjkl}
is nearly zero for some reason, no many-body effects will be ob-
servable in spite of the quasi-degeneracy of p^{-1} and $k^{-1} 1^{-1} j$. The

second factor which enters is the density of the $k^{-1}1^{-1}j$ configurations in energy space. If the density is low in the energy region of the single hole configuration, p^{-1}, the breakdown phenomenon will only occur if there is an accidental quasidegeneracy.
The intensity is then distributed only over a few lines. If the
density is high on the other hand the intensity becomes distributed over numerous lines each having only a few percent of the
total intensity. Since larger molecules have a high density of
configurations the breakdown phenomenon will dominate their
ionization spectra in the inner valence shell.

As the first example we discuss the CS molecule [27] . The PES of
CS (Fig.12) contains below 20 eV instead of three bands (due to
ionization from 7σ, 2π and 6σ; ionization of 5σ electrons
occurs above 20 eV) four bands [28] ; i.e. there is one band too
many.

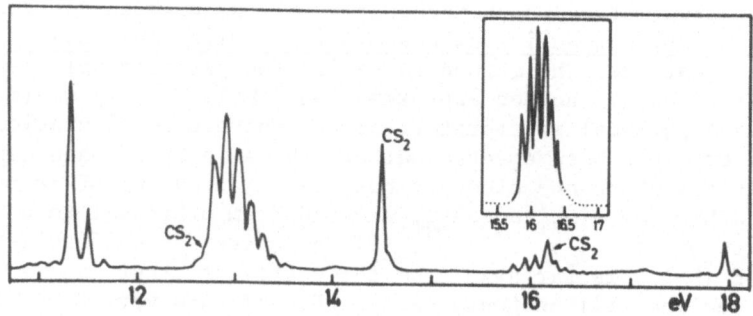

Fig.12: The PES of CS [28]

This phenomenon can be explained by looking at the schematic self-
energy potentials in Fig.'s 13 - 15. The solution for the 7σ ionization energy occurs in the main interval and leads to a large
pole strength (Fig.13); the same is the case for 2π ionization.

In the case of the 6σ orbital (Fig.14) one solution is still
found in the main interval but close to the first pole and another
solution of approximately equal pole strength in the first interval. These two solutions explains the experimental finding. The
6σ line is split into two lines at about 16 and 18 eV. by final
state correlation effects. None of these lines corresponds to
simple ionization from the 6σ orbital.

The 5σ orbital energy lies in the midst of many poles of the
self-energy (Fig.15). Here we find many solutions of about equal
intensity. The orbital picture of ionization thus breaks down for

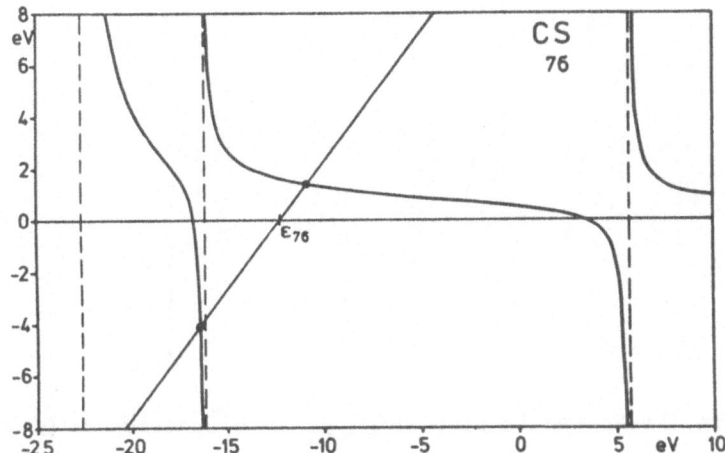

Fig.13: Schematic plot of $\Sigma_{76\,76}$ as a function of ω for CS

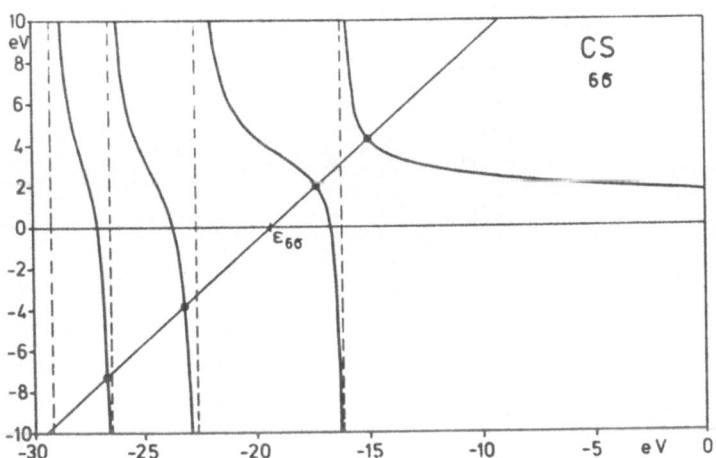

Fig.14: Schematic plot of $\Sigma_{66\,66}$ as a function of ω for CS

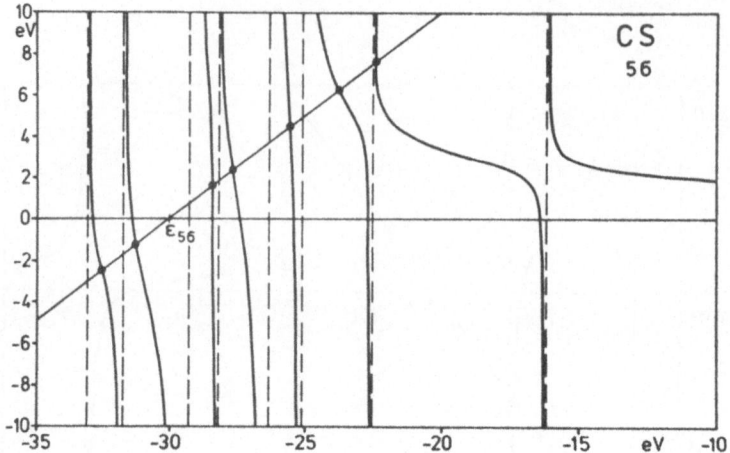

Fig.15: Schematic plot of $\Sigma_{5\sigma5\sigma}$ as a function of ω for CS

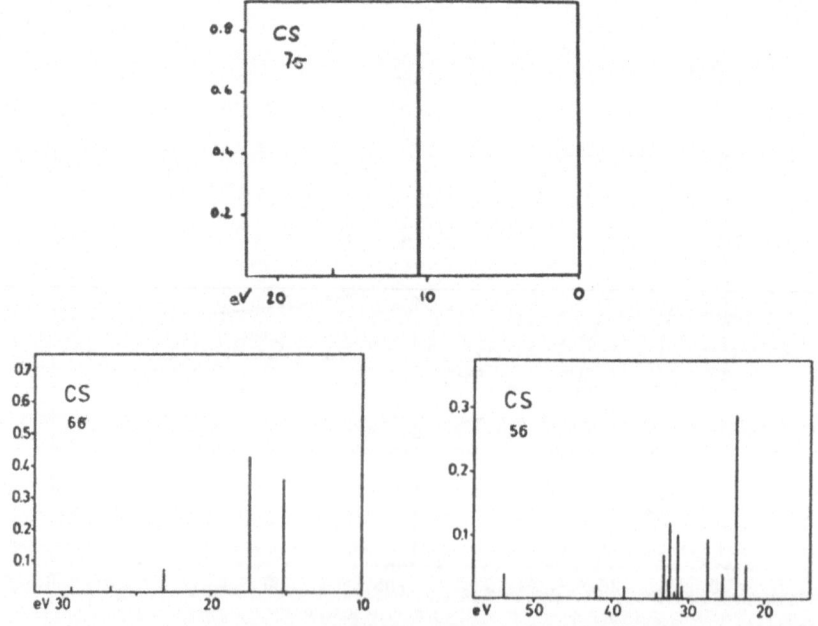

Fig.16: The calculated ionization spectra of CS (7σ, 6σ, and 5σ orbitals). Note the different ordinate scale for the 5σ spectrum.

ionization out of the 6σ and 5σ orbitals. The line spectra for 7σ, 6σ and 5σ ionization are given in Fig.16.

As the next example we consider CS_2 ; the PES of this molecule contains two intense satellite lines at very low energy (about 14.1 and 17.0 eV) [10,29] . Both arise from $2\pi_u$ ionization, but their intensity (about 4 % and 16 %) is still small compared to the main line. We will not discuss this point in detail, but turn to the energy region above 20 eV. The ESCA spectrum [30] is reproduced in Fig.17. It shows a broad band extending over about 20 eV which contains little structure.

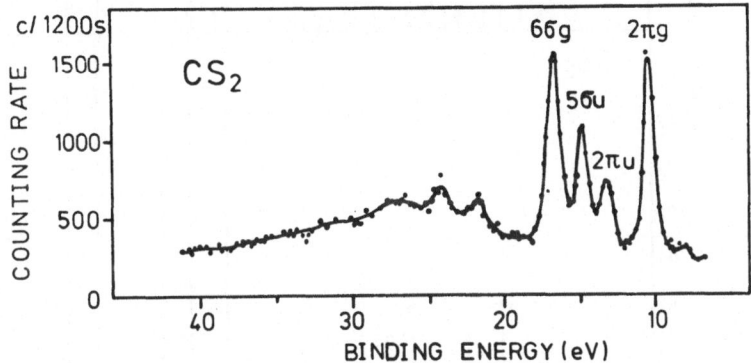

Fig.17: Mg K$_\alpha$ PES of CS_2 as recorded by Allan et al [30]

In the orbital model two bands ($4\sigma_u$ and $5\sigma_g$) should be found in this energy range. But this clearly cannot explain the observed structure. Vibrational broadening cannot account for it either. The explanation is given by the calculated spectrum (Fig.18). The $4\sigma_u$ and $5\sigma_g$ lines are smashed to pieces by many-body effects and numerous lines appear instead of the expected two lines. The two calculated spectra differ in some details which is to be expected, because of the complicated nature of these resonance states, but both explain the observed spectrum in a very satisfactory way.

This breakdown of the orbital model of ionization becomes more dramatic for larger molecules as mentioned above. The spectra about 20 eV begin to look like a continuum, but this breakdown also occurs at very low energies, i.e. for lines which are not of 2s character for first row atoms. The calculated spectrum for s-tetrazine has been given in Fig.7 for the energy range up to about 24 eV. Only the first six lines are interpretable in the orbital model, but here Koopmans' approximation fails badly.

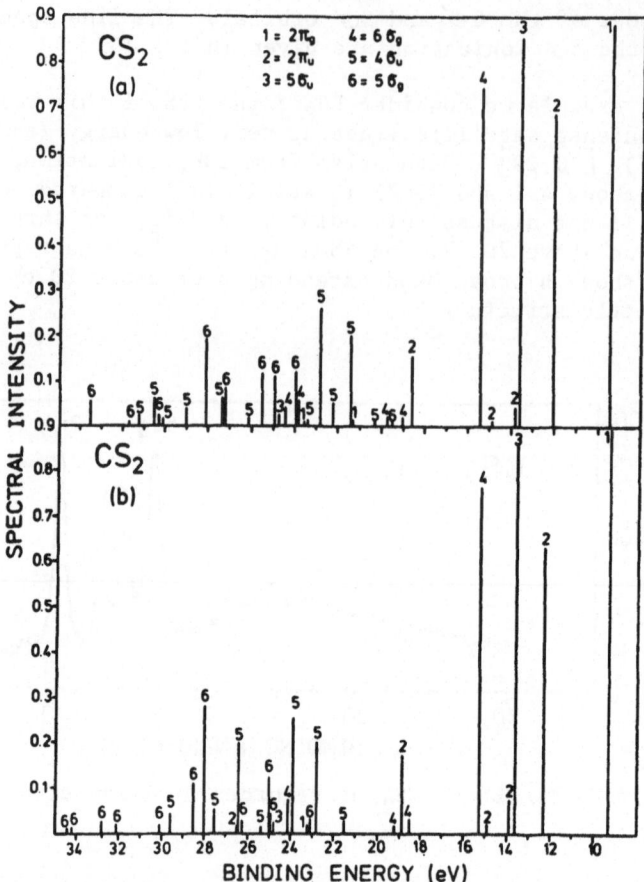

Fig.18: Calculated positions and relative intensities for the
        ionic states arising from valence ionization of CS_2
        a) basis set (12s 9p/9s 5p)
        b) basis set (12s 9p 2d/9s 5p 1d)

Above about 15 eV the orbital model breaks down and the lines
become split into two to four components. These orbitals are not
of 2s character. As a consequence the concept of orbital ordering
is inapplicable here. The spectrum in the energy range from 20 to
50 eV is given in Fig.19. The spectrum is considered only as a
qualitative one due to basis set limitations. A huge number of
lines is found instead of the six lines postulated by Koopmans'
hypothesis. The maximum pole strength is as small as 0.1 in some
cases. This corresponds in a CI language to the case that the
"dominant" configuration has only a coefficient of 0.3!

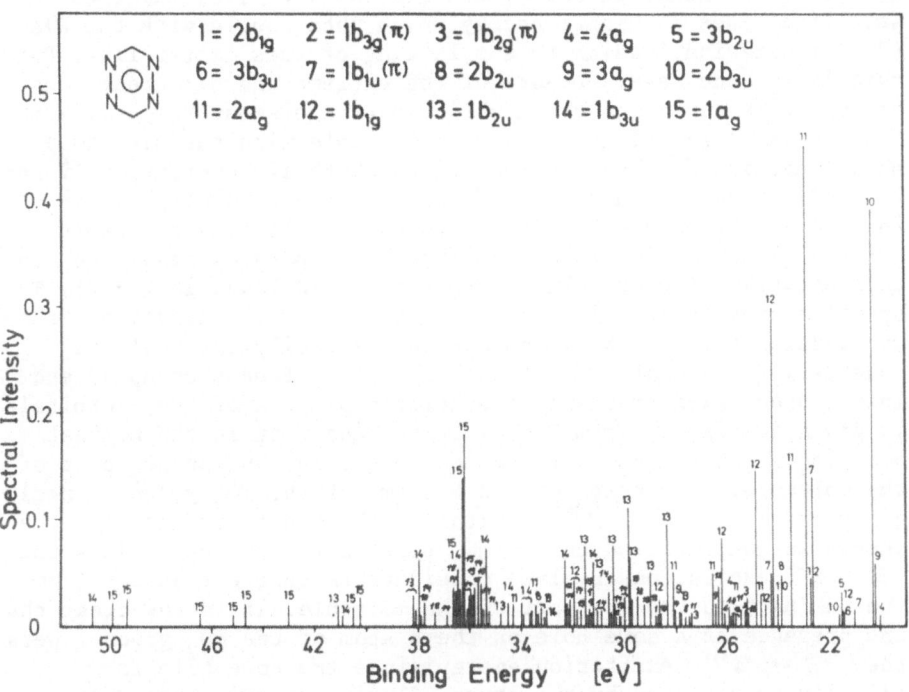

Fig.19: Ionization spectrum of s-tetrazine from 20 to about 50 eV

As a final example we would like to turn to ionization of core
electrons. In most cases one observes a main line with satellite
lines of much smaller intensity. But in some cases one encoun-
ters a different situation, e.g. for para-nitroaniline [31] .
This molecule has been treated in detail in Ref. 32. The N 1s
energy region of the Al Kα ESCA spectrum of this molecule is
shown in Fig.20. The N 1s lines resulting from nitrogen atoms
in the NH_2 and NO_2 groups, respectively, are clearly separated
because of the large chemical shift in this highly polar molecule.
The N1s (NO_2) line shows a double peak structure, the N1s (NH_2)
line, on the other hand, appears as a single line. One might now
assume that ionization out of the N1s (NH_2) orbital leads to
satellite lines at higher energy which then couple with the N1s
(NO_2) ionization leading to a splitting of this latter line. But
this is not the case. Because of the extreme and different localization characteristics the coupling matrix element V_{ijkl} is zero.
The N1s (NH_2) satellite lines do not couple with the N1s (NO_2)
main line. The $k^{-1} l^{-1} j$ configurations which lie energetically below and interact with the N1s (NO_2) single hole configuration
involve a hole in the N1s (NO_2) orbital itself plus a valence
$\pi \rightarrow \pi^*$ excitation. The shake-up energies in para-nitroaniline are
thus negative. The calculated spectra are included in Fig.21. The
specific properties which are responsible for the negative shake-
up energies for N1s (NO_2) ionization are easily rationalized
considering the charge distribution of the highest occupied and
lowest unoccupied orbitals. A schematic picture of these orbitals
is given in Fig.21. From Fig.21 it is seen that in the highest
occupied π orbital most of the charge resides on the NH_2 part of
the molecule, the charge on the N atom of the NO_2 group is negli-
gibly small. In the π^* orbital on the other hand most of the
charge is concentrated on the NO_2 part of the molecule. Thus the
$\pi \rightarrow \pi^*$ excitation involves considerable charge transfer from
the NH_2 part to the NO_2 part of the molecule. It is now clear that
the presence of a core hole on the N atom of the NO_2 group lowers
the $\pi \rightarrow \pi^*$ excitation energy since the core hole is
effectively screened by the charge flowing towards this N atom.
A core hole on the N atom of the NH_2 group, on the other hand,
must increase the $\pi \rightarrow \pi^*$ excitation energy since negative
charge is pulled away from the positive core hole in the
excitation. The energy gain through the screening of the N1s
(NO_2) core hole is large enough to make the $\pi \rightarrow \pi^*$ shake-up
energy negative.

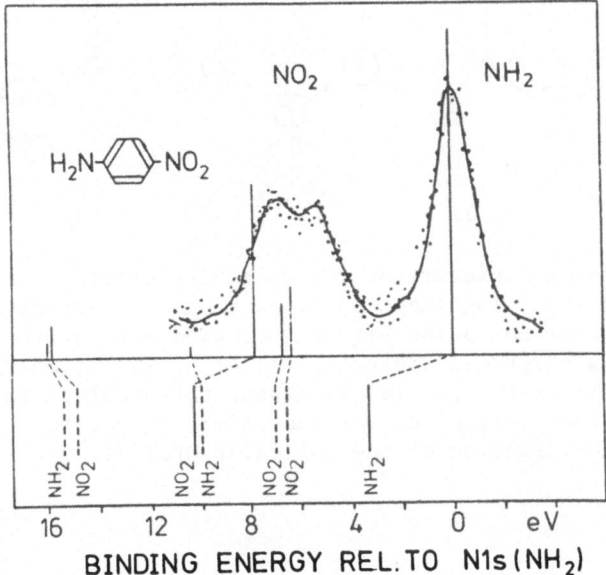

Fig.20: N1s energy region of the Al K_{α} ESCA spectrum of paranitro-
aniline (from Ref.31) together with the calculated ionization spectrum. The energies of the corresponding unper-
turbed states are shown in the lower panel.

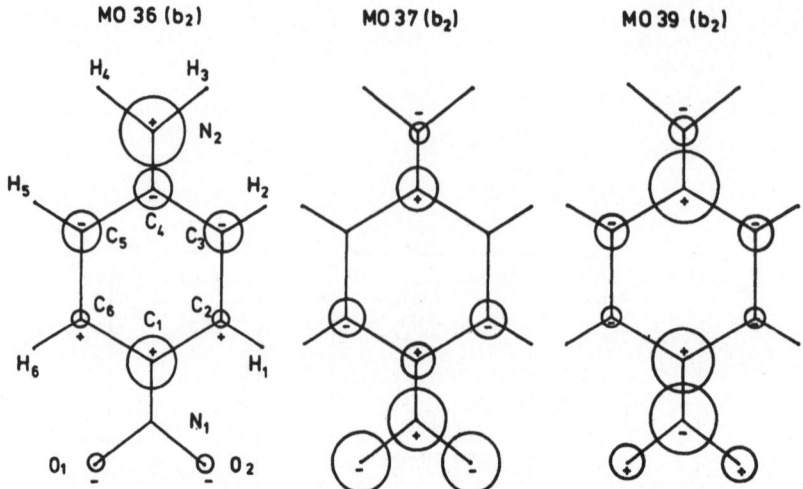

Fig.21: Schematic picture of the molecular orbitals involved in
the low-lying shake-up transitions of para-nitroaniline.
Orbital 36 is the highest occupied one, orbitals 37 and
39 are virtual orbitals, all of b_2 (π) symmetry. The
radii of the circles represent the absolute magnitudes
of the coefficients of the atomic p-type functions in the
LCAO expansion of the molecular orbital.

We have presented a number of applications of the Green's function
method to phenomena in the field of molecular photoelectron spectroscopy and have demonstrated that even some of the complicated
effects can be calculated with the help of this Green's function
method. With higher resolution and new types of measurements
being made in the field of photoelectron spectroscopy more information on molecular structure will become available. The Green's
function method is a powerful tool for the interpretation of
measurements as well as for predictions.

Acknowledgement

The authors would like to acknowledge the close cooperation with
Dr. G. Diercksen and Dr. W.P. Kraemer and the permission to use
the MUNICH SCF and transformation program of these authors and
the cooperation with Dr. H. Köppel.

References

1) T. Koopmans, Physica 1, 104 (1933)

2) L.S. Cederbaum, Theor.Chim.Acta 31, 239 (1973); J. Phys. B8, 290 (1975)

3) L.S. Cederbaum and W. Domcke, Adv. Chem. Phys. 36, 205 (1977)

4) J. Schirmer and L.S. Cederbaum, J. Phys. B11, 1889 (1978)

5) L.S. Cederbaum and W. Domcke, J. Chem. Phys. 60, 2878 (1974); 64, 603, 612 (1976)

6) W. von Niessen, L.S. Cederbaum and W. Domcke, "Excited States in Quantum Chemistry", Proceedings of the NATO Advanced Study Institute, Kos, Greece, 1978, ed. C.A. Nicolaides and D.R. Beck, D. Reidel, Publishing Company, Dordrecht 1978, p 183

7) W. Domcke, L.S. Cederbaum, H. Köppel and W. von Niessen, Mol. Phys. 34, 1759 (1977)

8) J.W. Rabalais, L. Karlsson, L.O. Werme, T. Bergmark and K. Siegbahn, J. Chem. Phys. 58, 3370 (1973)

9) W. von Niessen, L.S. Cederbaum and G.H.F. Diercksen, J. Chem. Phys. 67, 4124 (1977)

10) D.W. Turner, C. Baker, A.D. Baker and C.R. Brundle, "Molecular Photoelectron Spectroscopy", Wiley-Interscience, New York 1970

11) L.S. Cederbaum, W. Domcke and W. von Niessen, Chem. Phys. 10, 459 (1975)

12) G. Bieri, E. Heilbronner, V. Hornung, E. Kloster-Jensen, J.P. Maier, F. Thommen and W. von Niessen, Chem. Phys. 36, 1 (1979)

13) L.S. Cederbaum, Chem. Phys. Lett. 25, 562 (1974)

14) D.P. Chong, F.G. Herring and D. Mc Williams, J. Electron Spectry 7, 445 (1975)

15) W. von Niessen, W.P. Kraemer and G.H.F. Diercksen, Chem. Phys. 41, 113 (1979)

16) C. Fridh, L. Åsbrink, B.-ö. Jonsson and E. Lindholm, Inter. J. Mass Spectrom. Ion Phys. 9, 485 (1972)

17) R. Gleiter, E. Heilbronner and V. Hornung, Helv. Chim. Acta 55, 255 (1970)

18) W. von Niessen and G.H.F. Diercksen, J. Electron Spectry 16, 351 (1979)

19) P. Bischof, J.A. Hashmall, E. Heilbronner and V. Hornung, Helv. Chim. Acta 52, 1745 (1969)

20) G. Bieri, F. Burger, E. Heilbronner and J.P. Maier, Helv. Chim. Acta 60, 2213 (1977)

21) E. Heilbronner and H.-D. Martin, Helv. Chim. Acta, 55, 1490 (1972)

22) L.S. Cederbaum, W. Domcke, H. Köppel and W. von Niessen, Chem. Phys. 26, 169 (1977)

23) H. Köppel, W. Domcke, L.S. Cederbaum and W. von Niessen, J. Chem. Phys. 69, 4252 (1978)

24) H. Köppel, L.S. Cederbaum, W. Domcke and W. von Niessen, Chem. Phys. 37, 303 (1979)

25) D.C. Frost, S.T. Lee and C.A. McDowell, Chem. Phys. Lett. 23, 472 (1973)

26) C. Fridh and L. Åsbrink, J. Electron Spectry 7, 129 (1975)

27) J. Schirmer, W. Domcke, L.S. Cederbaum and W. von Niessen, J. Phys. B11, 1901 (1978)

28) N. Jonathan, A. Morris, M. Okuda, D. Smith and K.J. Ross, Chem. Phys. Lett. 13, 334 (1972)

29) J. Schirmer, W. Domcke, L.S. Cederbaum, W. von Niessen and L. Åsbrink, Chem. Phys. Lett. 61, 30 (1979)

30) C.J. Allan, U. Gelius, D.A. Allison, G. Johansson, H. Siegbahn and K. Siegbahn, J. Electron Spectry 1, 131 (1972)

31) S. Tsuchiya and M. Senō, Chem. Phys. Lett. 54, 132 (1978)

32) W. Domcke, L.S. Cederbaum, J. Schirmer and W. von Niessen, Phys. Rev. Letters 42, 1237 (1979); Chem. Phys. 39, 149 (1979)

NOVEL RADICAL IONS: GENERATION AND PROPERTIES

AN INTERIM REPORT ON PES AND ESR INVESTIGATIONS BY THE FRANKFURT GROUP*

H. Bock, G. Brähler, W. Kaim, M. Kira,
B. Roth, A. Semkow, U. Stein, A. Tabatabai

Institute of Inorganic Chemistry
Johann Wolfgang Goethe University
Niederurseler Hang, D-60 Frankfurt am Main 50
West Germany

The earliest tempestuous years of Electron Spin Resonance spectroscopy[1,2] - comprising development of the method, commercial production of high-resolution spectrometers complemented by the ENDOR technique and measurement of the hyperfine structures of thousands of radicals - should now be superseded by a time of reflection about practical application. Based on the experience gathered, it is now feasible that the preparative chemists participate by designing novel oxidizable compounds, inventing new redox systems, finding thermal or irradiative pathways for radical generation, studying their prospective properties and trying to isolate them. Thus in the future, new organic conductors may become available, although presently other areas of interest still dominate. To point out just one of the general auspices of ESR applications: on oxidation or on reduction, molecules M form new species M^{n+} or M^{n-}, and the loss or the acquisition of electrons is connected not only to the energy difference between the molecular ground state and the state of the ion generation but as well accompanied by a charge redistribution, i.e. structural changes. Therefore, in view of the current interest in detailed pathways of reactions, redox processes may attract some attention.

Along these lines, the Frankfurt Photoelectron Spectroscopy Group had started in 1976 to explore particularly the combination of PES and ESR techniques. Some of the results reported in publications VIII to XXXVII - listed in the special appendix - or still unpublished, shall serve as illustrations for information or application of interest to the chemist as outlined above.

---

* Part XXXVIII of the series on Radical Ions; for preceding publications cf. the special appendix. For the preceding 6th Essay on Molecular Properties and Models see Ref. 3.

STARTING POINT: THE RADICAL CATION OF A TETRASILYL-SUBSTITUTED ETHENE

In Spring 1976, we determined and compared the PE-spectroscopic data [XV] for planar bicyclic tetrasilyl-ethene derivatives (Fig. 1). The octamethyl derivative exhibited a low first ionization energy of only 7.98 eV. Therefore, a new oxidizing system, which had been "rediscovered" (cf.[4]) in a systematic search to replace nitromethane in the widely used $AlCl_3/H_3CNO_2$ oxidizing combination by an oxygen-free substitute was applied. It consisted of $AlCl_3$ in H_2CCl_2 — a solvent of low melting point, low viscosity even at low temperatures (important for ESR resolution), but of still high enough dielectricity constant to stabilize ions — , proved to be a most effective and selective oxygen-free oxidizing system, and produced the first organosilicon radical cation (Fig. 1)[VII]

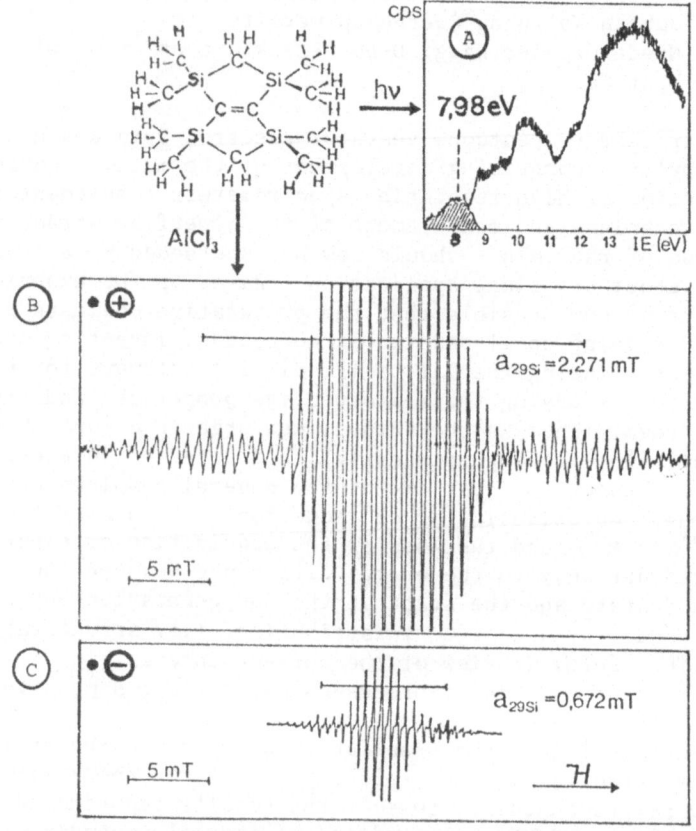

Figure 1. The 'bivalve'-shaped 2,2,4,4,6,6,8,8-octamethyl-2,4,6,8-tetrasila-[3,3,0]-oct-1(5)ene: (A) PE spectrum between 7 eV and 15 eV exhibiting IE_1=7.98 eV, (B) ESR spectrum of the radical cation generated with $AlCl_3$ in H_2CCl_2, and (C) ESR spectrum of the radical anion after reduction with K in DME.[VIII]

Reduction with K in dimethoxyethane yielded a radical anion (Fig.1: Ⓒ) persistent[5] even at room temperature[VIII,XV] which can be rationalized both by extensive charge delocalization as well as by the external methyl group shielding in the 'bivalve'-shaped radical anion.

Comparison of the $M^{\cdot \oplus}$ and $M^{\cdot \ominus}$ ESR spectra (Fig. 1: Ⓑ and Ⓒ) reveals that both show a hyperfine-structured center due to the coupling of the 24 methyl and the 4 methylene protons and two low-intensity satellites which can be traced back to ^{29}Si coupling (nuclear spin $I = 1/2$, natural abundance 4.7%). Their total widths, however, differ considerably.

In order to comparatively discuss the $M^{\cdot \oplus}$ and $M^{\cdot \ominus}$ ground states at all, one has to assume that structural changes in the bicyclic skeleton are negligible, that the ESR coupling constants of both species can be correlated in approximately the same way to the respective spin population, and that each of these will reflect the charge distribution. According to the higher effective nuclear charge of C relative to Si, it should be more difficult to delocalize the negative charge $>C\underset{=}{\overset{\ominus}{=}}C<$ onto the $Si(CH_3)_2$ substituents than to fill the positive hole $>C\underset{=}{\overset{\oplus}{=}}C<$ from the adjacent Si centers. Due to this $Si \rightarrow C$ electron donation, positive charge accumulates at silicon leading eventually to the large ^{29}Si coupling observed in the radical cation (Fig. 1: Ⓑ). The above interpretation is supported by the slightly smaller coupling constant a_{29Si} found for the analogous and isoelectronic hydrazine radical cation $>N\underset{=}{\overset{\oplus}{=}}N<$ [IX,XVI,XXXVI] (cf. Fig. 3), which would correspond to the larger effective nuclear charge N.

In general, the spin population in chemically related radical ions is of interest to the chemist both with respect to the individual species at hand as well as concerning the rationalization of the molecular properties in terms of charge distribution.[3]

DESIGN OF OXIDIZABLE ORGANOSILICON COMPOUNDS

Starting with the tetrasilylethene radical cation (Fig. 1), an effective general procedure has been developed to discover novel radical cations: the first ionization potential of the parent molecule M can be determined by photoelectron(PE) spectroscopy within ca. one hour (Fig. 1: Ⓐ), and provided IE_1 is below 8 eV, M is oxidized to $M^{\cdot \oplus}$ by simply adding $AlCl_3$ to its H_2CCl_2 solution in a tube within the ESR cavity. Thus most of the problems usually accompanying the search for new radical cations are reduced to the following two questions: will M be ionized below 8 eV, and will $M^{\cdot \oplus}$ persist?

As concerns the first ionization potential, some predictions are evident:[3] molecules M containing e.g. lone pairs n_E or bonds σ_{EE} of elements E exhibiting low effective nuclear charge, i.e. regions of high electron density, will be ionized more easily. Delocalization of the positive charge produced on electron expulsion will stabilize $M^{\cdot \oplus}$. For estimates, KOOPMANS' theorem $IE_1 = -\varepsilon_1^{SCF}$ proves to be helpful, although, among other shortcomings, it neglects electronic reorganisation.

Most valuable estimates of first ionisation potentials are based on the radical cation state comparison of chemically closely related molecules applying first and/or second order perturbation models.[3] For organosilicon substituted π systems, parametrization of 1st (cf. e.g.[XIX]) or 2nd order (cf. e.g.[XX]) perturbation models using known PES ionization energies turned out to be superior to the usual semi-empirical calculations both in practicability and in reliability of the results. For illustration, IE_1 of 1,4-bis(trimethylsilylmethyl)-benzene is approximated using known PES data from analogously substituted toluene derivatives (Fig. 2).

STEP I: DREAM UP A SYMMETRY-ADAPTED PERTURBATION MODEL

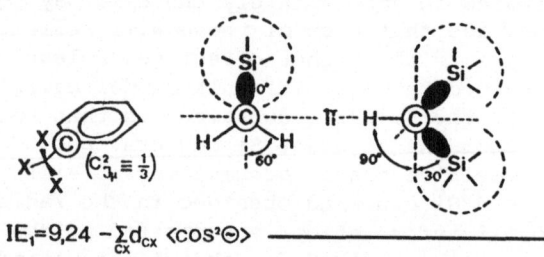

$$IE_1 = 9{,}24 - \sum_{cx} d_{cx} \langle \cos^2 \Theta \rangle$$

STEP II: PARAMETERIZE SUBSTITUENT EFFECTS USING KNOWN PES DATA

PES	X^1	X^2	X^3	IE (eV)	
$X^1\text{-}\underset{X^2}{\overset{X^2}{C}}\text{-}X^3$	H	H	H	8,84	$\left.\vphantom{\begin{matrix}a\\b\end{matrix}}\right\}$ $\overline{d_{CH}} = 0{,}27$ eV
	SiR_3	SiR_3	H	8,10	
	SiR_3	SiR_3	SiR_3	8,10	$\left.\vphantom{\begin{matrix}a\\b\end{matrix}}\right\}$ $d_{CSi} = 0{,}68$ eV

STEP III: FEED BACK PERTURBATION PARAMETERS INTO MODEL

H_2CSiR_3

?

R_3SiCH_2

$$IE_1^{calc.} = 9{,}24 - \left[2(0{,}68\cdot1) - 4(0{,}27\cdot0{,}25)\right] = 7{,}61 \text{ eV}$$

$$IE_1 = 7{,}75 \text{ eV}$$

Figure 2. The benzene π perturbation by bulky $[(H_3C)_3Si]_n H_{3-n}C$-substituents is angle-dependent (STEP I).[XX] Assuming sterically preferred conformations, bond interaction parameters are straightforwardly extracted from PES ionizations of analogously substituted toluene derivatives (STEP II). For 1,4-bis(trimethylsilylmethyl)benzene IE_1 = 7.61 eV is predicted, i.e. a value below 8 eV. Therefore, the compound has been synthesized, the experimental IE_1 determined and - according to the general procedure - the radical cation generated.

The radical cation state comparison of trimethylsilylmethyl-substituted benzenes (Fig. 2) displays some features of general interest. Thus both the α,α-bis- and the α,α,α-tris(trimethylsilyl)toluenes exhibit identical first ionization energies (Fig.2: Table in STEP II), which in turn support the assumption of sterically preferred conformations as well as the hyperconjugation model[6] employed, and also give some clues on how the spin might be distributed in the radical cations generated. It also should be pointed out explicitely that 1,4-substitution of benzene by $(H_3C)_3SiH_2C-$ groups lowers its first ionization potential by 1.5 eV ! Therefore, the powerful donor effect of ß trimethylsilyl substituents on π systems[7] as well as on lone pairs[8] constitutes one of the useful principles to design organosilicon compounds with low first ionization potentials (Fig. 3).

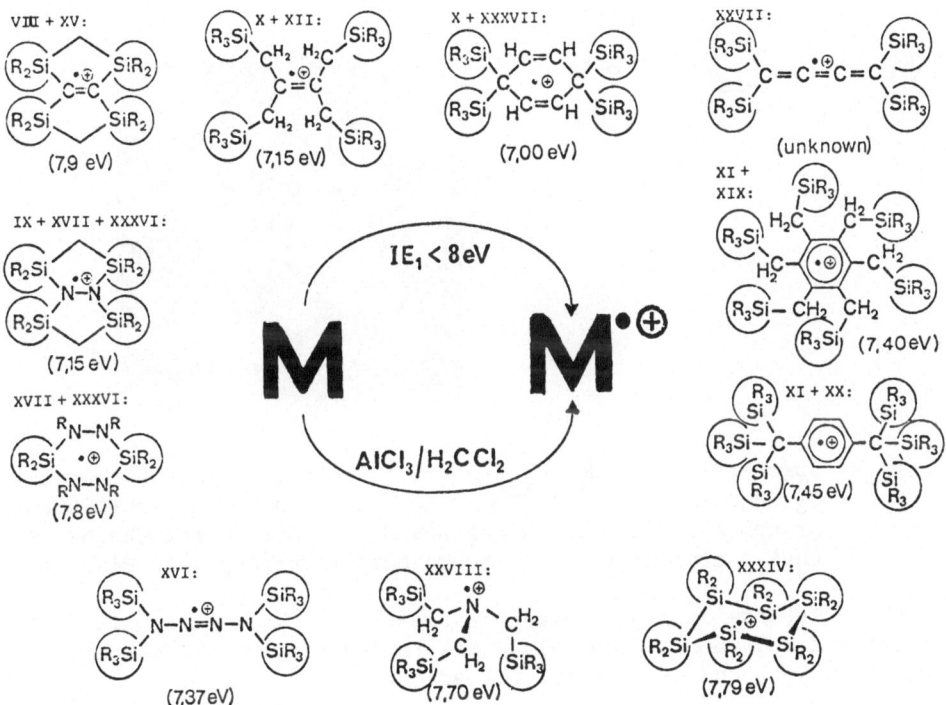

Figure 3. Examples of novel organosilicon radical cations, which have been generated by the selective oxidizing system $AlCl_3/H_2CCl_2$[4] and characterized by their ESR spectra. From a comparison of the numerous data collected, the oxidation potential of $AlCl_3/H_2CCl_2$ corresponds to approximately a first ionization potential of 8 eV. The PE spectroscopically determined vertical IE_1 are given in brackets; the parent compound of the tetrasilylbutatriene radical cation $(R_3Si)_2C=C=C=C(SiR_3)_2^{\cdot\oplus}$, which forms on oxidation of acetylene derivatives[XXVII], is still unknown. Roman numerals refer to the publications listed in the special appendix.

The novel organosilicon radical cations generated (Fig. 3) range from ethene derivatives[VII,X,XII,XV,XXXVII] through amines[XXVIII] to σ systems like the $Si_6(CH_3)_{12}^{\cdot\oplus}$ ring[XXXIV], which no longer contains π bonds or lone pairs. The selectivity of the oxidizing system $AlCl_3$ in H_2CCl_2 is nicely illustrated by the generation of tetrazene radical cations[XVI] without decomposition under $-N_2$ elimination.

The second question posed at the outset of this chapter – will the radical cations generated persist? – can be commented on partly as follows: the bulky methylsilyl substituents in α or β position to π systems or lone pairs stabilize the species $M^{\cdot\oplus}$ by charge delocalization and also enhance their persistance[5] by extensive shielding. Due to their 'bivalve'- or 'ball'-like shape with hydrogens on the surface, the radical cations (Fig. 4) are at least kinetically stable enough to be characterized by their ESR spectra.

$(H_3C)_3SiH_2C$ and $CH_2Si(CH_3)_3$
$(H_3C)_3SiH_2C$ $CH_2Si(CH_3)_3$
$(H_3C)_3SiH_2C$ $CH_2Si(CH_3)_3$

Figure 4. Space-filling model of hexakis(trimethylsilylmethyl)benzene $C_6(CH_2Si(CH_3)_3)_6$ [XI,XIX] with each 3 of the bulky substituent groups above and below the six-membered ring and the 54 methyl hydrogens coating the molecular surface (cf. Fig. 3).

For the existence of a persistent[5] radical ion, the reversibility of the 1st and 2nd redox potentials as determined from cyclic voltammetry has been suggested[9] as a useful criterion.

ESR SPECTRA OF ORGANOSILICON AND OF ORGANOSULFUR RADICAL CATIONS

The radical cation precursors (cf. Fig.3) have been synthesized mostly via organometallic routes,i.e. employing Wurtz-Fittig reactions (cf. X,XIX,XX), in situ Grignard technique (cf. X,XX,XXXVII) or reductive silylation using potassium (cf. X,XXI,XXXVII), possibly in "one pot" as in the preparation of the benzene derivative shown in Fig.4:

$$BrH_2C \; \; CH_2Br \quad \xrightarrow[-6\,NaCl\,-6\,NaBr]{12\,Na\,+\,6\,R_3SiCl} \quad R_3SiH_2C \; \; CH_2SiR_3 \qquad (1)$$

BrH_2C—⬡—CH_2Br R_3SiH_2C—⬡—CH_2SiR_3

$BrH_2C \; \; CH_2Br \qquad \qquad R_3SiH_2C \; \; CH_2SiR_3$

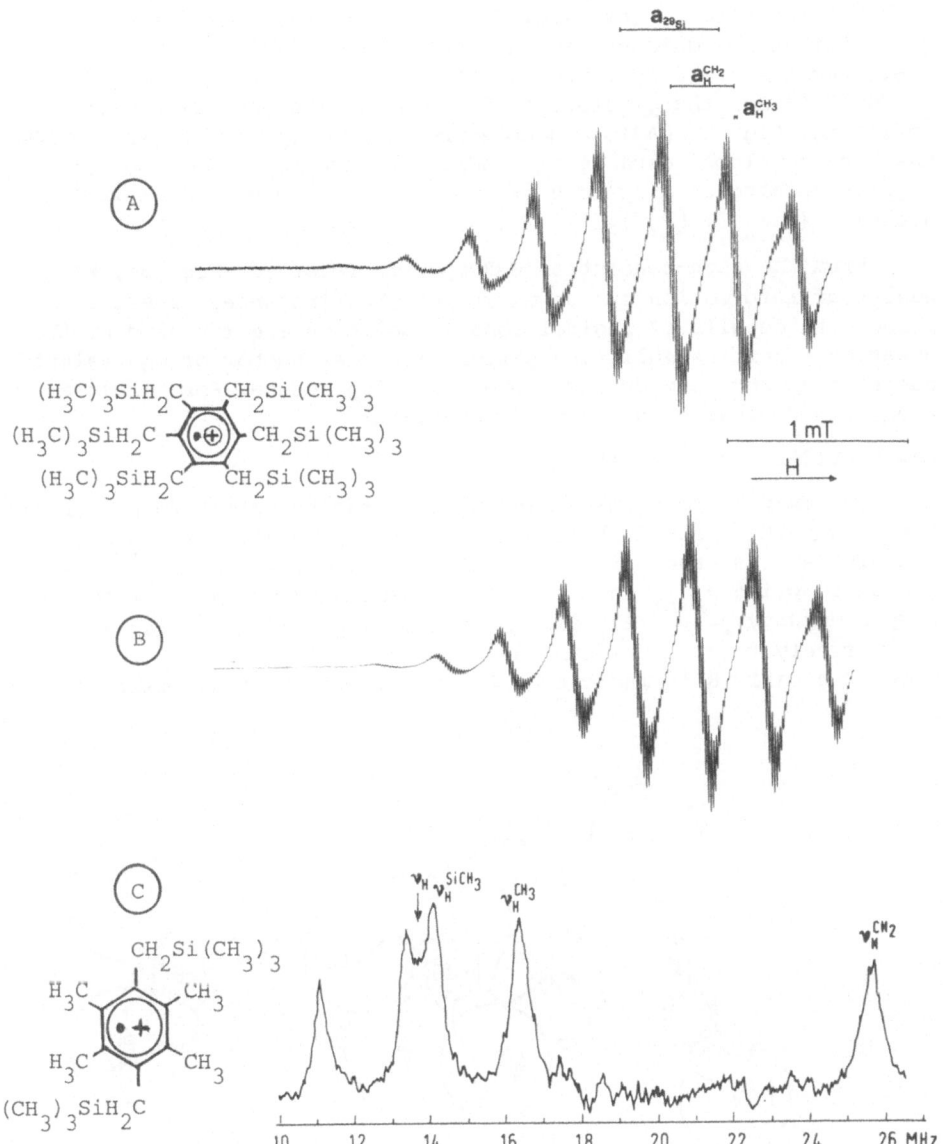

Figure 5. ESR spectra and their assignment: (A) The high-resolution
spectrum of $C_6(CH_2Si(CH_3)_3)_6^{\oplus}$ (cf. Fig. 4) displaying a
total of 2145 signals including the 55tet of the 54 methyl
protons with an intensity ratio between outer and central
line of 1:1946939425648112, (B) its simulation using the
ESPLOT program[XIX]. (C) The ENDOR spectrum of the corresponding tetramethyl-substituted radical cation: in spite of
the cis/trans isomerism[XIX], only 3 different hydrogen couplings are recorded.

All ESR spectra (cf. e.g. Figs. 1 or 5) have been recorded using Varian E9 equipment and are calibrated with Fremy's salt.[10] Their temperature dependance has been studied from 18o K up to 300 K.[11] For the simulation of the sometimes unusual hyperfine splitting (Fig. 5: (A)), a double-precision computer program ESPLOT has been developed capable of handling a total of 10000 theoretical signals arising from up to 8 groups of a maximum of 100 equivalent nuclei[11] (Fig. 5: (B)).

From the high-resolution ESR spectra recorded (Fig. 5), a wealth of information can be extracted. For instance, numerous structural details of radical ions in solution are revealed by the observed hyperfine splitting pattern i.e. the number of equivalent nuclei coupling, and by its temperature dependence. For illustration, the following topics are emphasized:

⟹ RADICAL ION FLEXIBILITY:

The example presented (Fig. 6) concerns the rotation of substituents π-C-CH$_2$X around their bond axis to a π radical ion center. The argument is based: i. on the HELLER/McCONNELL equation[12] for ß proton coupling $a_{H(\beta)}$ and its angle-dependent correlation with the π spin density ϱ_π (2), ii. on the value $\langle\cos^2\Theta\rangle = 0.5$ for a methyl group freely rotating within the ESR time scale, and iii. the defition of a ratio R [13] indicating the degree of sterical fixation (3).

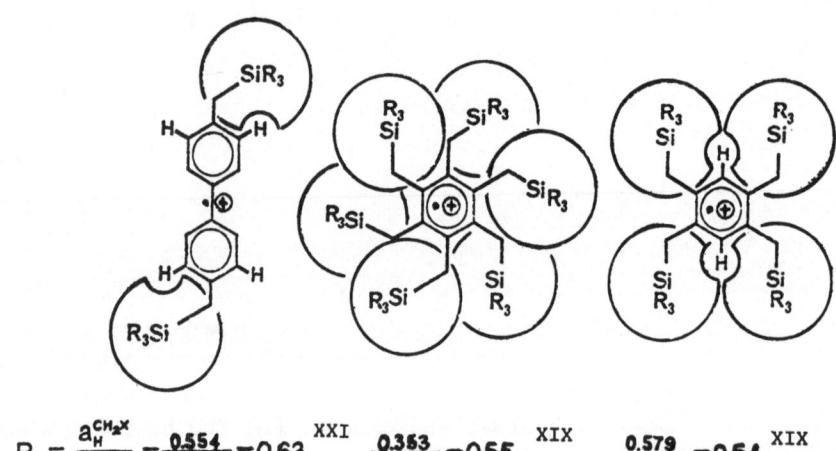

$$R = \frac{a_H^{CH_2X}}{a_H^{CH_3}} = \frac{0.554}{0.879} = 0.63 \quad ^{XXI} \quad \frac{0.353}{0.654} = 0.55 \quad ^{XIX} \quad \frac{0.579}{1.070} = 0.54 \quad ^{XIX}$$

Figure 6. R-values (see text) for benzene derivatives containing bulky (H$_3$C)$_3$SiH$_2$C-groups, indicating their steric interference with the ortho-hydrogens even in 4,4´-disubstituted biphenyl or their "blocked" optimum conformation in alternating "up and down" positions around the ring in sterically overcrowded radical ions (cf. Fig. 5) [XI,XIX,XXI].

$$a_{H(\beta)} = (B_o + B_2 \cdot \cos^2 \Theta) \cdot \varrho_\pi \qquad (2)$$

$$R = \frac{a_H^{CH_2X}}{a_H^{CH_3}} = \frac{\langle \cos^2 \Theta \rangle}{0.5} = \begin{cases} 1 : \text{"rotating"} \\ 0.5 : \text{"blocked"} \end{cases} \qquad (3)$$

Looking at the low R values for benzene derivatives with bulky $(H_3C)_3SiH_2C-$ substitutents (Fig. 6), their steric interference with ortho hydrogens or - more pronounced - with each other becomes evident. The resulting fixed conformations with the electron-rich SiC bonds in the π plane allow an optimum delocalization of the positive charge and thus considerably stabilize the radical cation generated.

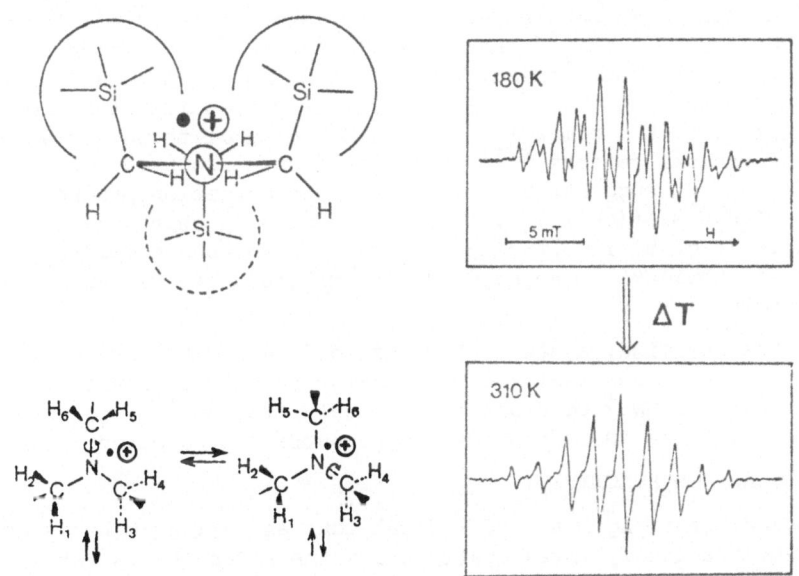

Figure 7. The 180 K ESR spectrum of tris(trimethylsilylmethyl)aminium radical cation exhibits 25 lines due to overlapping signals caused by a coupling of 1 N and 3 sets of each 2 H. The relatively small nitrogen coupling constant suggests a planar NC_3-skeleton and the low value R = 0.82 (3) an almost negligible steric interference of the 3 $(H_3C)_3SiH_2C$-groups. On warming of the H_2CCl_2 solution, selective line widths broadening starts due to interconversion between 6 individual conformations of which 2 are shown. The 310 K ESR spectrum already displays only 9 lines, although complete averaging within the ESR time scale is achieved only above room temperature[XXVIII].

➡ RADICAL ION FLUCTUATION

Pursuing the possible temperature dependence of ESR hyperfine splitting patterns may provide information on internal motion processes in radical ions. If e.g. the radical cation $\cdot^{\oplus}N(CH_2Si(CH_3)_3)_3$ (fig. 7), generated at 180 K by $AlCl_3$ in H_2CCl_2 solution, is warmed up to room temperature, a characteristic "flip-flop" starts between the 6 formal conformers containing each 2 or 1 $(H_3C)_3Si$-groups above and below the planar NC_3-skeleton. Ultimately, this fluctuation leads to proton equivalence within the ESR time scale [XXVIII].

➡ STRUCTURAL CHANGES ACCOMPANYING REDOX REACTIONS

As pointed out already in the introductory remarks, loss or acquisition of an electron by a molecule M is connected not only to an energy difference but also to potential structural changes in $M^{n\oplus}$ or $M^{n\ominus}$. An excellent example is provided by the one-electron oxidation of the twisted tetrasubstituted hydrazine derivatives, which yields radical cations $R_2N\overset{\cdot\oplus}{-}NR_2$ stabilized by planarisation[14,15,XVII,XXXVI]. For unsubstituted $H_2NNH_2^{\cdot\oplus}$ this is reflected by an INDO open shell hypersurface, the total energy minimum of which is calculated for a completely planar structure with a considerably shortened NN bond distance (cf. Fig. 8). Comparison of the ESR data of tetrasilylated hydrazine radical cations[XVII,XXXVI] based on calculated coupling constants a_X^{INDO}, suggests for the open-chain oxidation product , $((H_3C)_3Si)_2\overset{\cdot\oplus}{N}-N(Si(CH_3)_3)_2$, a structure with an only reduced dihedral angle (Fig. 8); i.e., the energetically favorable complete planarization is obviously prevented by the bulkiness of the $(H_3C)_3Si$-substituents.

In the meantime, a sterically shielded tetraalkyl hydrazine radical cation has been isolated. Its X-ray structure proves perfect planarity and an $NN\cdot^{\oplus}$ bond distance of only 126.9 pm [15] - in excellent agreement with the INDO open shell hypersurface prediction[XVII].

➡ RADICAL ION REACTIVITY

Thermal rearrangements of radical ions as well as other reactions can be advantageously investigated using the ESR probe. A rather spectacular case is the one of tetrakis(trimethylsilyl)butatriene [XXVII], of which both the radical cation as well as the radical anion are known so far but not the neutral parent molecule:

$$\tag{4}$$

The butatriene radical ions have been characterized unambigously by high-resolution ESR spectra exhibiting numerous ^{13}C and ^{29}Si satellites. Speculation about possible reaction intermediates included R_3Si-group transfer as well as the generation of the diacetylene dianion,

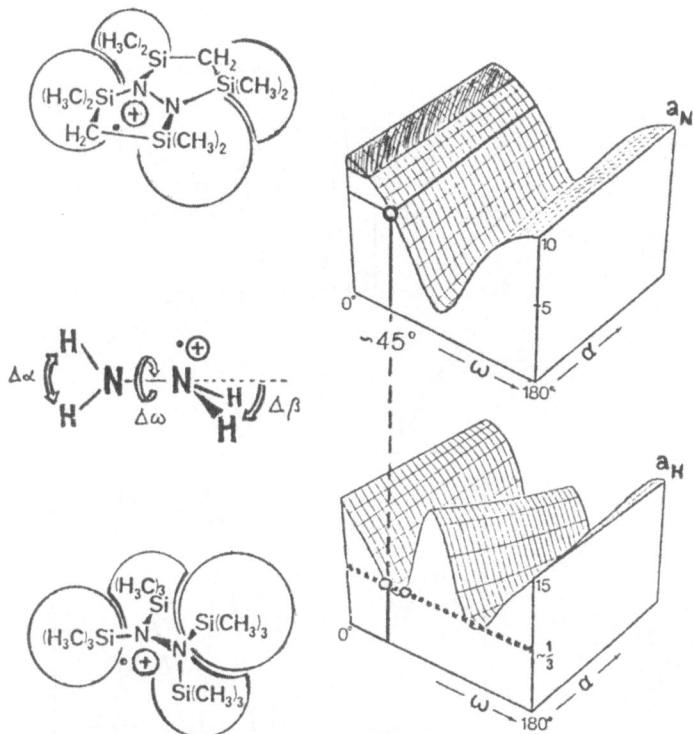

<u>Figure 8.</u> The hydrazine radical cation $H_2N\overset{\cdot\oplus}{—}NH_2$ is predicted by an
INDO open shell total energy hypersurface [XVII,XXXVI] to
be planar with the NN bond shortened to 128 pm (!). Addition-
ally calculated coupling constants depend predominantly
on the dihedral angle: twisting to $\omega = 90°$ leads to minimum
nitrogen and maximum hydrogen coupling constants. Inser-
tion of the appropriate ESR data into the INDO hypersurfaces for a_N or a_H and assuming near planarity fo the bicyclic radical cation, a dihedral angle $\omega \sim 45°$ is deduced
for the open chain species; i.e., apparently the steric interference between the bulky $(H_3C)_3Si$-groups prevents the
energetically favored full planarization of the $Si_2NNSi_2^{\cdot\oplus}$
skeleton [XVII,XXXVI].

$K^{\oplus\ominus}C\equiv C—C\equiv C^{\ominus\oplus}K$. Subsequent reductive silylation of bis(trimethylsilyl)diacetylene on a preparative scale expectedly allowed to iso-
late both starting material as well as hexakis(trimethylsilyl)-2-butyne:

$$2 \, R_3SiC=C-C=CSiR_3 + 4 \, K \longrightarrow \left[K^\oplus (R_3Si)_2C\overset{\ominus}{=}C=C\overset{\ominus}{=}C(SiR_3)_2\overset{\oplus}{K} + K^\oplus C\overset{\ominus}{\equiv}C-C\overset{\ominus}{\equiv}C^\oplus K\right]$$

$$\downarrow + 4 \, R_3SiCl \qquad\qquad \downarrow \qquad (5)$$

$$(R_3Si)_3C-C=C-C(SiR_3)_3 + R_3SiC\equiv C-C\equiv CSiR_3$$

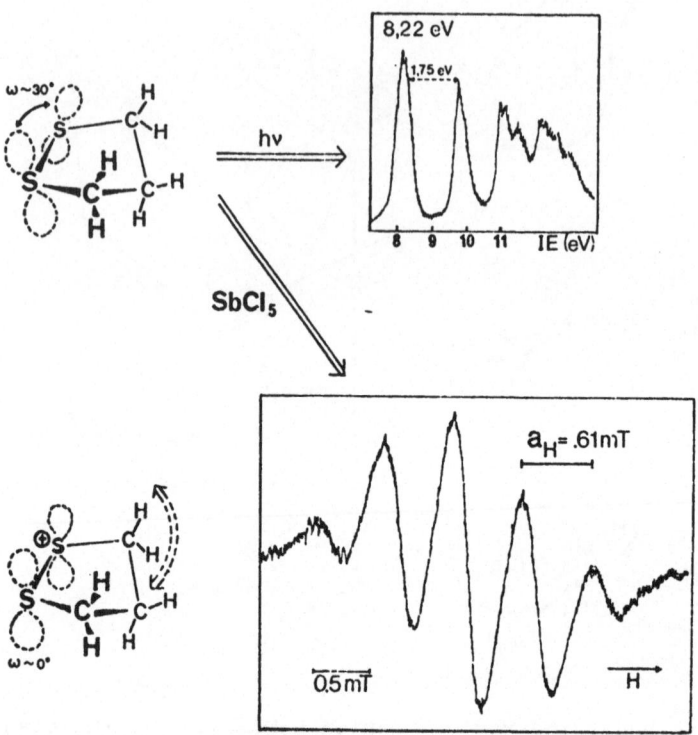

<u>Figure 9.</u> The unstable five-membered ring disulfide 1.2-dithiolane,
by analogy to known structures of derivatives, should also
exhibit a rather small dihedral angle of only ~30°. Accordingly, the split between the two sulfur lone pair ionization energies amounts to 1.75 eV (!), and its first ionization occurs already at 8.22 eV. Subsequent oxidation using
the more powerful SbCl$_5$ (cf. Figure 3) generates the first
radical cation of a saturated disulfide. Its temperature-
dependent ESR spectrum displays at 180 K a hyperfine splitting of 9 lines due to 2 pairs of equivalent hydrogens [16],
which changes into the quintet, shown above, at room temperature. For the obviously activated flickering of the external
methylene group within the ESR time scale, a barrier of
~8 kJ/mole is estimated from line-shape analysis. The unusually high g value of 2.0183 suggests, in addition, that
most of the positive charge must have been accomodated by
the electron-rich SS bond which – as predicted by INDO
open shell calculations – should turn planar on electron removal and also decrease slightly in length [16].

So far, the novel organosilicon radical cations have served exclusively to illustrate the chemists' approach using ESR spectroscopy as a useful tool to generate new species in new ways, to investigate their stability as well as other molecular properties, and to gather information, e.g., on structural changes during oxidation. However, the valuable correspondence between the first ionization potential, so easily determined by photoelectron spectroscopy and the oxidizability of a compound in solution, can be extended to other classes of compounds as well as to other oxidizing systems.

For instance, the coenzyme α-lipoic acid contains as its biochemically active group a saturated five-membered ring disulfide. Due to the unusually small dihedral angle of these heterocycles enforced by their ring size, the SS bond becomes especially electron-rich[16]: based on the low first ionization potential of e.g. 8.22 eV for 1,2-dithiolane (Figure 9), the first radical cation of a satured disulfide could be generated subsequently, employing $SbCl_5$ as a more powerful oxidizing reagent. Again, the ESR spectrum not only allowed to characterize the new species, but via its temperature dependence, an internal fluctuation could be detected as well as the activation barrier determined. The high g value measured for the disulfide radical cation together with additional open shell SCF calculations suggest localization of the positive charge in the (C)S$\overset{\oplus}{\equiv}$S(C) moiety accompanied by planarization and bond shortening[16].

RADICAL ANIONS, TRIANIONS AND "ISOMOLECULAR" RADICAL IONS

In general, if energy is supplied to or withdrawn from a molecule, it will turn into another state. If, in addition, electrons are acquired or released during this transition, then molecular ions, which may also be radicals, are formed as new particles which can themselves exist in numerous states of various energies[3] (Figure 10). The energy differences recorded by a variety of measuring techniques show that the total energy of the individual molecular states increases on going from those of a stable radical anion $M^{\cdot\ominus}$ via those of the neutral molecule M to those of the radical cation $M^{\cdot\oplus}$.

Radical cation states of molecules are especially suitable for an introductory discourse[3]: they are characterized by a typical electron hole, they can be observed by straightforward techniques like photoelectron or charge transfer UV spectroscopy[XIX], and the vertical ionizations can be interpreted on the basis of simple models.

Radical anions, on the other hand, are more easily and in a careful way generated, e.g., by the reduction on alkaline metal films in ether solution at low temperatures and, therefore, up to now their number by far exceeds the one of known oxidized species[1]. Radical anions can be considerably stabilized by ion pair-formation with the counter metal cation as well as by solvation, and their negative (!) energy difference relative to M is usually measured as the reduction potential $E_{1/2}^{Red}$ by electroanalytical techniques or as electron affinity using the recently developed Electron Transmission Spectroscopy[17].

Figure 10. A manifold of molecular states is accessible from the ground state $\Gamma(M)$ of a neutral molecule: e.g., on absorption of UV radiation, M is transferred into one of the numerous electronically excited states $\chi_i^j(M)$. Larger amounts of energy may lead to loss of an electron, thereby producing a radical cation in its ground state $\Gamma(M^{\cdot\oplus})$ or one of its electronically excited states $\chi_i^j(M^{\cdot\oplus})$. If on acceptance of an electron a stable radical anion is formed, i.e., given a positive electron affinity, then the total energy becomes more negative. In general for radical ions, generation in the gasphase and in solution has to be distinguished due to the potential stabilization by ion pair-formation with the respective counterion as well as by solvation. Information on the spin density - and with some deliberation on an approximate charge distribution - in $M^{\cdot\oplus}$ or $M^{\cdot\ominus}$ can be gained from high-resolution ESR spectra.

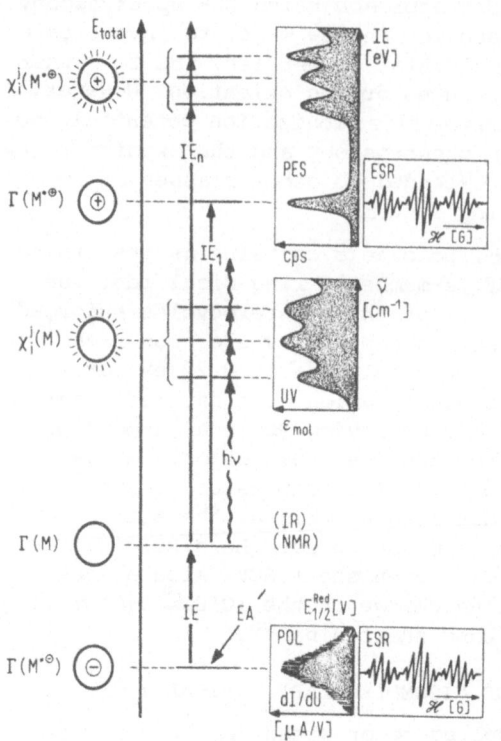

Radical anions and their ESR spectra are of interest to the chemist because their spin density as determined by the coupling constants allows to approximate - assuming some parallelism - the charge distribution, at least by correlation with the results of MO calculations. In this respect, most interesting are "isomolecular ions," i.e. radical cation $M^{\cdot\oplus}$ and radical anion $M^{\cdot\ominus}$ generated from the same molecule M. Therefore, in addition to the introductory organosilicon case (Figure 1), some other selected examples of the many radical anions investigated by the Frankfurt group - cf. lit.cit. VIII, XIII, XIV, XV, XXII, XXIII, XXIV, XXV, XXVI, XXIX, XXX, XXXII, XXXIII as well as unpublished results - are intended to enlarge the scope of this report.

An appropriate starting point is the chemically rather trivial improvement of radical anion generation by adding crown ethers to the solution, which via complexation of the alkali metal counterion also adds to the stabilization of the reduced system containing $M^{\cdot\ominus}$. For illustration, the puzzling two-step reduction of 1,4-bis(dimethylphosphino)benzene may serve (Figure 11): <u>with</u> dicyclohexyl-18-crown-6 a solution of the mono radical anion $M^{\cdot\ominus}$, stable even at room temperature, results; while <u>without</u> any complexing reagent and on pro-

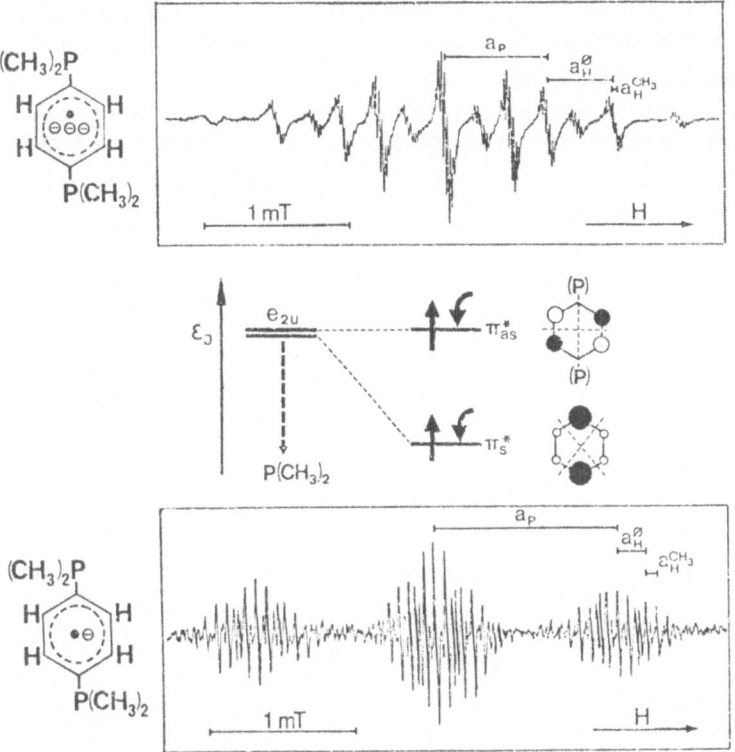

Figure 11. R_2P-Substituents are relatively poor electron donors but
reasonable electron acceptors: electrochemical reduction
in H_3CCN solution containing $R_4N^{\oplus}BF^{\ominus}$ as conducting electro-
lyte or chemical reduction using sodium in the presence
of dicyclohexyl-18-crown-6 as counterion complexing reagent yields a 195 line ESR spectrum due to a species $M^{\cdot\ominus}$
containing the expexcted coupling nuclei (2 ^{31}P + 12 H_{CH_3}
+ 4 H_{CH}). On prolonged contact with sodium film in THF
solution without any chelating additive, the trianion
forms. Its ESR spectrum exhibits - as anticipated from
benzene-$\pi(e_{2u})$-orbital perturbation arguments - a tremendous decrease of the spin population in the region of the
nodal plane, i.e., smaller ^{31}P coupling and an increase at
the C(H) ring centers, i.e., enlarged $a_H^{\emptyset}$ constant. For rationalization, one might tentatively anticipate that the P
lone pairs interact with the counterions and thereby di-
stribute the charge over the whole molecular skeleton (cf.
XXII and XXIII).

longed contact with the sodium film, the radical trianion $M^{\cdot \ominus \ominus \ominus}$ is detected and unambigously characterized by ESR [XXII,XXIII] (Figure 11). The multistep-reduction observed ESR-spectroscopically displays as another intriguing facet, that a benzene π perturbation MO model allows one to describe the spin populations of both a radical anion and trianion.

Bis(methylthio)naphthalenes belong to the few molecules which can both be oxidized to their radical cations $M^{\cdot \oplus}$ as well as reduced to their radical anions $M^{\cdot \ominus}$: the symmetrically disubstituted derivatives

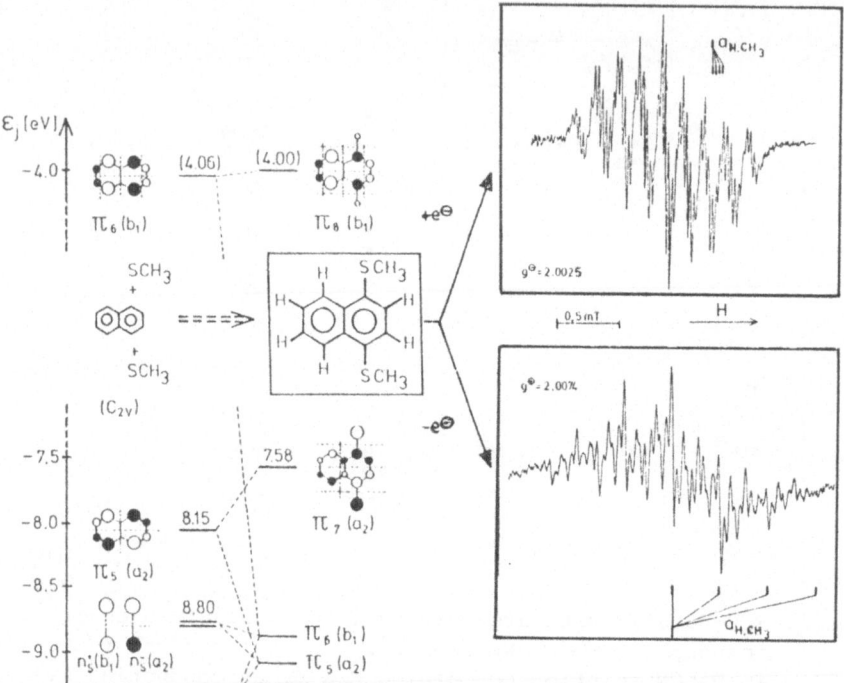

Figure 12. Effects of methylthio-substituents $-SCH_3$ on aromatic π systems can be satifactorily predicted by HMO calculations inserting PE-spectroscopically determined parameters α and β. Thus 1,4-disubstitution of naphthalene is predicted to raise the highest occupied orbital $\pi_7(a_2)$ by sulfur lone pair $n_S^-(a_2)$ admixture, while the more distant lowest unoccupied orbital $\pi_8(b_1)$ is hardly perturbed at all. Accordingly, the first ionization energy is lowered to 7.58 eV [XXXIII], and - in contrast to unsubstituted naphthalene - the radical cation can be easily generated as a species stable at room temperature. With the radical anion also being persistent, comparison of the ESR data of the isomolecular radical ions yields the interesting result: the sulfur atoms accommodate a significant portion of the spin population only in the radical cation [XXXIII] - as already suggested by the orbital diagrams.

investigated ^{XIV,XXXIII} (Figure 12) indicate in their photoelectron spectra that the first ionization potentials are lowered by at least 0.5 eV relative to the π-ionization of the parent hydrocarbon $C_{10}H_8$. On the contrary, only a negligible perturbation of the lowest un-occupied naphthalene-π-orbital by the H_3CS-substituents is calculated. Correspondingly, the spin distributions inferred from the ESR spectra (Figure 12) – exhibiting methyl proton coupling $a_H^{\cdot\oplus} \gg a_H^{\cdot\ominus}$ and values $g^{\cdot\oplus} \gg g^{\cdot\ominus}$ – demonstrate that the radical anions closely resemble the parent species $C_{10}H_8^{\cdot\ominus}$, whereas in the radical cation ground state of all isomers $(H_3CS)_2C_{10}H_6^{\cdot\oplus}$ the sulfur centers participate conside-rably ^{XIV,XXXIII}.

To rationalize the ESR data of "isomolecular" radical ions $M^{\cdot\oplus}$ and $M^{\cdot\ominus}$, quite often MO π perturbation arguments prove to be of help. To quote another example, for the 10 π electron system thieno[3,4-c]-thiophene the following "inner", i.e. highest occupied and lowest un-occupied, orbitals are calculated ^{XXIV}:

$$\tag{6}$$

$$\pi_5(a_u) \qquad \pi_6(b_{3g})$$

$$g^{\cdot\oplus} = 2.0020 \qquad g^{\cdot\ominus} = 2.0056 \qquad XXIV \tag{7}$$

Assuming as usual [1] that higher spin/orbit contribution of the hea-vier elements leads to an increase of g values, the ESR-spectroscopi-cally determined relation, $g^{\cdot\oplus} \ll g^{\cdot\ominus}$, fully agrees with the MO pre-diction: due to the nodal plane through the S centers in $\pi_5(a_u)$, no spin density at all is expected at sulfur in the radical cation, and correspondingly, its g value should not deviate much from the one for the free electron, $g = 2.0023$ ^{XXIV}.

"Isomolecular" radical ions $M^{\cdot\oplus}/M^{\cdot\ominus}$ have been repeatedly, yet never systematically, investigated; prototypes stretch from larger π compounds capable of adequate – and in alternate π systems even identical ^{3,18} – delocalization of both $\oplus$ and $\ominus$ charge to molecules which contain individual donor and acceptor regions, e.g.

$$\tag{8}$$

(cf.Fig.1) ^{XVIII,XV}

Internal donor/acceptor-type compounds should attract more attention because they often will form persistent radical ions which some-
times even might be isolated as salts $M^{\cdot\oplus}A^{\ominus}$ or $M^{\cdot\ominus}C^{\oplus}$ and might exhibit conducting properties. According to the obvious underlying princip·
les (8), their design is straightforward and limited rather by the fea·
sibility of their chemical synthesis.

INSTANCES OF SERENDIPITY: SPIN-LOCALIZED RADICAL IONS ?

In general, the spin of the unpaired electron in radical species
$M^{\cdot\oplus}$, $M^{\cdot}$ or $M^{\cdot\ominus}$ is distributed over the whole molecular skeleton. Therefore, every nucleus with non-zero spin momentum I and within a re-
gion of finite spin density will give rise to $(2I + 1)$ multiplicating

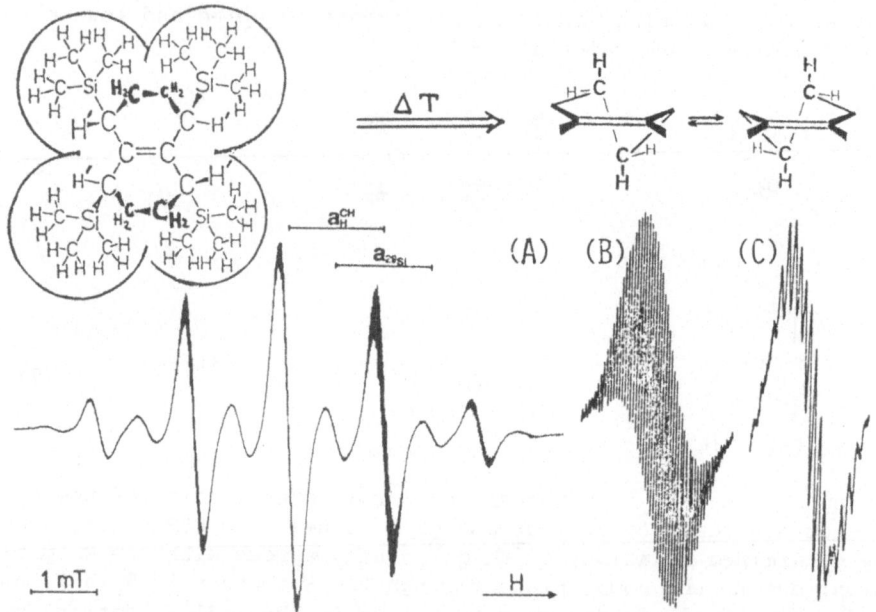

Figure 13. The radical cation of 1,4,5,8-tetrakis(trimethylsilyl)-
Δ^{4a}(8a)-octaline gives rise to an ESR spectrum for which
a total of 13875 lines is predicted: 2 of the 5 different spe
cies $M^{\cdot\oplus}$ dominate, one unlabelled and one containing 1 isotope ^{29}Si out of four (I = 1/2, 4,7% nat. abundance). The
unlabelled $M^{\cdot\oplus}$ is rigid at lower temperatures and then
displays sets of each 4, 4, 4 and 36 equivalent hydrogens.
To the resulting 5 x 5 x 5 x 37 lines, the ^{29}Si-labelled species adds another 2 x 4625 = 9250 lines. With increasing
temperature, equilibration of the $-H_2C-CH_2-$ bridge protons
begins. All individual coupling constants have been resolve
(A: full spectrum at 200 K, B and C: central line at 180 K
and 310 K)XXXVII.

signals. Thus from a radical with i groups of each n equivalent coupling nuclei a hyperfine splitting pattern will result, which comprises a total of $N = \prod_i (2n\,I + 1)$ lines. These are revealed in highly resolved ESR spectra if the respective coupling constants are not too small, and if the intensity ratios between central and outer lines do not grow too large (Figure 13, cf. also Figures 1, 5, 11, or 12).

The opposite — spin localization within the ESR time scale — as strange as it might seem, however, has been reported in the literature for radical species like:

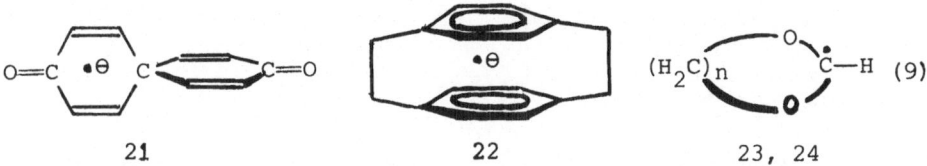

21   22   23, 24   (9)

The partial spin localization in radical anions of e.g. spiro[5.5]undeca-1.4.6.9-tetraene-3.8-dione [21] shown in (9), of spiro-bifluorenes or of [2.2]paracyclophane [22] has been carefully investigated including variation of reduction time or of solvent. The ESR results confirm that non-uniform spin distribution is governed by $M \cdot^{\ominus}/K^{\oplus}$ counter ion association at one π subsystem within the ESR time scale, and therefore favored by decreasing counter cation solvation.

In cyclic methyl radicals with the tervalent carbon bonded to two oxygens (9), the carbon proton coupling constants depend strongly on the ring-size [23,24]: they decrease from $a_H = 2.17\,mT$ for the five-membered radical to $a_H = \pm 0\,mT$ in the six-membered ring (9: n = 3), and amount to $a_H > |1\,mT|$ in open-chain compounds $(RO)_2C^\bullet - H$ [23]. In contrast, the analogous sulfur compounds show almost constant hydrogen coupling constants $a_H = 1.35\,mT$ to $a_H = 1.48\,mT$ [24]. The ESR results are interpreted in terms of differences in substituent perturbation $O \neq S$ [24], and for the alkoxy derivatives by induced pyramidality at the tervalent carbon center upon radical generation [23].

Among the numerous novel radical ions generated by the Frankfurt Group within the last three years, partly by chance but mostly by systematic screening, some more radical anions and even radical cations exhibiting "single line" ESR spectra due to spin localization have been discovered. Most of them can be subdivided into the following classes of compounds:

① Perphenylcyclopolysilane radical anions $(H_5C_6)_{2n}Si_n^{\bullet\ominus}$ (XXVI).

② Tetra(alkylthio)ethene radical cations $(RS)_2C \overset{\oplus}{=\!=} C(SR)_2$ (XXXV, 1o, 25).

③ Trithiocarbonate radical anions $(RS)_2C \overset{\ominus}{=\!=} S$ (XIII, XXIX, 10, 26–28).

Examples are presented in the following Figures 14 to 18.

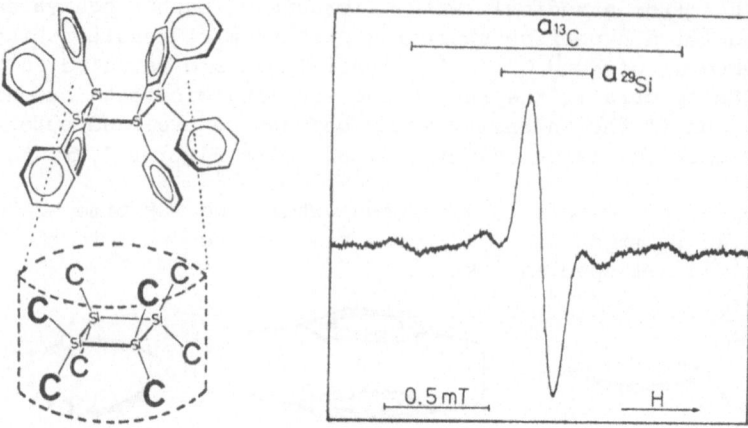

Figure 14. Perphenylcyclopolysilanes $[(H_5C_6)_2Si]_{4,5}$ are reduced by potassium in THF/DME mixture at low temperature to their radical anions **XXVI**. The resulting simple ESR spectra like the one shown for $(H_5C_6)_8Si_4^{\bullet\ominus}$ exhibit no visible H coupling fine structure and, therefore, suggest that the extra electron does not enter the phenyl rings. The observed ^{13}C and ^{29}Si isotope couplings further support a spin distribution confined to the Si_4 skeleton and the surrounding cylinder of the phenyl ring α carbons **XXVI**.

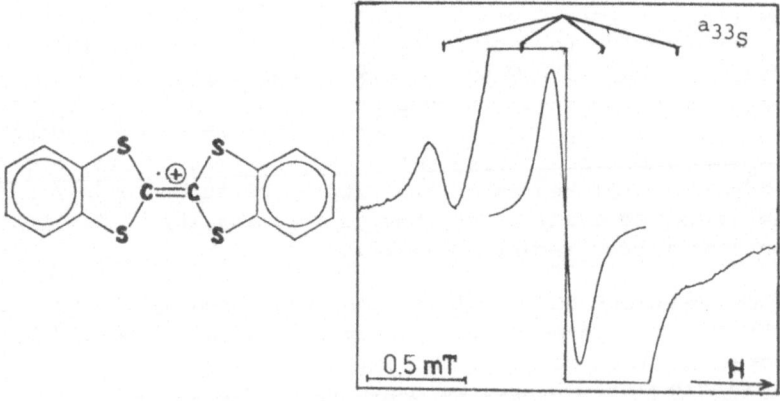

Figure 15. Dibenzotetrathiofulvalene is easily ionized in the gasphase $IE_1 = 6.81$ eV [29], and, correspondingly, can be oxidized in solution by $AlCl_3/H_2CCl_2$ [XXXV,25]. On extreme amplification of the single line of only 0.17 mT half-band width, a ^{33}S isotope coupling ($I = 3/2$, nat.abundance 0.7%) of $a_{33_S} = 0.413$ mT appears. The relatively high value $g = 2.0077$ also suggests that the spin density accumulates preferentially at the S centers i.e. within the $S_2C \stackrel{\oplus}{=\!=\!=} CS_2$ moiety.

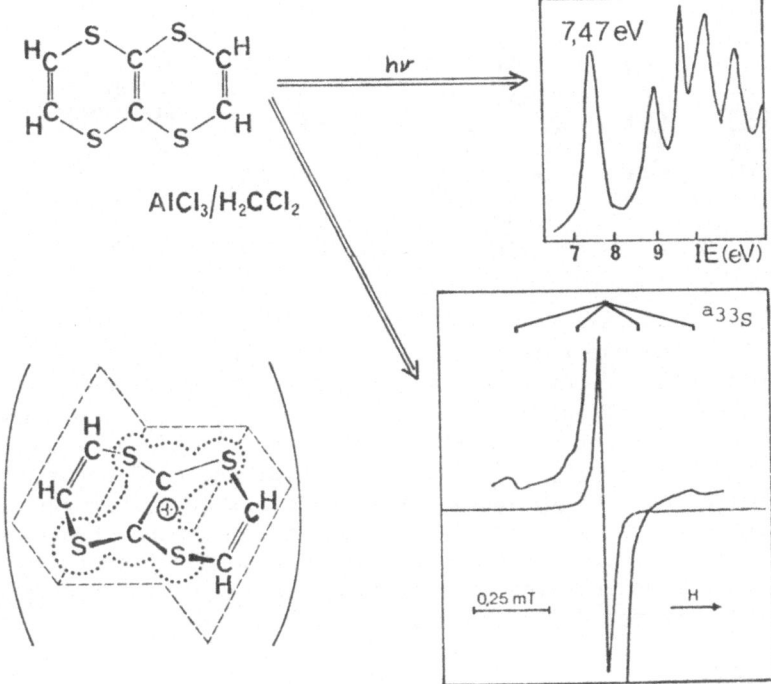

Figure 16. 1,4,5,8-Tetrathianaphthalene shows five separated bands
in the low-energy region of its PE spectrum, and the radical cation ground state is 7.47 eV above the molecular
ground state. Oxidation employing the selective reagent
$AlCl_3/H_2CCl_2$ generates again a species $M^{\cdot\oplus}$ exhibiting only
a single-line ESR spectrum, which on amplification reveals
a ^{33}S coupling of 0.414 mT [25]. A high spin density at the
sulfur centers also follows from a relatively high value
g = 2.0079. An INDO open shell calculation including geometry optimization via a Davidson-Fletcher-Powell subrou-
tine yields bending angles along the S···S axes of ca.
30°. Therefore, in addition to the spin localization
within the $S_2C\overset{\oplus}{=}CS_2$ moiety a radical cation "folding" as
indicated above in brackets helps to prevent the other-
wise expected hyperfine coupling of the 4 hydrogens present. The valence isomer, tetrathiafulvalene radical cation, exhibits a hydrogen quintet in its ESR spectrum [30].

        The third and most spectacular class of novel paramagnetic spe-
cies giving rise to single-line ESR spectra are trithiocarbonate radical anions [XIII,10, 25-28]. They have been discovered by comparing
the hydrogen coupling constants of the following series of thiathione
radical anions [XIII]:

$\overline{a_H}$:

$$(10)$$

The isotrithione radical anion generated electrochemically at a mercury cathode [10], exhibits in its ESR spectrum the rather small hydrogen coupling of only .o56 mT (10). This corresponds, via the McCONNELL relation for planar π radical anions [1], $a_{H,\mu}^{\pi} = |2.5\,mT|\,\varrho_\mu^\pi$, to the low π spin density $\varrho_C^\pi \sim 0.02$ at each of the ethene carbons i.e. the difference of $\Delta\varrho^\pi \sim 0.96$ must be localized within the trithiocarbonate group. The above rationalization is supported by a comparison of INDO closed and open shell charges [10], according to which the electron acquired on reduction should be almost completely "trapped" within the $S_2C=S$ moiety:

$$(11)$$

Obviously, the highly polar thiocarbonyl group $\overset{}{\underset{}{>}}C^\oplus = S^\ominus$ — a statement, experimentally confirmed by the vibrationyl fine structure of the ionization in the PE spectrum of $F_2C=S$ [31] — can act as an "electron trap" XIII,[32]. This reasoning based on the ESR results for thiathiones (10) as well as on the INDO charges (11), subsequently prompted a "trithiocarbonate synthesis and reduction rush", which yielded numerous radical anions with "single-line" ESR spectra (Figures 17 and 18, cf. also (12)).

The g values of the trithiocarbonate radical anions generated electrochemically using a mercury cathode[10] average an astonishing 2.00⁻ indicating a considerable sulfur contribution to the $M^\ominus$ ground state. Although some of them are rather short-lived even at low temperatures, for many of them carbon coupling constants a_{13C} could be determined in an especially designed electrochemical cell within the ESR cavity [10]:

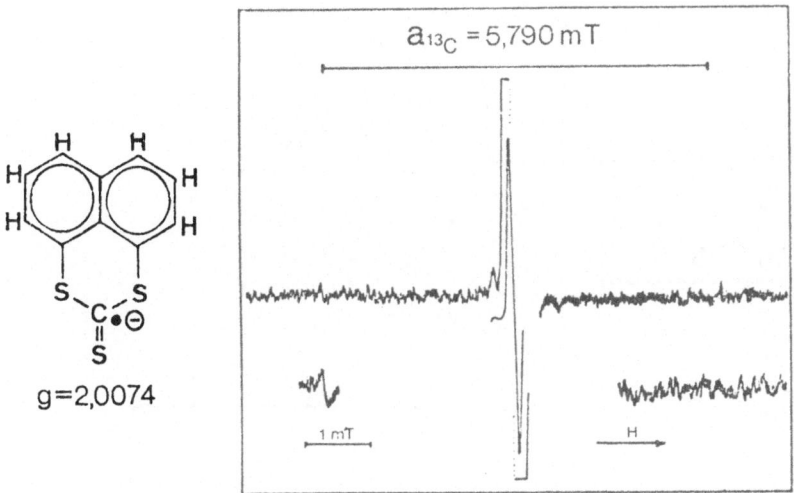

$a_{13C} = 5{,}790 \, mT$

g=2,0074

Figure 17. Electrochemical reduction of the peri-bridged naphtho-
1.8-trithiocarbonate in H_3CCN/DMF at 200 K yields a radical anion with a high g value and an ESR spectrum without
any hydrogen coupling. On extreme amplification a low-
field satellite is resolved, which on the following arguments is assigned to one ^{13}C isotope coupling: the observed dublet signal intensity of ca 0.5% compares well
the expectation of 0.55% for a radical anion containing
one ^{13}C isotope $(I = 1/2$, nat. abundance 1.1%). The high-
field line is broadened as would correspond to a positive
sign of the coupling constant a_{13C} [1]. The rather large
value of the ^{13}C coupling measured, suggests a pyramida-
lization of the tervalent thiocarbonyl carbon [33].

	$R:$	HC⁄ HC⟍	RS–C⁄ RS–C⟍	H_2C⁄ H_2C⟍			(12)
a_{13C} (mT)		---	5.054	---	5.571	5.790	
g value		2.0073	2.0072	2.0070	2.0073	2.0074	

Looking for even far-flung analogies, in radicals $F_nC^\cdot H_{3-n}$, the
^{13}C coupling increases considerably with increasing fluorine substitution [34] relative to the planar $H_3C^\cdot$ radical [1] — a finding, attributed [1,34] to an increase in height of the carbon pyramide. For the
isotrithione radical anion (cf. (10), (11) and (12)), an INDO open
shell hypersurface (cf. Figure 8) shows a total energy minimum for
a structure with the terminal S bent out of the S_2C plane by almost

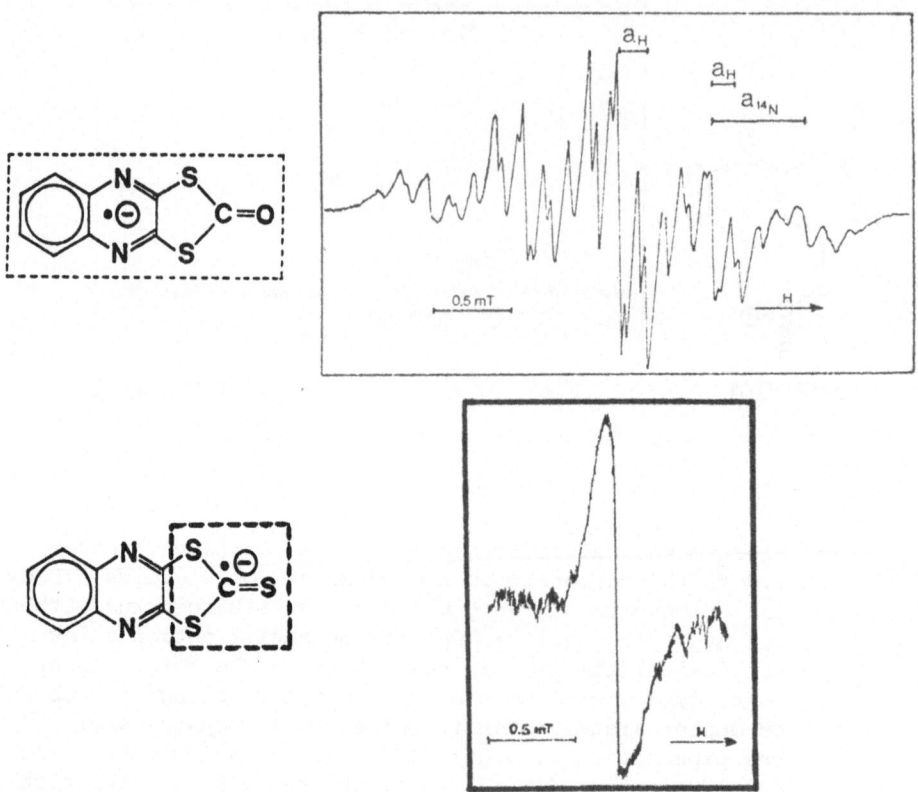

<u>Figure 18.</u> The acaricides, 2-oxo- and 2-thiono-1.3-dithiolo[4.5-b]-
chinoxaline [35] (ERADEX [R] and MORESTAN [R] by Bayer AG)
strikingly demonstrate the spin localization within trithiocarbonate groups [28]: the ESR spectrum of the oxo-radi-
cal anion displays the hyperfine structure expected for
complete delocalization over the molecular skeleton, i.e.
a splitting of two hydrogen triplets and one nitrogen
quintet. In contrast, the iso(valence)electronic sulfur
compound shows in its ESR spectrum a single line of only
~ .25 mT half-band width [28].

30°, and the missing hyperfine coupling could then be attributed to
changing sign (- to +) calculated for the coupling constant a_H^{INDO}
around 25° (S_2C)-S bending angle. Whether or not the trithiocarbonate
radical anions are best described by a "singly-occupied carbon lone
pair", the effect of the spin-localization within a $S_2C\dot{-}\overset{\ominus}{S}$ moiety
becomes especially eye-catching if the oxygen pan-handle of a large
and completely delocalized π radical anion is exchanged for sulfur
(Figure 18) [28].

OUTLOOK: ESR AS A USEFUL TOOL FOR THE CHEMIST

The onset of this interim report - which intentionally has been confined predominantly to the activities of the Frankfurt Group within the last three years - included the statement: "Based on the (ESR) experience gathered, it is now feasible that the preparative chemists participate by designing novel oxidizable compounds, inventing new redox systems ... for radical generation, studying their prospective properties and trying to isolate them". In this respect, the otherwise limited scope of this review may even have its merits - demonstrating the use of a useful tool by a team of chemists.

And as time passes by, radical ions which are persistent enough become isolated in increasing numbers. In addition to the illuminating X-ray structure of a sterically shielded hydrazine radical cation already quoted [15], we would also like to present an example of our own XXXV: on addition of equimolar amounts of iodine to a solution of hexakis(alkylthio)ethane in acetonitrile, first a white deposit of polymeric alkylsulfide precipitates, followed on careful solvent evaporation by the crystallization of the metal-black needles of tetrahydro-tetrathiafulvalene radical cation triiodide:

$$\left[\begin{array}{c} S \\ S \end{array}\!\!>\!\!C\!\!-\!\!C\!\!<\!\!\begin{array}{c} S \\ S \end{array}\right] + \frac{3}{2} I_2 \Rightarrow \left[\begin{array}{c} S \\ S \end{array}\!\!>\!\!C\!\!\overset{\cdot\oplus}{=}\!\!C\!\!<\!\!\begin{array}{c} S \\ S \end{array}\right] I_3^{\ominus} + (S\text{-}CH_2\text{-}CH_2\text{-}S)_n \tag{13}$$

Found % C 12.2 H 1.3   C 27.2 H 4.4
         12.2   1.4   26.1   4.3

What are the future prospects of the powerful ESR technique at the disposal of chemists? The answer can only be: "numerous and excellent!" But to preferentially replace speculation by novel observation, our report shall be concluded as follows: since the stimulating IBM Symposium in Bad Neuenahr four weeks ago, the oxidation of the simplest cluster available, i.e., tetrakis(tertbutyl)tetrahedrane containing 12 electrons in its C_4 skeleton, has been investigated by ESR assisted by extensive MNDO open and closed shell hypersurface calculations. On electron removal, the delicate charge balance governed by EULER´s equation for convex polyhedra, i.e., faces + vertices = edges + 2, breaks down and the cluster opens to form tetrakis(tert. butyl)cyclobutadiene radical cation [36]:

$$\xrightarrow[{-e^{\ominus}}]{AlCl_3/H_2CCl_2} \tag{14}$$

PUBLICATIONS ON RADICAL IONS 1976 - 1979 BY THE FRANKFURT GROUP

VIII. Radical Cation and Radical Anion of a Tetrasilyl-Substituted
      Ethylene.
      H. BOCK, G. BRÄHLER, G. FRITZ and E. MATERN, Angew. Chem. 88,
      765 (1976); Int. Edit. 15, 669 (1976).

  IX. Radical Cations of Tetrasilyl-Substituted Hydrazines.
      H. BOCK, W. KAIM and J. W. CONNOLLY, Angew. Chem. 88, 766
      (1976); Int. Edit. 15, 7oo (1976).

   X. Electron-Rich Alkyl Olefines and Their Radical Cations.
      H. BOCK and W. KAIM, Tetrah. Lett. 1977, 2343.

  XI. One-Electron Oxidation of Silylalkyl Benzenes in the Gas Phase
      and Solution.
      H. BOCK and W. KAIM, J. Organomet. Chem. 135, C14 (1977).

 XII. To what extend do bulky alkyl groups rotate?
      H. BOCK and W. KAIM, Nachr. Chem. Techn. Lab. 25, 3o6 (1977).

XIII. Thiathion Radical Anions: Are $(-S)_2C=S$ Groups Electron Traps?
      H. BOCK, G. BRÄHLER, A. TABATABAI, A. M. SEMKOW and
      R. GLEITER, Angew. Chem. 89, 745 (1977); Int. Edit. 16,
      724 (1977).

 XIV. Oxidation and Reduction of Methylthio-Substituted π Systems
      and the Electron Distribution in Their Radical Ions.
      H. BOCK and G. BRÄHLER, Angew. Chem. 89, 893 (1977); Int.
      Edit. 16, 855 (1977).

  XV. 2,4,6,8-Tetrasilabicyclo[3,3,o]-oct-1(5)enes: Formation of
      Radical Cations and Radical Anions of the Derivatives -SiH,
      -SiF and $-Si(CH_3)_3$.
      G. FRITZ, E. MATERN, H. BOCK and G. BRÄHLER, Z. anorg. allg.
      Chem. 439, 173 (1978).

 XVI. One-Electron Oxidation of Silyl-Substituted Tetrazenes and
      Pentazenes.
      H. BOCK, W. KAIM, N. WIBERG and G. ZIEGLEDER, Chem. Ber. 111,
      315o (1978).

XVII. Structural Changes during the Oxidation of B-, C-, Si- and
      P-Substituted Hydrazines.
      H. BOCK, W. KAIM, A. M. SEMKOW and H. NÖTH, Angew. Chem. 9o,
      3o8 (1978), Int. Edit. 17, 286 (1978).

XVIII. Note: Do Persistent Radical Cations of Phosphines Exist?
W. KAIM, H. BOCK and H. NÖTH, Chem. Ber. <u>111</u>, 3276 (1978).

XIX. One-Electron Oxidation of R_3SiCH_2-Substituted Benzenes in the Gasphase and in Solution.
H. BOCK and W. KAIM, Chem. Ber. <u>111</u>, 3552 (1978).

XX. The Hyperconjugative Stabilization of p-Xylene Radical Cations by R_3Si-Substituents.
H. BOCK, W. KAIM and H.-E. ROHWER, Chem. Ber. <u>111</u>, 3573 (1978).

XXI. ESR-Investigation of R_3SiCH_2- or R_3Si-Substituted Radical Cations of Anthracene, Biphenyl and 9,1o-Dihydroanthracene.
W. KAIM and H. BOCK, Chem. Ber. <u>111</u>, 3585 (1978).

XXII. Radical Anion and Radical Trianion of 1,4-Bis(dimethylphosphino)benzene.
W. KAIM and H. BOCK, J. Amer. Chem. Soc. <u>1oo</u>, 65o4 (1978).

XXIII. R_2P- and R_2N-Substituted Benzenes: The Charge Distribution in Their Cations, Anions and Trianions.
W. KAIM and H. BOCK, Chem. Ber. <u>111</u>, 3843 (1978).

XXIV. The Electronic Structure of Tetraphenylthieno[3.4-c]thiophene: Photoelectron , Electron Spin Resonance and Electronic Absorption Spectra.
R. GLEITER, R. BARTETZKO, G. BRÄHLER and H. BOCK, J. org. Chem. <u>43</u>, 3893 (1978).

XXV. Radical Anions of Tin-Substituted Naphthalenes.
H BOCK, W. KAIM and H. TESMANN, Z. Naturforschg. <u>33b</u>,1223 (1978).

XXVI. Radical Anions of Perphenylcyclopolysilanes.
M. KIRA, E. HENGGE and H. BOCK, J. organomet. Chem. <u>164</u>, 277 (1979).

XXVII. Radical Ions of Tetrakis(trimethylsilyl)butatriene.
W. KAIM and H. BOCK, J. organomet. Chem. <u>164</u>, 281 (1979)

XXVIII. $\cdot^{\oplus}N(CH_2Si(CH_3)_3)_3$ - A Stable Fluctuating Aminium Radical Cation.
H. BOCK, W. KAIM, M. KIRA, H. OSAWA and H. SAKURAI, J. organomet. Chem. <u>164</u>, 295 (1979).

XXIX. Thioformaldehyde Radical Anion: A PNO-CEPA ab initio Study.
P. ROSMUS and H. BOCK. JCS Chem. Comm. <u>1979</u>, 334.

XXX. Reduction of R_3SiO-Substituted Benzene Derivatives.
H. BOCK and W. KAIM, Z. anorg. allg. Chem. <u>443</u>, (1979).

XXXI.   NMR- and ESR-Spectroscopic Investigation of the Isomerization
         of Five- and Six-Membered B/N-Heterocycles.
         H. NÖTH, W. WINTERSTEIN, W. KAIM and H. BOCK, Chem. Ber. 112,
         2494 (1979).

XXXII.   Alkyl Ligand Exchange in Bis(cyclopentadienyl)dialkyltitanium
         (III) Radical Anions.
         M. KIRA, H. BOCK, H. UMINO and H. SAKURAI, J. organomet.
         Chem. 173, 39 (1979).

XXXIII. Oxidation and Reduction of Methylthio-Substituted Naphtha-
         lenes - A Molecular State Comparison.
         H. BOCK and G. BRÄHLER, Chem. Ber. 112,3o81(1979).

XXXIV.   Novel Organosilicon Radical Cations: The One-Electron Oxi-
         dation of Permethylpolysilanes.
         H. BOCK, W. KAIM, M. KIRA and R. WEST, J. Amer. Chem. Soc.
         1o1,   (1979).

XXXV.   The Oxidation of Hexa(alkylthio)-substituted Ethanes to
         Tetra(alkylthio)ethane Radical Cations.
         H. BOCK, G. BRÄHLER, U. HENKEL, R. SCHLECKER and D. SEEBACH,
         Chem. Ber. 112,   (1979).

XXXVI.   Structural Changes Accompanying the One-Electron Oxidation
         of Hydrazine and its Silyl Derivatives.
         H. BOCK, W. KAIM, H. NÖTH and A. SEMKOW, J. Am. Chem. Soc.
         1o2,   (198o).

XXXVII. Ionisation and One-Electron Oxidation of Electron-Rich Olefins.
         H. BOCK and W. KAIM, J. Am. Chem. Soc. 1o2,   (198o).

GENERAL REFERENCES

(1) Cf. e.g. K. SCHEFFLER and H. B. STEGMANN "Elektronen-Spinreso-
     nanz", Springer Verlag 197o, or J. E. WERTZ and J. R. BALTON
     "Electron Spin Resonance", McGraw-Hill 1972.

(2) Cf. e.g. the annual Specialist Periodical Reports of the Che-
     mical Society "Electron Spin Resonance" Vol. 1 (1973) and
     following volumes.

(3) Cf. e.g. the lecture in honor of Erich Hückel's 8oth birthday,
     H. BOCK, Angew. Chem. 89, 631 (1977); Angew. Chem. Int. Ed.
     Engl., 16, 613 (1977), and the literature quoted therein.

(4) We only learned later on that $AlCl_3$ had been used before e.g. by D. H. GESKE and M. MERRITT, J. Am. Chem. Soc. 91, 6921 (1969). Speculation on the unkown course of the redox reaction as well as on the resulting reduction product include chloride abstraction from H_2CCl_2 forming an $AlCl_4^{\ominus}$ counter ion and subsequent oxidation by the carbonium ion $H_2C^{\oplus}Cl$, or the direct formation of Al-Al bonds.

(5) D. GRILLER and K. U. INGOLD, Acc. Chem. Res. 9, 13 (1976).

(6) Cf. e.g. the review by C. PITT, J. Organomet. Chem. 61, 49 (1973) and the literature quoted therein.

(7) Cf. e.g. H. BOCK and H. SEIDL, J. Organomet. Chem. 13, 87 (1968), Chem. Ber. 1o1, 2815 (19o8) (together with M. FOCHLER) or J. Chem. Soc. B 1968, 1158; or H. BOCK and H. ALT, J. Am. Chem. Soc. 92, 1569 (197o); and literature quoted.

(8) Cf. e.g. E. HEILBRONNER, V. HORNUNG, H. BOCK and H. ALT, Angew. Chem. 81, 357 (1969), Int. Ed. Engl. 8, 524 (1969).

(9) S. F. NELSON and C. R. KESSEL, J. Am. Chem. Soc. 99, 2392 (1977).

(1o) Thesis G. BRÄHLER, University of Frankfurt 1978.

(11) Thesis W. KAIM, University of Frankfurt 1977.

(12) C. HELLER and H. McCONNELL, J. Chem. Phys. 32, 1535 (196o).

(13) T. M. McKINNEY and D. H. GESKE, J. Am. Chem. Soc. 89, 2806 (1967). Cf. also E. W. STONE and A. H. MAKI, J. Chem. Phys. 37, 1326 (1962).

(14) Cf. the recent review by S. F. NELSON, Acc. Chem. Res. 11, 14 (1978) and literature quoted.

(15) S. F. NELSON, W. C. HOLLINSED, C. R. KESSEL and J. C. CALABRESE, J. Chem. Soc. 1oo, 7876 (1978).

(16) H. BOCK, U. STEIN and A. SEMKOW, chem. Ber. in print: Part XXXXI of Radical Ions.

(17) Cf. eg. K. JORDAN, J. A. MICHEJDA and P. D. BURROW, J. Am. Chem. Soc. 98, 1295, 7189 (1976) and lit. cit.

(18) Cf. F. GERSON: Hochauflösende ESR-Spektroskopie. Verlag Chemie Weinheim 1967.

(19) F. GERSON, CH. WYDLER, F. KLUGE, J.Magn. Reson. 26, 271 (1977).

(2o) D. H. GESKE, J.L. RAGLE, M.A. BAMBENEK, A.L.BALCH, J. Am. Chem. Soc. 86, 987 (1964); B. M.Latte, R.W. TAFT, ibid.89, 5172 (1967).

(21) F. GERSON, R. GLEITER, G. MOSHUK and A. S. DREIDING, J. Am.
Chem. Soc. 94, 2919 (1972) or F. GERSON, B. KOWERT and B. M.
PEAKE, ibid. 96, 118 (1974).

(22) F. GERSON, W. B. MARTIN and CH. WYDLER, J. Am. Chem. Soc. 98,
1318 (1976).

(23) A. J. DOBBS, B. C. GILBERT and R. O. C. NORMAN, J. Chem. Soc.
(A) 1971, 124.

(24) E. A. C. LUCKEN and B. PONCIONI, JCS Perkin II 1976, 777.

(25) H. BOCK, G. BRÄHLER, M. CAVA and L. LAKSHMIKANTHAM, unpublished
results, cf. [10].

(26) Master Thesis A. TABATABAI, University of Frankfurt, 1976.

(27) H. BOCK, G. BRÄHLER, J. MEINWALD and D. DAUPLAISE, unpublished
results, cf.[10].

(28) H. BOCK, G. BRÄHLER and K. H. BÜCHEL, unpublished results.

(29) J. SPANGET-LARSEN, R. GLEITER and S. HÜNIG, Chem. Phys. Lett.
37, 29 (1976).

(3o) F. B. BRAMWELL, R. C. HADDON, F. WUDL, M. L. KAPLAN and J. H.
MARSHALL, J. Am. Chem. Soc. 100, 4612 (1978).

(31) K. WITTEL, A. HAAS and H. BOCK, Chem. Ber. 1o5, 3877 (1972).

(32) Cf. also the ESR studies by J. VOSS and coworkers, e.g. C.-P.
KLAGES and J. VOSS, Angew. Chem. 89, 744(1977); Angew. Chem.
Int. Ed. Engl. 16, 723 (1977).

(33) Cf. e.g. Specialist Periodical Report "ESR", Vol.3, The Chemical Society London 1976, p.155 and literature quoted.

(34) R. W. FESSENDEN and R. H. SCHULER, J.Chem.Phys. 43, 27o4 (1965).

(35) Cf. K. H. BÜCHEL "Pflanzenschutz und Schädlingsbekämpfung",
Thieme Verlag Stuttgart 1977, p. 87.

(36) Radical Ions XXXIX: H. BOCK, B. ROTH and G. MAIER, Angew. Chem.,
in print.

POTENTIAL SURFACE STUDIES OF OPEN SHELL SYSTEMS

Peter Bischof

Institut für Organische Chemie der THD

Petersenstrasse 22, D 6100 Darmstadt

The high reactivity of radicals makes it extremely difficult to study individual, elementary reaction steps by conventional experimental techniques. There are very often a number of competing reaction pathways which sometimes lead to ambiguous, in some cases even contradictive, interpretations of the obtained results.

"Experimental computer chemistry"[1] provides an alternative course to study such processes, because each elementary reaction step can be simulated. Furthermore, such methods enable the chemist to investigate reactions which do not take place and to understand why they don't.

The aim of such calculations must be to provide a link between experiment and understanding which, if correct, is independent of the used model.

In this sense, the most commonly applied approach involves applying Eyring's theory of the activated complex[2] to the study of the corresponding potential energy surfaces.

In order to proceed with this approach, we need some means to meet the following two conditions:

1) A model which describes the potential surface as accurate and easily accessible as possible and

2) A mathematical procedure which allows location of the essential structures with the least amount of numerical calculations.

Clearly, the first point calls for some reasonable compromise between reliability and computational effort. While ab initio methods beyond the Hartree Fock level have proven to be as accurate as experimental techniques to calculate intramolecular forces, these methods still are limited to rather small molecules. For systems of moderate or large size one is therefore forced to use some semiempirical approach to describe the potential energy surface.

Being aware of its limitations, MINDO/3 [3)] still seems to be the most reliable of these methods. Its main advantage as compared with other semiempirical models lies in the fact that reasonable heats of formations as well as satisfactory geometries have been obtained for a large number of widely different molecules [4)].

Since the parameters used in this scheme hopefully balance the electronic correlation energy, its adequate application to open shell systems should be made using a restricted Hartree Fock version, such as the half electron method [5)]. However, this method leads to serious problems in the gradient calculations which is a grave handicap in view of the second requirement mentioned above. We have therefore extended MINDO/3 to an unrestricted Hartree Fock version using the same set of parameters and formalism [6)].

A first important check for its suitability is that given by the comparison of the calculated properties of radicals in their ground state with the available experimental data or results of ab initio calculations. The latter are especially helpful in connection with equilibrium geometries, because experimental data are scarce.

Fig. 1 shows the correlation between calculated and experimental heats of formation of a large series of radicals. It should be kept in mind that the data represented in this plot have not been involved in the parametrization procedure. Nevertheless, they are in reasonably good accord with experiment [6)].

The largest deviations from the theoretical line with unit slope were found for radicals containing heteroatoms. Quite generally, the radicals are calculated to be too stable by an average of 40 kJ/mole. This tendency is not a serious shortcoming, because it only shifts the whole energy potential surface toward lower values. Our calculations showed on the other hand [6)] that the relative stabilities of isomer radicals are in most cases predicted quite correctly.

Survey of the ground state geometries shows that the method provides reasonably good estimates of the structural parameters.

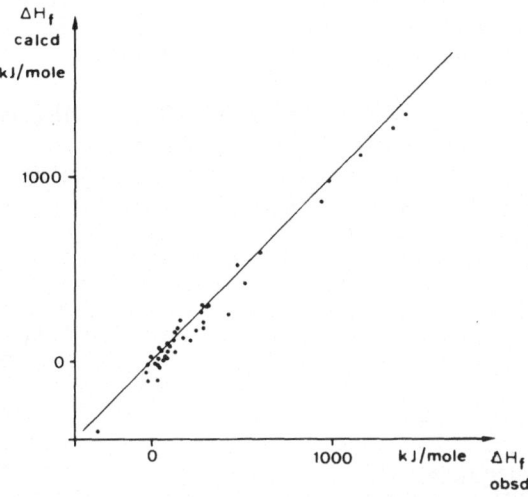

Fig. 1: Correlation of calculated and experimental ΔH_f of radicals

Table 1 summarises some calculated and spectroscopic data of diatomic radicals [7]. Although calculation and experiment compare reasonably well, the data disclose a general shortcoming of our method: All bondlengths are consistently too short, all stretching vibration wavenumbers too high [8].

Tab. 1: Calculated and experimental data of Diatomic Radicals. (Bond lengths in Å, wavenumbers in cm^{-1}.)

R·	state	r_o(calc)	r_o(exp)	v_o(calc)	v_o(exp)
CC	$^3\Pi_u$	1.261	1.312	1993	1641
CCl	$^2\Pi$	1.641	1.642	903	846
CH	$^2\Pi$	1.119	1.120	3347	2862
CN	$^2\Sigma^+$	1.132	1.172	2403	2069
CO$^+$	$^2\Sigma^+$	1.102	1.115	2396	2214
CP	$^2\Sigma^+$	1.482	1.562	1590	1240
HCl$^+$	$^2\Pi$	1.275	1.315	2627	2675
NN$^+$	$^2\Sigma_g^+$	1.050	1.116	2688	2207
NO	$^2\Pi$	1.164	1.151	2117	1904
OO	$^3\Sigma_g^-$	1.206	1.207	2107	1580
OO$^+$	$^2\Pi_g$	1.137	1.123	2426	1876
OH	$^2\Pi$	0.947	0.971	4027	3735

Diatomic radicals are definitely not the stronghold of any
semiempirical treatment. However, the shown comparison proves
that the MINDO/3 results are at least realistic.

Fig. 2 shows a few calculated MINDO/3 equilibrium geometries
which are compared with results of ab initio calculations [9]. The
CC bondlengths again are predicted to be shorter, but the general
features are reproduced quite well.

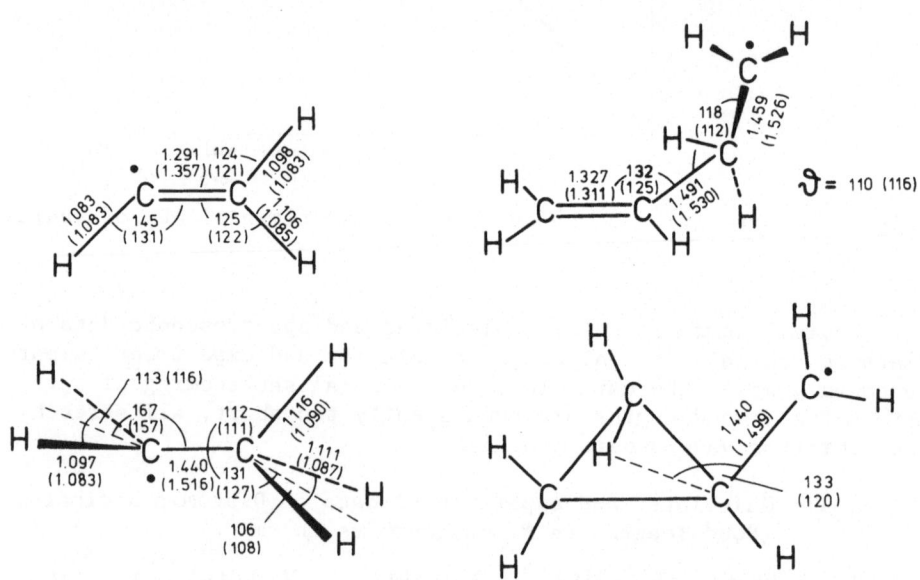

Fig. 2:   Comparison of predicted MINDO/3-UHF equilibrium geometries
          with results obtained by single configuration ab initio
          calculations. Only the optimised value of the latter are
          given in parantheses. (Ref. 32))

Judging from the results discussed so far, we conclude that
the potential energy surface properties, at least in the vicinity
of predicted minima, are realistic and useful. As will be shown
later, the energies of activated complexes are reasonable too.

We shall now turn to the second requirement mentioned above.
As already anticipated, we wish to locate minima and saddle
points on the potential surface of interest.

Optimizing a certain function or localizing a minimum in a multidimensional space is a problem common to all fields of science. Accordingly, there is a large number of available procedures [10]. We have chosen the method developped by Fletcher and Powell [11], because the calculation of the gradients in the MINDO/3-UHF scheme is trivial, if the bond order matrix is kept constant.

In this way, the equilibrium structures were located. They were verified to be true minima by the standard procedure of calculating and diagonalising the force constant matrix A. This was done in mass weighted cartesian coordinates which directly lead to normal vibrations or principal directions $|q>$ in the way proposed by Wilson, Decius and Cross [12].

Pulay has shown [13] that in the calculation of second derivatives, the so called wave function forces cannot be neglected. This leads to the necessity of computing the matrix A by using a grid of points, each calculated by a full SCF calculation.

It turns out that this verification needs about as much computer time as the search for the minima. However, the obtained data together with the located structures allows to estimate the standard entropies of the species by using statistical thermodynamics [14].

Obviously the contributions of internal rotors in this model are naively replaced by contributions of torsional vibrations with very low vibrational frequences. For radicals containing a large number of internal rotors the method is therefore bound to fail.

Table 2 summarizes some calculated entropies and compares them with "experimental" values. The term "experimental" is somewhat misleading, since the standard entropy of a radical cannot be measured. O'Neal and Benson have tabulated values obtained by the difference method (DM), based on the experimental values measured for the parent molecules and corrected for the missing hydrogen atom [15].

The tabulated standard entropies and their reduced internal contributions show that the observed errors have about the magnitude of the calculated vibrational parts. Nevertheless, they might be used as a further corroboration that the size of the radicals (and therefore their geometries) are quite accurate.

In addition, we used these standard entropies to roughly estimate the effects geometrical changes might have on the rate constants of radical rearrangement reactions [16].

Tab. 2: Comparison of calculated with "experimental" standard
         entropies. S^i is calculated by subtracting the trans-
         lational part from S^o. (All values in J/Kmole.)

	S^o_{298}		S^i_{298}		
	calc	"exp"	calc	"exp"	Δ
CH_3	199	194	57	52	5
C_2H_5	253	248	103	98	5
C_3H_7	284	285	128	129	-1
$i-C_3H_7$	294	283	138	127	11
$t-C_4H_9$	317	312	158	153	5
CH_2CHCH_2	262	260	105	103	2
$CH_2CH(CH_2)_2$	323	312	165	154	11

     The localization of saddle points is a much more severe prob-
lem than localizing minima.   A minimum on the potential energy
surface E in most cases corresponds to a structure, which can be
described by a classical Kekulé formula. This in turn is a useful
guide to make a good starting point choice. On the other hand
such chemical intuition does not help much in estimating the
structure of an activated complex.

     McIver and Komornicky have developped a procedure which at
first sight seems to be the method of choice [17]. Their theory
is based on the fact that minima as well as saddle points are
stationary points on the potential energy surface E. Hence they
are minima on the gradient square surface σ. The value of σ is
given by:

$$\sigma = \sum_{i=1}^{3N-6} (\frac{\delta E}{\delta S_i})^2 = \langle g | g \rangle$$

This way the localization of an activated complex is thus reduced
to a "simple" optimization of the value of σ. This can in principle again be performed using the Fletcher-Powell procedure.
However, the calculation of the derivatives of σ involves the
determination of the force constant matrix $\mathbf{A}$, because

$$\frac{\delta \sigma}{\delta S_i} = 2 \sum_{j=1}^{3N-6} A_{ij} \frac{\delta E}{\delta S_j}$$

or in vector notation

$$| \sigma ' \rangle = 2 \mathbf{A} | g \rangle$$

     As already indicated, the calculation of $\mathbf{A}$ is a tedious piece
of work. Since this has to be done at each point reached in the

optimization procedure, the localization of the desired saddle
points became hopelessly slow.

We therefore turned to an alternative procedure, which proved
to be superior as far as the computer time consumption is concerned.
The method can be described as follows: Find the saddle point
as economically as possible and spend more time in verifying its
nature to be the true saddle point of interest.

This is performed in five steps:

1) The reactant structure is optimized. From the geometry defin-
    ing variables a_i (bondlengths, bondangles and dihedral angles)
    the one is chosen as a trial reaction coordinate α, which
    apparently changes its value the most as the reactant conti-
    nuously turns into the product. In most cases, α is the
    length of the bond being broken or formed in the process.

2) This variable is then varied in small increments at which
    points the energy is optimized with respect to all other vari-
    ables a_i thus making all first derivatives zero

    $$\left(\frac{\delta E}{\delta a_i}\right)_\alpha = 0$$

    within a specified accuracy. At this geometry $|a\rangle$ the deri-
    vative of E with respect to α is calculated.

    This "step search" is repeated, until the geometry is located,
    at which $$\left(\frac{\delta E}{\delta \alpha}\right)_{a_i \neq \alpha} = 0$$

    The geometry obtained this way shall be denoted $|t\rangle$. McIver
    has pointed out [17] that this structure does not necessarily
    have to be a saddle point. We therefore applied the following
    tests:

3) The force constant matrix A was calculated and the 3N-6 prin-
    cipal directions $|q\rangle$ obtained as its eigenvectors. It was
    checked that one and only one calculated eigenvalue was nega-
    tive. The principal direction $|q^-\rangle$ belonging to this negative
    eigenvalue is in fact "the way over the hill" and has been
    called the transition vector [17] or the intrinsic reaction
    coordinate (IRC) [18].

4) The first derivatives of E with respect to the principle di-
    rections $|q\rangle$ were calculated by finite differences insuring
    that they all were below the convergence criteria.

    Taking for granted that these checks are sufficient to prove
    that $|t\rangle$ really is a saddle point, there still remains the
    problem to find whether or not it is the only activated

complex along the reaction path. This was done in the final step:

5) The structure $|t\rangle$ was modified by the transition vector $|q^-\rangle$
    which led to two nearly identical structures $|t\rangle^+$ and $|t\rangle^-$ as
$$|t\rangle^+ = |t\rangle + \lambda|q^-\rangle$$
$$\text{and } |t\rangle^- = |t\rangle - \lambda|q^-\rangle.$$

The factor λ was chosen as 0.01. The obtained geometries were
then transformed back to the internal coordinate system $|a\rangle$
and the structures were optimized. $|t\rangle^+$ converged to structu-
re of the product (if the reaction was a one step reaction),
$|t\rangle^-$ to the starting point reactant.

Sometimes, however, the relaxed structure of $|t\rangle^-$ was quite
different from that of the starting point. In these cases,
the starting structure and the relocated reactant structure
were always found to be local minima separated by small activation energies such as rotational barriers.

The program system which we just described is, apart from
the spin unrestricted modification of MINDO/3, identical to the
one used to study reactions on a singlet surface.

It's application to a reaction of interest leads to a scheme
as the one shown in Fig. 3. This picture shows as an example the
rearrangement Norbornenyl-Nortricyclyl radical, which has been
studied experimentally by Giese [19]. The arrows shown in the plot
of the activated complex correspond to the nuclear displacements
along the intrinsic reaction coordinate.

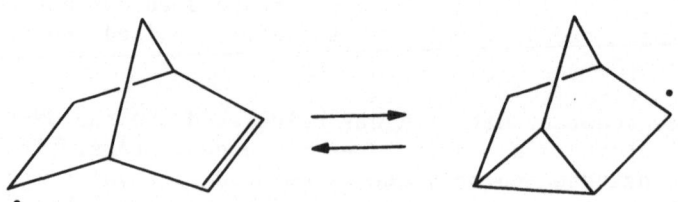

Giese has concluded from his measured kinetic data that

1) both reactants have about the same energy, the norbornenyl
    radical being at most [20] 8kJ/mole more stable than the nor-
    tricyclyl radical, and

2) the activation energy amounts to 48 kJ/mole for the forward,
    to 40 kJ/mole for the reverse reaction.

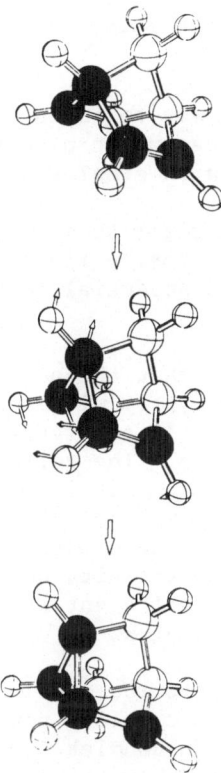

Fig. 3: Calculated structures of the Norbornenyl/Nortricyclyl
         radical rearrangement.

MINDO/3-UHF predicts both radicals to have the same heat of
formation, the activation energy is calculated to be 48 kJ/mole.

This reaction is an example of a widely observed type of
radical rearrangement: It is a homoallyl/cyclopropyl carbinyl
rearrangement. The shaded atoms in Fig. 3 indicate the carbon
atoms which form the parent system. We shall return to a detailed
study of the parent reaction later on.

An important aspect we have not mentioned so far might serve
as a further justification for the application of "experimental
computer chemistry" to radical rearrangements.

For singlet molecules, the Woodward Hoffmann rules [21] provide a very useful way to predict the ease with which a reaction
will proceed, and what stereochemical consequences are to be ex-

pected. Unfortunately, these rules are not conclusive for doublet systems [22].

However, the analogue way to investigate reactions could explain expected differences in the potential surface behaviour of open shell as compared with singlet systems.

Such differences are thought to be especially pronounced for so called "forbidden" reactions, while for "allowed" reaction paths, smooth, well behaved potential surfaces are expected for either type of reaction.

It is well understood that the activated complex of a for-bidden reaction path for a system with an even number of electrons cannot be described by a single configuration. This seems to be different for radical systems. The correlation diagrams for both are compared in Fig. 4.

In this simple picture the crossing of the frontorbitals is thought to take place in the vicinity of the activated complex, only if these orbitals have different symmetry behaviour with respect to the symmetry element R.

A crude estimate leads to the conclusion that in the singlet case three singlet (and one triplet) configurations are of com-parable energy, while in both doublet cases only two configura-tions of different symmetries have to be considered. Hence in the former configuration interaction might lead to an avoided crossing, while in the latter the surfaces should cross indeed [23].

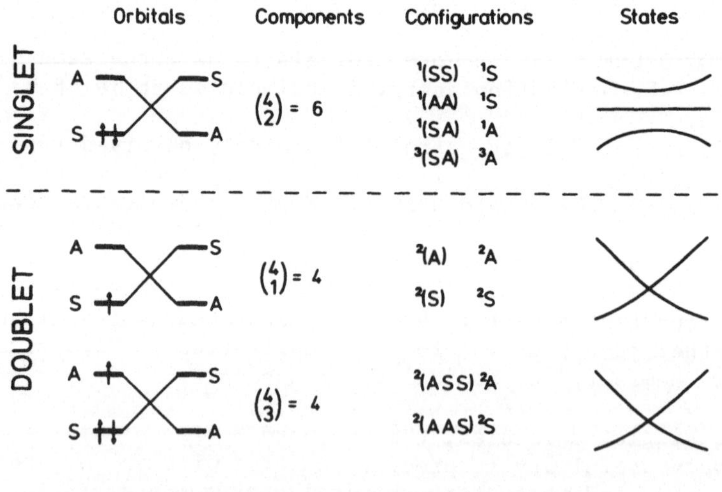

Fig. 4: Correlation diagrams of "forbidden" reaction pathways.

Evidently the expected electronic structure of the activated
complex in the doublet case resembles that of a Jahn Teller state
radical, and detailed calculations of such systems have led us to
expect a surface behaviour as the one shown in Fig. 5 [23]. In
this plot Q_1 corresponds to a symmetrical reaction coordinate,
Q_2 to the Jahn Teller distortion, which is necessarily antisymmetric with respect to the symmetry element R.

This qualitative scheme therefore predicts that a "forbidden"
disrotatory radical rearrangement proceeds via two enantiomer
reaction paths.

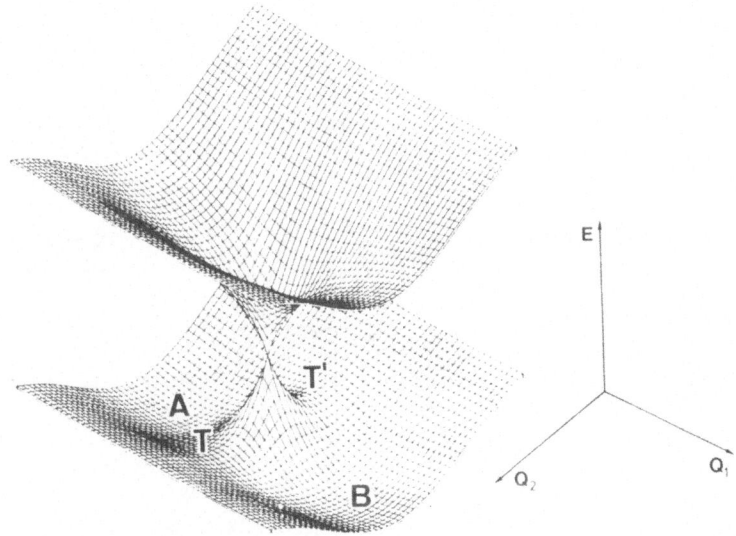

Fig. 5: Model potential surface of a "forbidden" doublet reaction.

We wish to illustrate this point by a specific example, for
which experimental data are available.

Sustmann [24] has investigated the rearrangement of the
bicyclo-[3,2,0]-heptadienyl radical, which leads to the tropyl radical in an adamantane matrix by ESR, and found a free activation
energy of 90 kJ/mole at 323K.

Since the search for the activated complex led to some problem, the calculated reaction profile is shown in Fig. 6. As the
trial reaction coordinate we have chosen the bondlength C^1C^5.
Enforcing C_S symmetry, the search did not lead to a saddle point,
since the derivative with respect to α was steadily increasing
until no SCF convergence was achieved anymore. By the independent optimization of the 41 remaining variables, the deviation from
C_S symmetry became appreciable for $\alpha > 1.9$ Å, and led to a real
saddle point which has no symmetry.

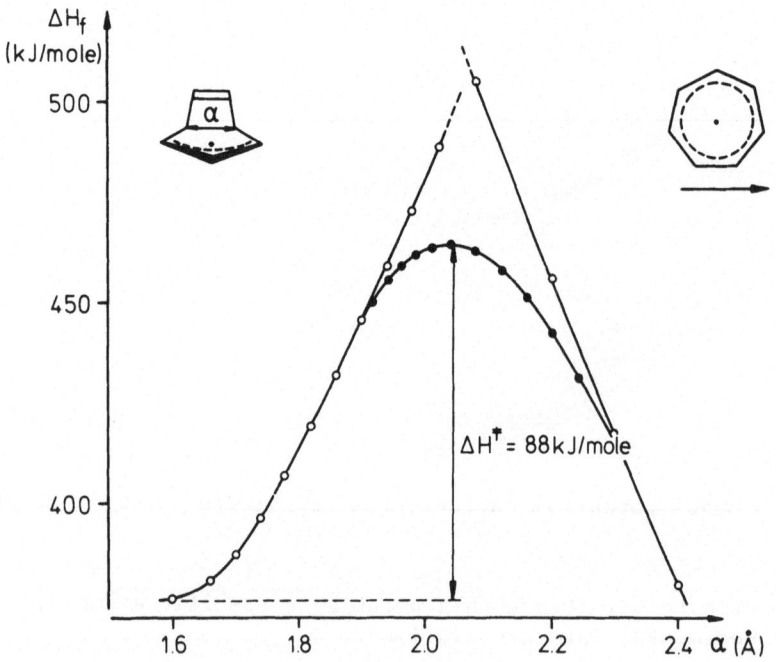

Fig. 6: Calculated reaction profile for the ringenlargement of
         bicyclo-[3,2,0]-heptadienyl radical.

The calculated activation enthalpy amounts to 88 kJ/mole [23],
the free activation enthalpy $\Delta G^{\ddagger}_{323}$ (which includes the estimated
small change in entropy) to 86 kJ/mole [25]. The calculated
structures are shown in the usual way in Fig. 7.

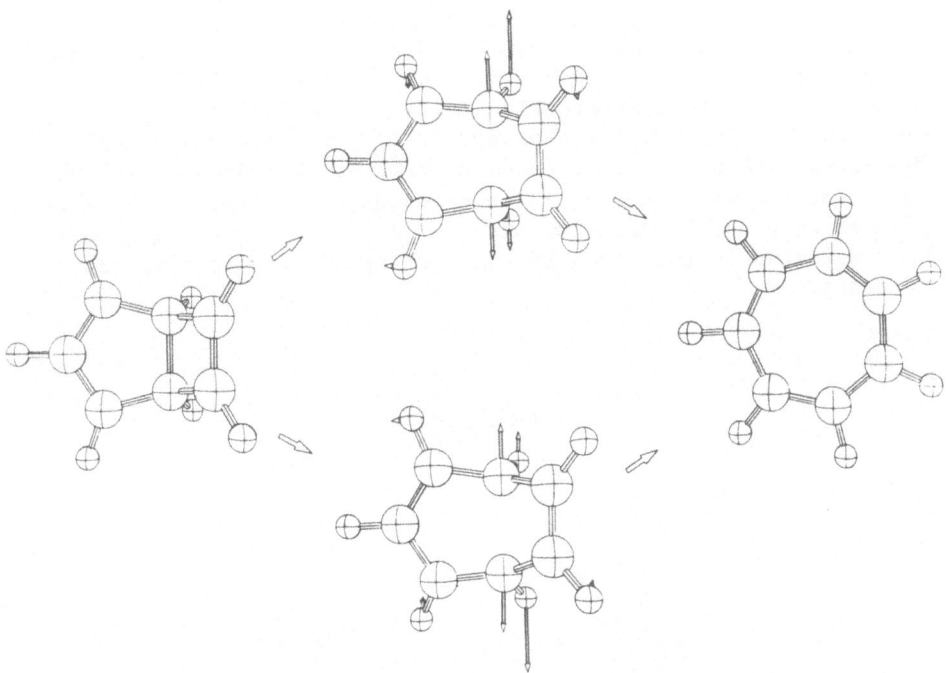

Fig. 7: Calculated structures of the ringenlargement of
bicyclo-[3,2,0]-heptadienyl radical.

The given examples show that "allowed" as well as "forbidden" reaction pathways can be treated reasonably well with
MINDO/3-UHF. We are aware that the tremendous numerical accuracy encountered above is probably due to a fortuitous cancellation
of errors, because the standard deviation of radicals' heats of
formation in their equilibrium structure is much larger as already anticipated in Fig. 1.

We shall now turn to a series of related reactions which
have attracted interest of many experimentalists and have given
birth to a vast number of papers [26). Their main appeal lies in
the fact that the observed reaction pathways follow a course
which is unexpected.

Radicals add very rapidly to molecules containing a CC
double bond. This intermolecular reaction is known to involve a
small activation barrier in the order of twenty kJ/mole. Theory [27) is on line with experiment [28), that in case of relatively
nonpolar additions, this reaction takes the Markownikow route
which leads to the thermodynamically more stable products.

If the double bond and the radical center are part of the same molecule, the addition (which in this case is a cyclization) proceeds exclusively along the alternative anti-Markownikow course if the molecule is sufficiently small [29], while for larger systems, the "normal" Markownikow cyclization is observed as well [30]. The parent system is illustrated in Fig. 8. For each number of methylene groups n, there are three types of reactants and likewise three possible reaction pathways A, M, and S, which have to be considered in the study of the corresponding potential energy surfaces.

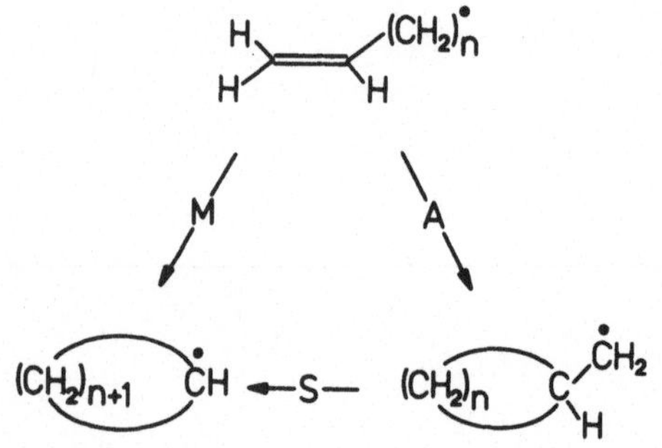

Fig. 8: Parent system of ω-alkenyl radical cyclizations.

Since in this notation the number n includes the radical center, the smallest member of the series is the homoallyl radical (n = 2), where the two functional groups of the open chain ω-alkenyl radical are still separated by one linking methylene group.

We have studied the four homologue systems with n = 2, 3, 4, and n = 5 [16,31]. First, we located the structures of the reactants, and we shall look at the obtained results. Fig. 9 summarizes the four lowest minima of the ω-alkenyl radicals. All other rotamers were found to be slightly higher in energy.

Apparently the structures have very much in common: They consist of an all trans-methylene-chain substituting the terminal double bond in gauche conformation, while the radical center plane at the other end in all radicals intersect the adjacent methylene group.

As already shown in Fig. 2, the structure of the smallest member of the series has been calculated by ab initio [32], Hehre's result being essentially identical with ours except for the conformation of the radical site, where the two results contradict.

Our prediction is strongly supported by experimental findings based on ESR experiments [29].

Fig. 9: Equilibrium structures of ω-alkenyl radicals.

The charge distribution at the reaction sites is worth to be mentioned: The double bond is strongly polarized, the terminal carbon atom being the negative end of the local dipole. The radical center carbon atom is negatively charged too.

Both effects can be explained by simple perturbational models, the former has been analyzed in detail by Hoffmann [33].

Fig. 10 shows the calculated structures of the observed cyclo-alkyl carbinyl radicals. In the cyclopropyl carbinyl radical, the exocyclic methylene group assumes the perpendicular conformation, while in all other members of the series, the equilibrium confor-mation is parallel with the ring of the cycloalkyl substituent.

In the cyclopropyl carbinyl radical, the cyclopropyl substi-tuent provides a very effective orbital to conjugate with the localized ρ orbital at the radical center: The antisymmetric Walsh orbital [34].

The resulting conjugation leads to a marked stabilization of the radical which more than outbalances the repulsion between the two eclipsed hydrogenatoms. For any higher homologue such an effect is much smaller because the corresponding Walsh orbitals are too low in energy.

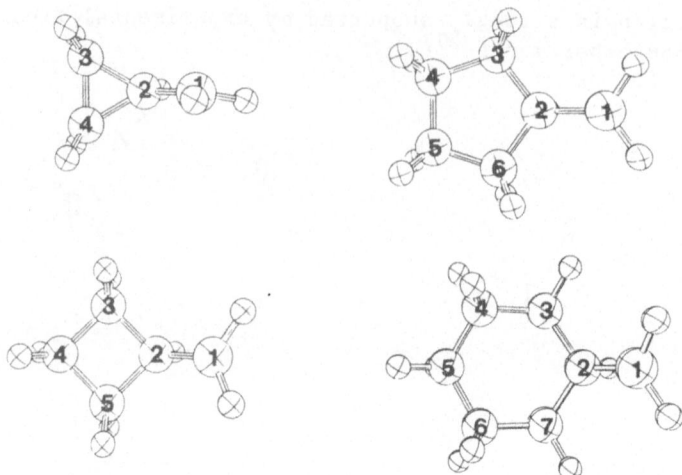

Fig. 10:   Equilibrium structures of cycloalkyl carbinyl radicals.

This leads to a relatively high rotational barrier around the exocyclic CC bond of 7 kJ/mole for the cyclopropyl carbinyl radical, while in all other cases the corresponding values were less than 2 kJ/mole.

To complete the summary of reactant structures, the calculated geometries of the cycloalkyl radicals are shown in Fig. 11. They show no peculiar aspects worthwhile to be discussed in this context.

The calculated heats of formation of the reactants predict that in all cases the cycloalkyl radical is the most stable. The ω-alkenyl radical is the least stable with the exception of the smallest system, where the ringstrain of the three-membered ring reverses the ordering of the ω-butenyl and the cyclopropyl carbinyl radical.

This is on line with experiment [15], although the absolute heats of formation are all too low as anticipated earlier.

The outcome of the cyclization reaction is not governed by the stability of the products. This is a fact which follows from the experimental findings and which is reproduced by our calculations.

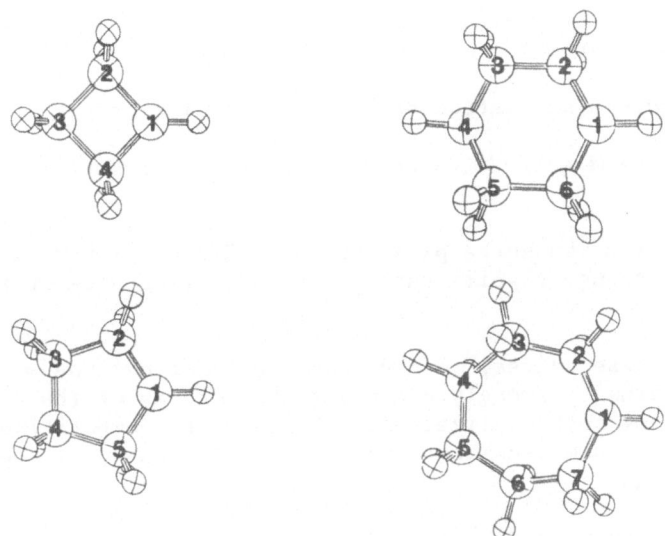

Fig. 11: Equilibrium structures of cycloalkyl radicals.

The calculated activation parameters are summarized in Tab. 3. For the activation entropies of route A and route M cyclizations, we needed to estimate the standard entropies of the open chain ω-alkenyl radicals, which for the reasons given above should not be obtained by the direct calculations.

Tab. 3: Comparison of energies, entropies and free energies of activation for the three investigated types of rearrangements. ($\Delta H^{\neq}$ and $\Delta G^{\neq}_{298}$ in kJ/mole, $\Delta S^{\neq}_{298}$ in J/Kmole.)

n	Route M			Route A			Route S		
	$\Delta H^{\neq}$	$\Delta S^{\neq}_{298}$	$\Delta G^{\neq}_{298}$	$\Delta H^{\neq}$	$\Delta S^{\neq}_{298}$	$\Delta G^{\neq}_{298}$	$\Delta H^{\neq}$	$\Delta S^{\neq}_{298}$	$\Delta G^{\neq}_{298}$
2	124	−39	135	52	−34	62	136	−15	140
3	91	−46	105	76	−43	88	139	−22	146
4	67	−47	81	67	−34	77	136	−22	142
5	65	−60	82	77	−60	95	126	−19	132

Looking at the standard entropies of linear alkanes [35], it can be found that the lengthening of the chain by one methylene group raises the standard entropy consistently by 40 J/Kmole. Therefore, we calculated the standard entropy of the smallest member of the series, adding 40 J/Kmole to each next higher homologue.

From the calculated activation enthalpies the following trends are apparent:

1) The route A activation enthalpy slowly increases,

2) The route M activation enthalpy at first strongly decreases, while

3) The activation enthalpies for route S are all very high and do not change significantly with increasing size of the system.

From these values, we conclude, that for n<4 the anti-Markownikow pathway strongly dominates the outcome of the cyclization, while for n⩾4 both cyclization mechanisms are competitive, the Markownikow route becoming more and more dominant, as the system becomes large.

Furthermore, a direct rearrangement via route S is highly unlikely. This pathway corresponds to a 1,2-alkyl shift, which indeed has never been observed in saturated radical systems [30].

All these results have been fully corroborated by experiment [26]. The reasons for these trends can be found in the structures of the corresponding saddle points.

We shall begin the discussion with the route A activated complexes. The obtained structures are shown in Fig. 12, where the arrows display the intrinsic reaction coordinates.

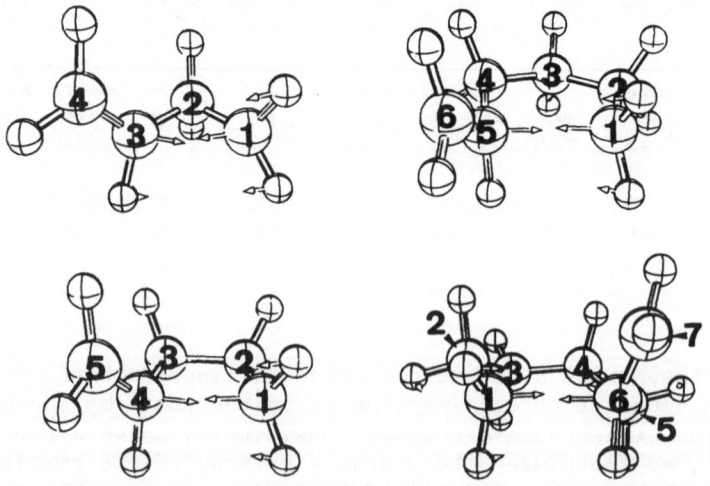

Fig. 12: Activated complexes for route A cyclizations.

Note, that in all cases, the terminal double bonds are essentially intact and that the transition vectors resemble a simple stretch vibration.

This is quite different in the saddle points associated with route M shown in Fig. 13.

The formerly simple intrinsic reaction coordinates in these cases are strongly coupled with a rotation of the methylene group which in the ω-alkenyl radical was part of the terminal double bond.   The rotation diminishes, as the size of molecule increases.

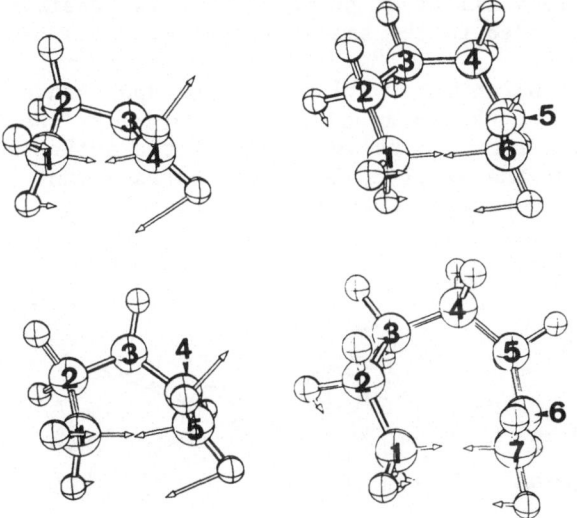

Fig. 13:   Activated complexes for route M cyclizations.

This rotation indicates a dominant decoupling effect of the double bond in the activated complex of small systems.   Such a torsional twist is associated with a large increase in energy.

The amount of the double bond decoupling is given by the corresponding dihedral angle between the p orbitals which participate in the π bond.

These angles α were calculated to be 28° for n=2, 18° for n=3, 9° for n=4, and 5° for n=5.

In a simple perturbational picture the associated raise in energy ΔE can be estimated, if we assume that the π/π^{*} -split is proportional to the p/p overlap, and that the complete decoupling of ethylene amounts to 260 kJ/mole [36].

The corresponding results are 31 kJ/mole for n=2, 13 kJ/mole for n=3, 3 kJ/mole for n=4, and is negligible for any higher member of the series.

This indeed accounts for the major part of the trend calculated for the route M activation energies. The decoupling angle α decreases as n gets larger, because the radical gains more dynamic freedom to attack the double bond in a better fashion.

An entirely different point of view is possible, if the series is extended to the smallest system, where n=1.

This extension was initiated by the appearance of the activated complex of the Markownikow cyclization of the ω-butenyl radical, which reminded us very much of the activated complex we calculated for the allyl/cyclopropyl rearrangement [37]. The two calculated structures are compared in Fig. 14.

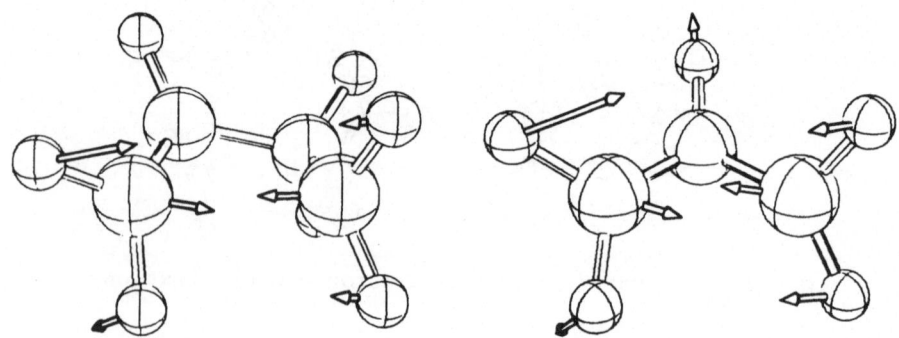

Fig. 14:   Comparison of activated complexes of route M cyclization
            of ω-butenyl radical with the allyl/cyclopropyl rearrange-
            ment.

If the route A and route M cyclizations are formulated accordingly, it turns out that the former is simply a rotation of the terminal methylene group, thus leading to the "two-membered" ring, while the latter corresponds to the classical case of a forbidden radical rearrangement: The allyl/cyclopropyl rearrangement. Updating the calculated activation energies finally leads to Fig. 15, which turns the formerly unexpected outcome of ω-alkenyl radical cyclizations into the most natural, well understood experimental fact.

This extension, which in the backward glance is trivial, has to our knowledge never been made so far. We definitely believe that such insights are the real destinations which should be reached by "experimental computer chemistry".

Numerical accuracy of the method in this view is not a crucial point, but helps to trace the reasons for the observed trends to the correct origins.

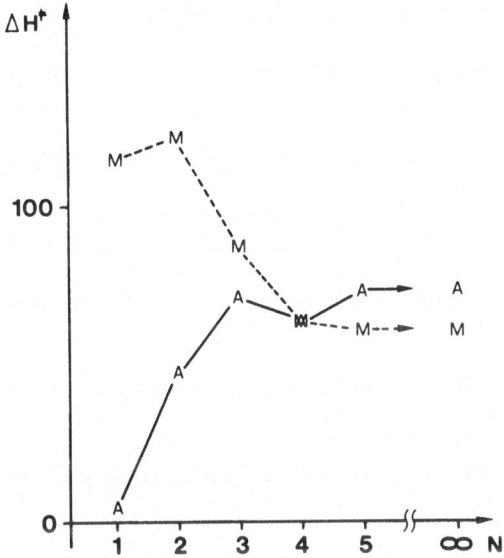

Fig. 15: Correlation of activation enthalpies of route A and route M cyclizations with increasing number of methylene groups n. n=∞ corresponds to the intermolecular radical addition.

Acknowledgment. The author is grateful to Professor C. Rüchardt, who initiated the study of ω-alkenyl radical cyclization mechanisms by a motivating discussion, Mrs. P. Hofmann for patiently typing the manuscript and Dr. M. Gold who prepared the slides.

References:

1) M. J. S. Dewar, Chemistry in Britain, 11, 97 (1975).

2) H. Eyring, J. Chem. Phys., 3, 107 (1935).

3) R. C. Bingham, M. J. S. Dewar and D. H. Lo, J. Am. Chem. Soc.,
   97, 1285 (1975).

4) R. C. Bingham, M. J. S. Dewar and D. H. Lo, J. Am. Chem. Soc.,
   97, 1294, 1302, 1307 (1975); M. J. S. Dewar, D. H. Lo and
   C. A. Ramsden, ibid, 97, 1311 (1975).

5) M. J. S. Dewar, J. A. Hashmall and C. G. Venier, J. Am. Chem.
   Soc., 90, 1953 (1968).

6) P. Bischof, J. Am. Chem. Soc., 98, 6844 (1976).

7) Experimental results taken from G. Herzberg, "Spectra of
   Diatomic Molecules", Van Nostrand, Princeton, N. J., 1950.

8) M. J. S. Dewar and G. P. Ford, J. Am. Chem. Soc., 99, 1685
   (1977).

9) W. A. Lathan, W. J. Hehre and J. A. Pople, J. Am. Chem. Soc.,
   93, 808 (1971).

10) See for example: R. Fletcher, "Optimization", Acad. Press,
    London (1969).

11) R. Fletcher and M. J. D. Powell, Comput. J., 6, 163 (1963).

12) E. B. Wilson, J. C. Decius and P. C. Cross, "Molecular Vibra-
    tions", McGraw Hill, N.Y. 1955.

13) P. Pulay, Molec. Phys., 17, 197 (1969).

14) F. N. Godnew, "Berechnungen thermodynamischer Funktionen aus
    Moleküldaten", Deutscher Verlag der Wissenschaften, Berlin
    (1963).

15) H. E. O'Neal and S. W. Benson in J. K. Kochi, "Free Radicals",
    Vol. II, 275 (1973), J. Wiley & Sons, N.Y. 1973.

16) P. Bischof, Tetrahedron Lett., 15, 1291 (1979).

17) J. W. McIver Jr. and A. Komornicki, J. Am. Chem. Soc., 94,
    2625 (1972).

18) K. Fukui, S. Kato and H. Fujimoto, J. Am. Chem. Soc., 97, 1 (1975).

19) B. Giese and K. Jay, Chem. Ber., 110, 1364 (1977).

20) B. Giese, private communication.

21) R. B. Woodward and R. Hoffmann, "The Conservation of Orbital Symmetry", Verlag Chemie, Weinheim, 1970.

22) H. C. Longuet-Higgins and E. W. Abrahamson, J. Am. Chem. Soc., 87, 2045 (1965).

23) P. Bischof, J. Am. Chem. Soc., 99, 8145 (1977).

24) R. Sustmann and D. Brandes, Tetrahedron Lett., 1791 (1976).

25) P. Bischof, unpublished results.

26) M. Julia, Pure Appl. Chem., 15, 167 (1967); J. Julia, Accounts Chem. Res., 4, 386 (1971); C. Walling, J. H. Cooley, A. A. Ponaras, and E. J. Racah, J. Am. Chem. Soc., 88, 5361 (1966), and references cited therein

27) V. Bonacić-Koutecky, J. Koutecky and L. Salem, J. Am. Chem. Soc., 99, 842 (1977); M. J. S. Dewar and S. Olivella, J. Am. Chem. Soc., 100, 5290 (1978), and references cited therein.

28) See for example B. Giese and J. Meixner, Tetrahedron Lett., 32, 2779 (1977).

29) D. J. Edge and J. K. Kochi, J. Am. Chem. Soc., 94, 7695 (1972).

30) J. W. Wilt in J. K. Kochi "Free Radicals", Vol. I, J. Wiley & Sons, New York (1973).

31) P. Bischof, submitted for publication.

32) W. J. Hehre, J. Am. Chem. Soc., 95, 2643 (1973).

33) L. Libit and R. Hoffmann, J. Am. Chem. Soc., 96, 1370 (1974).

34) A. D. Walsh, Nature, 159, 167, 712 (1947).

35) A summary of related data are cited in ref. 15).

36) M. H. Wood, Chem. Phys. Lett., 24, 239 (1974).

37) G. Friedrich and P. Bischof, to be published.

COMPUTED PHYSICAL PROPERTIES OF SMALL MOLECULES

Wilfried Meyer and Peter Botschwina

Fachbereich Chemie der Universität Kaiserslautern

D-6750 Kaiserslautern, Germany

Pavel Rosmus and Hans-Joachim Werner

Fachbereich Chemie der Universität Frankfurt

D-6000 Frankfurt, Germany

INTRODUCTION

There is no reliable intrinsic criterion which allows one to judge the quality of a computed quantum mechanical wavefunction with respect to a particular property. Even if one uses a good approximate solution of the many-electron Schrödinger equation, the numerical values of physical properties of molecules may exhibit errors which are hardly predictable from a priori arguments. This is not surprising if one considers the numerous approximations that are made even in calculations accounting for large portions of the electron correlation. The only practicable way to establish the degree of reliability of quantum chemical calculations seems to consist in systematic investigations of homologous or related molecules, where comparison with accurate experiments is possible for at least several members of a series.

In this contribution, we will give a brief summary of a few

representative studies performed in our laboratories, dealing with the calculation of various properties of small molecules from high-ly correlated electronic wavefunctions. In particular the following topics will be treated: potential energy functions and spectroscopic properties derived from them, dissociation energies, ionization energies, proton and electron affinities, dipole moment functions and static dipole polarizabilities.

In earlier applications, we have employed the PNO-CE method (Configuration Expansion by means of pseudonatural orbitals), which allows for a particular compact representation of the wavefunction without any substantial loss in correlation energy. More recent results have been obtained by the Self-Consistent Electron Pairs (SCEP) method, which yields an exact solution within the given con-figuration space of single and double substitutions with respect to the Hartree-Fock reference function. In addition to the tradi-tional treatment (CI with singles and doubles), both methods may employ the Coupled Electron Pair Approximation (CEPA), which modi-fies the Schrödinger equation so as to approximately account for the effect of the most important higher substitutions. For the purpose of this symposium, we feel that emphasis should be given to discus-sion of the quality of the numerical results, and, therefore, we re-fer to the literature[1-3] for details of our computational methods.

APPLICATIONS

1. Potential energy functions and derived spectroscopic properties

In our systematic investigation of the ground states of the diatomic hydrides LiH to HCl,[4] between 95 % (LiH) and 85 % (HCl) of the valence shell correlation energy have been accounted for. The derivation of the familiar spectroscopic constants (r_e, ω_e, α_e and $\omega_e x_e$) from the pointwise given potential functions by means of polynomial fitting has been carefully checked for numerical sig-nificance. The standard deviations in the spectroscopic constants amount to: r_e: 0.003 Å, ω_e: 14 cm^{-1}, α_e: 0.005 cm^{-1}, and $\omega_e x_e$: 1.5

cm^{-1}. Somewhat larger errors may result for diatomics with multiple
bonds, since higher substitutions play a larger role in these cases.
For polyatomic molecules, the size of the tractable basis set may
further restrict the achievable accuracy. Table 1 contains several
typical examples.

The good agreement obtained for the neutral diatomic hydrides[4]
allowed us to make rather accurate predictions of thus far unknown
spectroscopic constants for a larger number of positive[9] and negative[10] ions. Particularly for the negative ions, the observation
of gas-phase emission or absorption spectra seems to be very difficult. Only relatively crude information is available from photodetachment measurements. The Franck-Condon factors computed from
our potential curves of the AH and AH⁻ species fully support the
assignments of the photoelectron spectra of the negative ions (see
Ref. 10 for details). However, it could be demonstrated that erroneous spectroscopic constants had been derived from the spectra
in most cases due to an incorrect guess of the sign of the geometry
shift.

While the construction of potential curves from experimental
data is now almost a routine matter for diatomic molecules, the
situation is much less favorable for polyatomic molecules. Even
for stable triatomics, very few reliable "experimental" potential
energy functions (PEFs), which cover a larger region around the
equilibrium geometry, have been published. Present day ab initio
calculations employing highly correlated wavefunctions yield PEFs
from which the vibrational term energies are obtained with an error
of about 1 per cent. Such an accuracy is certainly sufficient to
make valuable predictions for unstable molecules and excited or
ionized states of stable molecules; calculations of this kind may be
of help to experimentalists involved in matrix spectroscopy, laser
Stark spectroscopy and electron or photoelectron spectroscopy. Vibrational band centers may be obtained by diagonalizing the vibration-
rotation Hamiltonian in a basis of products of one-dimensional oscil-

Table 1: Comparison of selected theoretical and experimental
spectroscopic constants

		R_e (Å)	α_e (cm^{-1})	ω_e (cm^{-1})	$\omega_e x_e$ (cm^{-1})
OH [a]	Theory	0.971	0.724	3743	85
	Experiment	0.971	0.714	3740	86
HF [a]	Theory	0.917	0.787	4169	90
	Experiment	0.917	0.795	4138	90
SiH [a]	Theory	1.526	0.216	2034	36
	Experiment	1.520	0.219	2041	35
HCl [a]	Theory	1.278	0.309	2977	53
	Experiment	1.275	0.307	2991	53
HBr [b]	Theory	1.418	0.234	2644	44
	Experiment	1.415	0.233	2649	45
BO [c]	Theory	1.210	0.017	1873	11.8
	Experiment	1.204	0.017	1886	11.8
PN [d]	Theory	1.499	0.004	1298	5.4
	Experiment	1.491	0.005	1337	6.9
CH_4 [e,f]	Theory	1.091		3037	13.6
	Experiment	1.086		3022	11.0
SiH_4 [g,f]	Theory	1.486		2265	8.7
	Experiment	1.473		2261	7.3

(a) Ref.4; (b) Ref. 5; (c) Ref. 6; (d) Ref. 7; (e) Ref. 1;
(f) A. Robiette, private communication; (g) Ref. 8

lator functions (for a review of different techniques see Ref. 11).
One example of recent unpublished work is given in Table 2
and will be discussed in some detail. The lowest singlet state of

the HPO molecule was observed in emission several years ago[12], but only recently were all three fundamentals observed in an Argon matrix at $4°K$.[13,14] A new analysis of the gas-phase emission spectrum of $HP^{16}O$, published just a few months ago[15], yielded improved values for ν_2, ν_3, $2\nu_2$ and $\nu_2 + \nu_3$, which are in very good agreement with the calculated values (see Table 2). Our calculations yield a strong anharmonic resonance between the first overtone of the "bending" mode and the "PH stretching" fundamental. The absorption at higher frequency, which is a bit more PH stretching in character, is calculated to be 2016 cm^{-1} and is thus smaller than the matrix value by as much as 79 cm^{-1}. The matrix value, however, is very probably blue-shifted by at least 30 cm^{-1} (for HNO, the corresponding matrix shift amounts to 31 cm^{-1} and, since the PH potential in HPO is more shallow than the NH potential in HNO, ν_1 of HPO is probably somewhat more strongly affected by matrix effects). On the other hand, the calculations probably underestimate ν_1 and simultaneously overestimate the PH equilibrium bond length to some extent.

Table 2: Theoretical and experimental vibrational frequencies for
the lowest singlet state of HPO and isotopes[a]

	$HP^{16}O$		$DP^{16}O$		$HP^{18}O$		$DP^{18}O$	
	theor.	exp.	theor.	exp.	theor.	exp.	theor.	exp.
ν_2	991	985	742	(750)	988	–	738	(746)
ν_3	1187	1188	1185	(1186)	1142	(1144)	1141	(1142)
ν_1	2016	(2095)	1452	(1530)	2013	(2095)	1451	(1530)
$\nu_2 + \nu_3$	2174	2169	1924	–	2127	–	1876	–
$2\nu_3$	2360	2362	2358	–	2273	–	2271	–

(a) P. Botschwina, unpublished. Experimental values are taken from Ref. 15; those values given in parentheses refer to matrix absorptions (see Ref. 13 and 14)

Table 3: Calculated and experimental vibrational band centers (in cm^{-1}) for water isotopes (a)

Quantum numbers n_1 n_2 n_3	H$_2^{16}$O		H$_2^{17}$O		H$_2^{18}$O		HD^{16}O	
	calc.	exp.	calc.	exp. or predicted	calc.	exp. or predicted	calc.	exp.
0 1 0	1594.6	1594.6	1591.2	(1591.2)	1588.2	1588.3	1403.3	1403.5
0 2 0	3149.7	3151.6	3143.3	(3145.2)	3137.4	(3139.3)	2781.0	2782.2
1 0 0	3657.1	3657.1	3653.2	3653.2	3649.8	3649.7	2723.0	2723.7
0 0 1	3755.9	3755.9	3748.3	3748.4	3741.5	3741.6	3707.3	3707.5
0 3 0	4662.7	4666.8	4653.3	(4657.4)	4644.8	(4648.9)	4145.5	4145.6
1 1 0	5232.6	5235.0	5225.7	(5228.1)	5218.9	5221.2	4097.1	4100.0
0 1 1	5335.5	5331.3	5324.5	5320.3	5314.7	5310.5	5090.4	5089.6
0 4 0	6128.5	6136.4	6116.4	(6124.3)	6105.3	(6113.2)	5506.7	-
1 2 0	6766.4	6775.0	6757.1	(6765.7)	6747.8	(6756.4)	5416.4	-
0 2 1	6877.0	6871.7	6862.9	(6857.6)	6850.2	(6844.9)	6451.4	6452.1
2 0 0	7202.0	7201.5	7194.7	(7194.2)	7187.2	7185.9	5364.5	5363.6
1 0 1	7249.7	7249.8	7239.3	7238.7	7229.3	7228.9	6415.1	6416.6
0 0 2	7443.6	7445.1	7431.2	(7432.7)	7418.9	(7420.4)	7250.3	-

(a) for details of the potential energy function and the experimental literature see Ref. 16

The calculated PEFs may be improved by making use of experimental information. A very simple approach, requiring only the fundamentals of one isotope (and, if available, the "experimental" equilibrium geometry), has recently been applied to H_2O, HNO, HOF and HOCl.[16] Results obtained for four different isotopes of water are given in Table 3. The standard deviation for the lowest 13 band centers of $H_2^{16}O$ is 4.2 cm^{-1}, the maximal error amounting to 8.6 cm^{-1}. Similar accuracy is obtained for the other 6 isotopes studied so far,[16] and many of the calculated values stand as predictions. Isotopic shifts in the fundamentals are reproduced within an accuracy of 0.5 cm^{-1}, and - since the mass effect is small - very accurate predictions should be possible for $H_2^{17}O$ and $H_2^{18}O$ by assuming the same errors as obtained for $H_2^{16}O$; these numbers are given in parentheses in table 3 when experimental values are missing.

2. Dissociation energies and proton affinities

The accurate determination of dissociation energies of molecules is of fundamental importance for thermochemistry. Since the heats of formation of polyatomic molecules are obtained by combining thermochemical data of smaller fragments, there is also a need for the knowledge of the dissociation energies of diatomic hydrides, which are so far rather uncertain in many cases. Unfortunately, the theoretical calculation of dissociation energies presents a rather serious problem, as well. If the electronic structure changes strongly in the dissociation products relative to the educts, one must expect large correlation contributions to the dissociation energy, which in principle cannot be fully accounted for by a quantum-chemical calculation involving a limited basis set. On the other hand, if the electronic structure of the products closely resembles that of the educts, the correlation contributions may be quite small. This is the reason, why proton affinities can be calculated very accurately.

In our study of the dissociation energies of the AH systems, we eliminated this difficulty by investigating a series of homologous molecules using a well-defined scheme for the choice of all parameters

entering the computation. As demonstrated in table 4, the theoretical
D_o values reveal a systematic error which steadily increases to
the right end of the row of the periodic system. Since the extra correlation caused by bond formation is to a large part of the high-energy type, the basis set deficiencies affect it to about the same degree
as the atomic correlation. E.g., about 82 per cent of the correlation
contribution to D_e $(CH_4 \rightarrow C + 4 H)$ are obtained from a PNO-CEPA calcula
tion accounting for about 89 per cent of the correlation energy.[1]

The systematic behavior of the remaining errors allowed us to recom
mend values for so far uncertain dissociation energies of diatomic hydrides.[4] E.g.,a value of 1.23±0.05 eV was recommended for MgH in full
agreement with an experimental value of 1.27±0.03 eV published three
years later.[17] For the reasons mentioned above, the calculated dissociation energies of the AH^+ systems and the proton affinities of the
first and second row atoms agree much better - within about 0.05 eV -
with the known experimental values. Thus the value calculated for D_o
of OH^+ (5.03 eV) is in excellent agreement with the recently published
value of 5.00±0.01 eV.[18]

For the negative ions AH^- we have compared our calculated dissociation energies with empirical ones obtained from the relation
$D_o(AH^-) = EA_o(AH) + D_o(AH) - E(A;H)$.[10] Again, the very systematic behavior of the errors suggests that the data used in the above cycle
are also consistent with the calculation of $D_o(AH^-)$. A summary of
calculated and experimental dissociation energies for the diatomic
hydrides of the first row and their ions is given in Table 4.

3. Ionization energies

During the last decade, much effort has been spent in the investigation of valence shell and core ionization processes. In this fiel
theoretical studies also made many valuable contributions to various
spectroscopic problems. The systematic work performed for the first
ionization energies of the diatomic hydrides allows well to demonstrat
the reliability of ionization energies calculated from highly correlated wavefunctions. In Table 5, theoretical values for ionization

Table 4: Dissociation energies (D_o, in eV) of the first-row hydrides

	AH		AH^+		AH^-	
	CEPA	exp.	CEPA	exp.	CEPA	exp.
LiH	2.40	2.429	0.13	0.09+0.06	1.96	–
BeH	2.01	1.99+0.01	3.04	3.10+0.08	1.80	1.94+0.1
BH	3.34	3.42+0.04	2.02	1.95+0.09	2.68	–
CH	3.29	3.45+0.01	4.03	4.04+0.02	3.32	3.43+0.1
NH	3.18	3.20+0.16	3.48	3.7+0.4	3.60	3.85+0.15
OH	4.11	4.4+0.01	5.03	4.82+0.15	4.62	4.77+0.01
HF	5.58	5.86+0.02	3.45	3.42+0.01	–	–

energies (calculated as energy differences $E(AH^+) - E(AH)$) are compared with the available experimental values. The PNO-CEPA calculations covered about the same portions of the valence shell correlation
energies as in the neutral hydrides; i.e., between 84 and 95 per
cent in the first row and between 80 and 93 per cent in the second
row. The calculated correlation contributions to the ionization energies reveal errors which closely parallel those of the valence correlation energies, i.e., they increase up to 16 per cent for HF and up
to 23 per cent for HCl. These results demonstrate that errors of a few
tenths of an electron volt must be expected for any calculation unless the computational method allows for compensating errors or,
accidentally, the correlation contribution to the ionization energy
is very small. Nevertheless, such an accuracy can be regarded as
sufficient for the assignments of the photoelectron spectra.

Since for many of the diatomic hydrides only rough estimates
of the ionization energies are available in the literature, our
recommended values represent a valuable help for experimentalists.
For instance, the ionization energy of the OH radical was predicted
as 13.0±0.1 eV, and the recently published photoionization experiments yielded 13.01 eV.[18]

Table 5: Adiabatic ionization energies of the first- and second-
         row diatomic hydrides (in eV; estimates in parentheses)

Hydride	SCF	CEPA	Exptl.
LiH	6.63	7.66	(6.5±0.5)
BeH	8.15	8.31	8.21±0.08
BH	8.44	9.63	9.77±0.05
CH	10.04	10.53	10.64±0.01
NH	12.75	13.30	13.10±0.2
OH	11.40	12.68	13.18±0.1
HF	14.31	15.74	16.044±0.003
NaH	5.76	6.88	(6.5)
MgH	6.76	6.92	(6.8)
AlH	7.34	8.35	(8.4)
SiH	7.36	7.84	8.01±0.08
PH	9.63	10.04	(9.5±0.5)
SH	9.23	10.02	10.40±0.03
HCl	11.66	12.49	12.748±0.005

4. Electron affinities

A large amount of experimental work involving techniques like
electron impact, ion cyclotron resonance, or laser photodetachment
measurements has been devoted to the investigation of negative ions,
partly due to their importance for processes in the ionosphere. In
this connection, it is of considerable interest to know if the
ground states of the negative ions are bound, i.e., if a molecule possesses a positive electron affinity. However, this is not
easy to decide by theoretical methods, since the changes in the electron correlation energies caused by the attachment of an electron
are of the same order of magnitude as the electron affinities themselves. One may differentiate between three main effects which com-
pensate each other partially. Firstly, the correlation increases by
the extra correlation of the added electron with all the others;
secondly, there is a decrease of the correlation among the other
electrons due to the fact that the added electron occupies orbital
space which was available for correlation in the neutral system;

thirdly, there is generally a slight increase in the correlation due
to orbital expansion resulting from the excess of negative charge.
In the case of the OH radical, these correlation effects have been
calculated as 2.0 eV, -0.6 eV and 0.2 eV, respectively, while the
observed electron affinity amounts to 1.83 eV. The deficiences of a
finite basis set mainly affect the first term and are therefore relatively enhanced by its partial compensation with the second term.

The electron affinities of the diatomic hydrides, calculated
by the PNO-CEPA method, are given together with the observed values
in Table 6. The expected defect varies between 0.16 eV (SiH) and

Hydride	CEPA	Experiment
LiH	0.26	–
BeH	0.48	0.7+0.1
BH	0.03	(0.$\overline{1}$5)
CH	1.04	1.25
NH	0.01	0.38
OH	1.51	1.83
NaH	0.31	–
MgH	0.83	1.05+0.06
AlH	0.03	(0.$\overline{1}$5)
SiH	1.13	1.29
PH	0.76	1.02
SH	2.12	2.33

Table 6: Adiabatic electron affinities of the first- and
second-row diatomic
hydrides (in eV; estimates
in parentheses)

0.37 eV(NH), but all trends of the observed affinities are well
reproduced. For instance, the calculations support the somewhat surprising finding that the electron affinities of CH and SiH are very
similar (1.25 and 1.29 eV, respectively) whereas those of NH and PH
are rather different (0.38 and 1.02, respectively). The decrease
in the electron affinity from the fourth to the fifth group may be
understood from the fact that the attached electron has to be placed
into an orbital which is already occupied by another electron, and
thus it experiences relatively strong Coulomb repulsion.

Studies of negative ions by photoionization techniques may

also yield valuable information concerning the neutral species. For instance, it should be possible to observe the transient species vinylidene[19], H_2CC, cyclopentadienylene, C_5H_4[20], and thioformaldehyde[21], H_2CS, by photoionization of their radical anions in their ground states which are calculated to be bound.

5. Dipole moment functions

The dipole moments as functions of the internuclear distances provide the basis for the calculation of vibration-rotation transition probabilities and are thus of crucial importance for the analysis of infrared chemiluminescence experiments (such experiments give information about reaction dynamics and chemical laser performance). There are serious problems in obtaining accurate dipole moment functions from experimental data and there is really a need for good functions from quantum-chemical calculations.

The most complete and reliable dipole moment information yet available from experiments for any molecule is the dipole moment function of hydrogen fluoride in its electronic ground state[22].

This function is well suited for a sensitive check of the quality of the computed wavefunction for a larger region of internuclear separation. We have performed a series of calculations for this molecule employing a large and flexible basis set for comparing the performance of different methods (SCEP, SCEP-CEPA and MC-SCF)[23]. The SCEP-CEPA wavefunction, which covers the largest portion of the correlation energy, is virtually identical with the empirical curve over the given region (see

Table 7: Dipole moment functions for HF

R (a.u.)	SCEP-CEPA (Debye)	Experiment
1.2	1.389	1.380
1.3	1.462	1.453
1.4	1.537	1.530
1.5	1.615	1.609
1.6	1.694	1.689
1.7331	1.801	1.796
1.8	1.855	1.850
1.9	1.933	1.929
2.1	2.081	2.078
2.3	2.208	2.209
2.5	2.302	2.312
2.7	2.351	2.375
2.9	2.345	2.378
3.1	2.276	2.295
3.2	2.218	2.209

Table 7). Thus the theoretical function most likely provides very precise transition probability coefficients for all transitions up to about the nineth vibrational level of HF.

We have also calculated the dipole moment functions for the ground states of the HCl and HBr molecules, which, especially in the case of HBr, represent a definite improvement over previous empirical curves. In Table 8 we compare several vibrational matrix elements of the dipole moments with experimental values.

Table 8: Vibrational matrix elements of the dipole moments for the HF, HCl and HBr molecules[23]

	HF Theory	HF Experiment[22]	HCl Theory	HCl Experiment[24]	HBr Theory	HBr Experiment[25]
μ_e	1.801–18	1.796–18	1.102–18	1.093–18	0.821–18	0.834–18
R_o^o	1.825–18	1.827–18	1.121–18	1.109–18	0.829–18	0.829–18
R_1^1	1.869–18	1.865–18	1.148–18	1.139–18	0.844–18	0.861–18
R_2^2	1.911–18	1.909–18	1.174–18	1.169–18	0.859–18	0.876–18
R_o^1	9.74–20	9.85–20	6.64–20	6.70–20	3.25–20	3.80–20
R_o^2	–1.28–20	–1.25–20	–7.65–21	–7.02–21	–2.29–21	–2.86–21
R_o^3	1.57–21	1.63–21	6.37–22	5.15–22	–2.92–22	–9.00–22

note: 1.801–18 is read as $1.801 \cdot 10^{-18}$ esu·cm; 1 Debye = 10^{-18} esu·cm

The dipole moment functions which were previously calculated for HF and HCl from PNO–CEPA wavefunctions[4] exhibit only slightly larger error bounds than the SCEP–CEPA functions. From the former dipole moment functions of the hydrides the rotationless Einstein coefficients of spontaneous emission have been derived and are listed in Table 9. Due to the systematic errors in the CEPA wavefunctions and the neglect of rotational effects, errors of 10 to 15 per cent may well occur in the tabulated values.

Table 9:   Rotationless Einstein coefficients of spontaneous emission for the $v=1 \to v=0$ transitions of diatomic hydride ground states (in sec^{-1})[26]

A =	Li	Be	B	C	N	O	F
AH	51.3	64.5	244.0	119.0	34.9	11.5	192.3
AD	16.5	20.1	73.5	34.8	9.7	3.6	56.1

A =	Na	Mg	Al	Si	P	S	Cl
AH	49.5	65.4	208.3	128.2	55.5	2.1	34.6
AD	13.4	18.6	56.2	35.2	15.2	0.5	9.5

6. Static dipole polarizabilities

Molecular polarizabilities are of relevance to phenomena like absorption, refraction, light scattering and, indirectly, inter-molecular forces. Depolarization ratios and Kerr constants may be evaluated from the polarizability anisotropies, which also play a role for the determination of quadrupole moments and anisotropies of magnetic susceptibilities.

Theoretical calculations of molecular polarizabilities have been restricted until 1975 almost exclusively to the coupled and uncoupled Hartree-Fock methods. Even for small molecules like H_2O, NH_3, or CH_4, the best ab initio calculations showed errors of more than 20 per cent.[27,28] These errors are due to both the use of inadequate basis sets and the neglect of electron correlation. Therefore, we found it worthwhile to investigate the requirements which have to be met for the calculation of molecular polarizabilities with an accuracy of a few per cent.[29]

Using the finite perturbation technique[30]   we have applied
the Hartree-Fock, PNO-CI and PNO-CEPA methods to calculate the polarizabilities of the molecules HF, H_2O, NH_3, CH_4, CO[29] and N_2.[31] Flexible Gaussian type basis sets with various optimized polarization
functions have been used such that the calculated polarizabilities
can be assumed to be very close to the limits of the respective
methods. Vibrational averaging, which has an effect of a few per
cent on the results, has been approximately taken into account.
The calculated mean polarizabilities are listed in Table 10 and
compared with the experimental values. As for other properties,

Table 10: Calculated and experimental static mean polarizabilities[a]

Method	HF	H_2O	NH_3	CH_4	CO	N_2
HF-SCF	4.98	8.68	13.61	16.69	12.40	11.67
PNO-CI	5.50	9.52	14.55	17.08	12.87	11.55
PNO-CEPA	5.67	9.86	14.96	17.22	13.13	11.55
Exp.	5.60	9.82	14.82	17.28	13.08	11.74

(a) in atomic units (1 a.u. = $1.48176 \cdot 10^{-25}$ cm^3)

the CEPA method yields the most accurate polarizabilities. The deviations from the experimental values are less than 2 per cent in all
cases.

The calculated polarizability anisotropies are presented in
Table 11 and compared with available experimental values. Polarizability anisotropies are small differences of the polarizability components and are therefore probably less accurate than the mean values.
The calculated anisotropies differ by 5 to 15 per cent from the experimental values of Muenter[32] for HF and Bridge and Buckingham[33] for
NH_3 and CO. The anisotropies obtained from Kerr measurements by Stuart
et al.[34] seem to be much less reliable. In particular their CO value,

which is larger by a factor of almost two, is ruled out by our calculations. For H_2O, experimental values of the polarizability component.
are obviously not yet known.

Table 11: Calculated and experimental polarizability anisotropies[a]

Method	HF $\alpha_{\parallel}-\alpha_{\perp}$	NH$_3$ $\alpha_{\parallel}-\alpha_{\perp}$	CO $\alpha_{\parallel}-\alpha_{\perp}$	N$_2$ $\alpha_{\parallel}-\alpha_{\perp}$	H$_2$O[b] α_{xx}	α_{yy}	α_{zz}
HF–SCF	1.45	0.80	3.19	5.22	9.37	8.06	8.62
PNO–CI	1.45	1.68	3.62	4.78	9.95	9.17	9.45
PNO–CEPA	1.44	2.24	3.91	4.48	10.14	9.66	9.79
Exp.	1.49	1.94	3.53	4.67			

(a) in atomic units, see Table 10; (b) C_2-axis identical with the
z-axis with the molecule lying in the xy-plane

For a detailed investigation of the basis set requirements and
correlation effects we refer to Ref. 29. A more detailed biblio-
graphy of previous theoretical and experimental work is also given
there.

The dependence of the mean polarizability on the internuclear
distances is of particular interest for the vibrational excitation
in molecular beam experiments with charged particles as well as for
Raman scattering spectroscopy. Such curves are very difficult to obtain from experimental information and theoretical ones are highly
required. As an example, the PNO–CEPA curve for N$_2$ in the range
$1.6 \leq R \leq 2.8$ (a.u.) is well approximated by the function[31]

$$\bar{\alpha}(R) = 11.51 + 5.3 (R - R_e) \quad \text{in a.u., with } R_e = 2.0744.$$

REFERENCES

1.   W. Meyer, Int. J. Quant. Chem. Symp. 5:341 (1971); J. Chem.
     Phys. 58: 1017 (1973); Theor. Chim. Acta 35:277 (1974)

2.   W. Meyer, "Configuration Expansion by Means of Psendonatural
     Orbitals", in: "Modern Theoretical Chemistry", Vol.3,
     H.F. Schaefer III, ed., Plenum Press, New York (1977); and
     references therein

3.   W. Meyer, J. Chem. Phys. 64:2901 (1976)

4.   W. Meyer and P. Rosmus, J. Chem. Phys. 63:2356 (1975)

5.   H.-J. Werner and P. Rosmus, to be published

6.   P. Botschwina, Chem. Phys. 28:231 (1978)

7.   P. Rosmus and W. Meyer, to be published

8.   A.M. Semkow, P. Rosmus, H. Bock and P. Botschwina, Chem. Phys.
     40:377 (1979)

9.   P. Rosmus and W. Meyer, J. Chem. Phys. 66:13 (1977)

10. P. Rosmus and W. Meyer, J. Chem. Phys. 69:2745 (1978)

11. G.D. Carney, L.L. Sprandel and C.W. Kern, "Vibrational
     Approaches to Vibration-Rotation Spectroscopy for Polyatomic
     Molecules", in: "Advan. Chem. Phys.", Vol. 37, I. Prigonine
     and Stuart A. Rice, eds., Wiley, New York (1978)

12. Lam Thanh My and M. Peyron, J. Chim. Phys. 59:688 (1962);
     60:1289 (1963); 61:1531 (1964)

13. M. Larzillière and M.E. Jacox, Proc. 10th Materials Research
     Symposium on Characterisation of High Temperature Vapor and
     Gases, J.W. Hastie, ed., Natl. Bur. Std, Washin

14. M. Larzillière and M.E. Jacox, "Infrared and ultraviolet
     absorption spectra of PO and HPO isolated in an argon matrix",
     to be published

15. M. Larzillière, N. Damany and Lam Than My, Canad. J. Phys.
     57:539 (1979)

16. P. Botschwina, Chem. Phys. 40:33 (1979)

17.   A. Balfour and B. Lindgren, Canad. J. Phys. 56:767 (1978)

18.   S. Katsumata and D. R. Lloyd, Chem. Phys. Letters 45:519 (1977)

19.   P. Rosmus, P. Botschwina and J. P. Maier, to be published

20.   P. Rosmus, unpublished work

21.   P. Rosmus and H. Bock, J. Chem. Soc., Chem. Commun. 334 (1979)

22.   R. N. Sileo and T. A. Cool, J. Chem. Phys. 65:117 (1976)

23.   H.-J. Werner and P. Rosmus, to be published

24.   F. G. Smith, J. Quant. Spectrosc. Radiat. Transfer 13:717 (1973)

25.   R. N. Stocker and A. Goldman, J. Quant. Spectrosc. Radiat. Trans-
      fer 16:335 (1976); F. A. van Dijk and A. Dymanus. Chem. Phys.
      Letters 5:330 (1970)

26.   P. Rosmus and W. Meyer, unpublished results

27.   G. P. Arrighini, C. Guidotti and O. Salvetti, J. Chem. Phys.
      52:1037 (1970)

28.   R. Moccia, Theor. chim. Acta (Berl.) 8:192 (1967); R. Moccia,
      J. Chem. Phys. 40:2164,2176,2186 (1964)

29.   H.-J. Werner and W. Meyer, Mol. Phys. 31:855 (1976)

30.   H. D. Cohen and C. C. J. Roothaan, J. Chem. Phys. 43:S34 (1965)

31.   P. Botschwina, in: Proceedings of the fourth seminar on computa-
      tional methods in quantum chemistry, B. Roos and G. H. F.
      Diercksen, eds., Max-Planck-Institut für Physik und Astrophysik,
      München (1978)

32.   J. S. Muenter, J. Chem. Phys. 56:5409 (1972)

33.   N. J. Bridge and A. D. Buckingham, Proc. R. Soc. A 295:334 (1966)

34.   H. A. Stuart, "Molekülstruktur", 3rd edition, Springer-Verlag,
      Berlin (1967)

CALCULATION OF ELECTRONICALLY EXCITED STATES IN MOLECULES: INTENSITY AND VIBRATIONAL STRUCTURE OF SPECTRA, PHOTOCHEMICAL IMPLICATIONS

S.D. Peyerimhoff

Lehrstuhl für Theoretische Chemie
Universität Bonn, 5300 Bonn 1, W. Germany

R.J. Buenker

Lehrstuhl für Theoretische Chemie
GesamthochschuleWuppertal, 5600 Wuppertal 1, W. Germany

ABSTRACT

The present contribution demonstrates the feasibility and reliability of present-day ab initio CI calculations for the prediction and interpretation of molecular spectra with respect to transition energy, vibrational structure and intensity. A short introduction is given into the general theoretical procedure which has been found to be applicable for the treatment of any excited or ionized state regardless of its spatial characteristics (valence-shell, Rydberg or inner-shell excitation) and multiplicity, as well as for any structural conformations (molecular equilibrium or transition state). Numerous examples are given and comparison with corresponding experimental data is made whenever possible. The applicability of the method for the study of photochemical processes and the required theoretical extension thereto are also discussed.

1. INTRODUCTION

Future chemistry will have to deal more and more with molecules in their electronically excited states as well as with short-lived species such as positive and negative ions and free radicals. While molecules possess only a single ground state they can appear in many excited states, each with its own chemical behavior, and hence regardless of how varied the scope of the ground-state chemistry might be, information concerning this one species can never be more than a small fraction of what is worthwhile knowing about the molecule as a whole. Consequently the study of the electronically excited states of molecules, their energetics, structure, life-time and possible interactions, is one of the fastest growing areas of research in modern-day chemistry and physics.

In the past virtually the only means of investigating excited and ionized states of molecules has been experimental spectroscopy. It is well-known, however, that in such work a wide variety of quite different techniques must be employed, depending on the energy and multiplicity of a given excited state, before a reasonably comprehensive description of this system can be said to have been achieved, whereby optical methods ranging from IR to UV, as well as electron impact, magnetic circular dichroism and photoelectron-spectroscopy are among the most common types of experimental procedures presently available for such objectives (1). Alternatively recently developed theoretical methods based on quantum theory can be employed effectively for the same purpose, even though they are not yet capable of the same degree of numerical accuracy as is obtainable in high-resolution spectroscopy. The reason for the latter assertion is simply that such purely theoretical approaches have the distinct advantage of being applicable for virtually any excitd or ionized state, regardless of its particular spatial or spin characteristics or the chemical stability of the system, and not only in the neighborhood of the ground state equilibrium structure but also for the entire range of nuclear conformations up to and including the various dissociation limits available to it (2). The present contribution will discuss the current status of such theoretical treatments and will illustrate their applicability in a number of examples, whereby the long-range goal in those investigations is to obtain sufficient understanding of molecular excited states to be able to accurately predict their behavior in chemical reactions.

2. THEORETICAL TREATMENT

A. General Considerations

The quantummechanical methods employed for spectroscopic investigations are generally divided into two categories, the non-empirical (or ab initio) treatments and the class of semiempirical procedures. The present work will deal with the first category only. Because of a variety of advances in the methodogical and numerical treatment of complicated mathematical systems along with parallel developments in the design of high-speed digital computers, the field of ab initio computations has expanded rapidly over the past ten years, so that calculations of this type are now used almost exclusively for the study of the electronic structure of small molecular systems and at the same time are finding even greater applicability in the treatment of much larger systems, previously thought to be the exculsive domain of semiempirical methods.

The basic elements of the theoretical instrument as shown in relationship to its experimental counterpart are presented schematically in Fig. 1. In the former information concerning the sample is contained

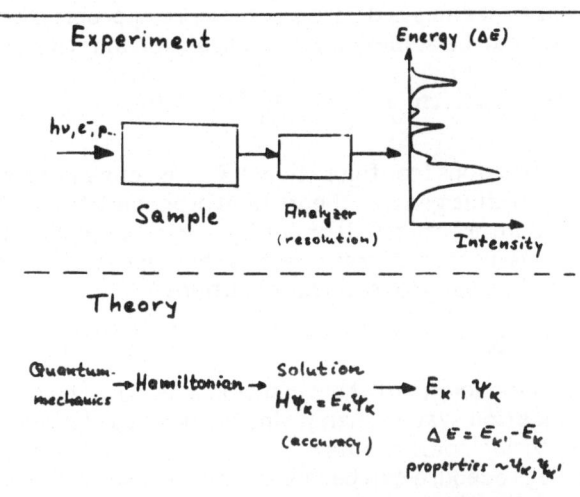

Fig. 1 Schematic comparison of an experimental and theoretical device for recording spectra

in the electronic Hamiltonian operator H in the form of nuclear charges and their locations (in cartesian coordinates, for example) as well as the number of electrons present in the system; it thus defines the molecule (or ion) for which spec-. troscopic investigation is sought. Routinely all electrostatic interactions between all electrons and all nuclei in the system are considered in the analytical form of H, whereby relativistic effects (including spin-orbit coupling, for example) are generally neglected, only to be included (usually at a later stage) when a higher degree of accuracy (resolution) is actually required. The basic theoretical tool (Fig. 1) is the solution of the Schrödinger equation $H\Psi_k = E_k \Psi_k$, (or some mathematically equivalent system), and just as in the case of a given piece of experimental equipment the level of its resolving power will depend to a large extent on how much attention to detail has been given to its design, as well as on the amount of time and energy (usually electrical) which can be expended in the actual running of the apparatus. Finally, while the form of the experimental results is ultimately a recording of the intensity of spectral lines versus energy, the output of the theoretical spectrometer ist he energy E_k and the wavefunction Ψ_k for a given state (at least when the Schrödinger equation is employed), whereby transition energies between states k and k' are obtained as energy differences $\Delta E = E_k - E_{k'}$, according to the Bohr principle, and properties of the individual states (or transition probabilities between them) can be derived directly from the corresponding wavefunctions.

B. Configuration Interaction Procedure

It is well-known that the single-configurational (or self-consistent field SCF) wavefunction is not adequate for the description of potential energy surfaces, relative stabilities and dissociation energies, for example, and in the present context it is important to note that the deficiencies of this type of theoretical approach are usually even more apparent when the description of excited or ionized states of molecules is desired. A far more appropriate representation of the wavefunction can be achieved via an expansion in terms of a suitable n-electron basis

$$\Psi_k (1 \dots n) = \sum_i c_{ik} \, \Phi_i (1 \dots n) \tag{1}$$

whereby the basis functions (configurations) Φ_i are conveniently chosen to be mutually orthonormal; this general theoretical procedure is usually referred to as the configuration interaction (CI) or configuration mixing (CM) or superposition of configurations method. A treatment of this type leads in practice to a set of simultaneous (homogeneous) linear equations

$$\sum_i c_{ik} (H_{ik} - E_k) = 0 \text{ with } H_{ik} = \int \Phi_i H \, \Phi_k d\tau \tag{2}$$

of the same order as the length of the expansion in eq. (1). The solution of the resulting (secular) equation system then yields the desired eigenvalues E_k and the corresponding expansion coefficients c_{ik} for the electronic wavefunctions. In essence then the CI procedure can be looked upon as involving a matrix representation of the Hamiltonian operator in the Schrödinger equation, the essential part of the theoretical spectrometer in Fig. 1. Practical solutions are obtained by diagonalizing the associated secular equations, a procedure which is especially well suited to the computational requirements of digital computers.

The most critical technical question which arises in this connection is the choice of a configuration set Φ_i with which the wavefunction Ψ_k of eq. (1) can closely approach the true solution of the Schrödinger equation. Generally a fixed set of one-electron functions ϕ_i (atomic orbitals AO's or molecular orbitals MO's or a convenient transformation thereof) is chosen at the beginning of a calculation and the configurations Φ_i are constructed therefrom, whereby the actual form of these basis functions is an antisymmetrized (because of the Pauli principle) product (Slater determinant, or a small sum thereof) of such one-electron species

$$\Phi_i (1 \dots) = \mathcal{A} \{ \phi_i (1) \, \phi_2 (2) \dots \phi_n (n) \}. \tag{3}$$

It is easy to show that from a fixed set of m such (spatial) one-electron basis functions and n electrons one can form a total of $\binom{2m}{n}$ distinct Slater-determinants, which is referred to as the full CI space associated with a particular one-electron (AO) basis. If the number of electrons is moderately large and a reasonable-sized AO basis is to be maintained, the value of $\binom{2m}{n}$ can become very large (for example, for m = 50 and 10 electrons, it is roughly 10^{13}) and because of this fact it has often been assumed that the attainment of reasonably converged CI expansions will be inpractical for the forseeable future. Nevertheless experience and theoretical arguments have shown that not all of these configurations are equally important and in the past five years a number of research groups dealing with large-scale CI calculations have agreed that for all practical purposes (at least for moderately sized molecular systems) it is sufficient to employ only those configurations which are related to the p most important configurations in the CI expansion by at most single or a double orbital substitution (excitation) (2-4). These p most important terms in the wavefunction of eq. (1) are generally denoted as reference or main (leading, generating) species and they contribute a certain minimal amount 0.3 % (i.e. $|c_{ik}|^2 > 0.003$) to the total CI expansion, say for concreteness, 0.3 % on a $|c_{ik}|^2$ basis. With such a choice for a multi-reference double-excitation CI (MRD-CI) the order of the requisite CI secular equation is roughly equal to $pn^2(2m-n)^2/4$ which, for p=3 reference species and the same values of m and n as before, is roughly 6×10^5; furthermore, when the effects of (spin and spatial) symmetry blocking are taken into account the maximum order of secular equation to be solved in this example easily may drop to around 10^4. Overall experience in the past years in the present research group shows that the CI space truncated in this manner typically ranges from 50 000 to 500 000 for AO basis sets containing 70-100 one-electron spatial functions with several reference configurations and a moderately large number (5-25) of correlated electrons.

Since present-day computer methods (5) for the solution of secular equations of general (albeit sparse) constitution are conveniently applied for orders up to only 5000-10 000 it is obvious, however, that for practical reasons additional simplifications have to be made. Our preferred route around this problem, which has been successfully applied in the past years (2,6) and is employed for all examples to be given in the present contribution, consists of a two-step process: first the various configurations of the MRD-CI space are ordered in a quantitative (though approximate) manner according to their predicted energy contribution to a given molecular state, with all those which appear to be capable of lowering the energy by more than a given threshold value (typically 10^{-5} hartree = 0.00027 eV) are included directly in the diagonalisation procedure (routinely the order of these secular equations runs between 2000 and 8000); this first step is then followed by a second in which the contribution of the remaining configuration (a large number of species which are only very weakly interacting with the reference configurations in the CI expansion) is taken into account by a simple extrapolation procedure. In the latter step the energy contribution of a configuration can be determinded, for example, by comparing secular equation results (7,8) with and without the individual configuration under investigation (together with the set of reference species) or with various equivalent versions for the individualized selection procedure employing perturbation methods (3,9). The energy extrapolation technique is conveniently undertahen by constructing the hamiltonian matrix for various selection threshold values T (for example T= 1×10^{-5}, 2×10^{-5}, 3×10^{-5}, 4×10^{-5} hartree) and by plotting the resulting E (T) curve along with similar curves formed from addition of the energy-lowering contribution of the unselected weakly interacting species $\Sigma_i \Delta E_i$ (T) with an arbitrary weighting factor: $E(T) + \lambda \Sigma_i \Delta E_i$ (T). Extrapolation of this family of curves to the common point at zero threshold which corresponds to the energy of the entire MRD-CI space is thereby easily achieved. More details concerning this selection and extrapolation scheme may be found elsewhere (7,8, 10).

The great advantage of the MRD-CI procedure including configuration selection and energy extrapolation from a theoretical point of view is that it is completely general, i.e. it has no restrictions either with respect to the type of configurations which can be included (as is the case in numerous other CI procecures)
or in the type of states which can be treated thereby, and additionaly it has the
nice feature that it can be extended in a natural way to higher accuracy (better
spectral resolution) simply by increasing the set of reference configurations or
by lowering the selection threshold. In practice it is found that this method allows
one to obtain the energy corresponding to the entire MRD-CI space with good
accuracy (as a rule of thumb the extrapolation error is 0.01 eV for a contribution
of $\Sigma_i \Delta E_i = 0.01$ hartree of the unselected species at the smallest threshold employed) at a computational expense which is only about 5 % or less of what would
be required to solve the entire CI problem by direct methods. Typical computer
times for various examples can be found in Refs. 2 and 8; a standard such calculation might take anywhere from 30 sec to 15 min (IBM 370/168) depending on its
complexity and the size of AO basis set employed (routinely between 50 and 90
functions).

Finally it should be pointed out that the effects of energy extrapolation to
zero threshold are important and that erroneous results, especially for potential
energy surfaces, can be obtained if the CI treatment is simply terminated after
the selection step (11). All other properties besides the energy can be calculated
with the wavefunction corresponding to the largest truncated MRD-CI space
(smallest energy threshold) and so far this level of sophistication has been found
to be quite satisfactory for one-electron properties such as dipole and quadrupole moments or for transition moments between electronic states. In water, for
example (12), relatively small differences have been observed for calculated
transition intensity values in going from a threshold of 100 μh to 20 μh, and similar experinece has been obtained recently for C_2 in evaluating life-times of the
Swan-bands (13). As a further example the property

$$\langle \Psi (\pi, \pi^*) | \Sigma_i \; x_i^2 | \Psi (\pi, \pi^*) \rangle$$

for ethylene changed by only 1.3 % (14) in going from a secular equation size of
roughly 2000 to that of the full MRD-CI space (in this case 11896). Thus the general experience to date has been that while extrapolation of the energy to the
zero threshold is often critical in obtaining a sufficient level of accuracy for this
important quantity, satisfactory results for the corresponding one-electron properties of the systems can be obtained by considering only the most important
terms in the corresponding CI expansion.

3. CALCULATION OF VERTICAL ELECTRONIC SPECTRA

A. Types of Excited States

In spectroscopy the various possible excited states are usually distinguished
(15) according to their characteristics as either a valence-shell or a Rydberg
species. Experimentally a Rydberg state is most easily identified by the fact that
it is part of a series of states which converges towards a given ionization limit;
its term value T_n (energy difference with respect to the ionization potential)

can be described according to a simple formula involving the ionization potential. Its charge distribution is quite expanded in space, a characteristics which is also often used to distinguish between Rydberg and valenc-shell states, since the behavior of the more diffuse charge distribution in condensed phase experiments is quite different (more affected) than that of the much more contracted valence-shell state. The relatively low intensiy in a Rydberg transition ($f \leq 0.08$ per degenerate component) and the small singlet-triplet splitting in Rydberg series can be looked upon as a direct consequence of the diffuse charge distribution. In MO theory the upper orbital in a Rydberg state is generally denoted in terms of AO's (or with the united-atom designation) because of its atomic-like (or one-center) charge distribution (at least in the major outer region of the MO). It can be looked upon in the main as being non-bonding, and hence the nuclear geometry of a molecule in a Rydberg state generally resembles very closely that of the respective positive ion.

By contrast valence-shell states are generally characterized simply by noting the compact MO's (constructed from the AO's of the constituent atoms) whose occupation is changed in going from ground to excited state. Their nature is much more dependent on the individual molecule than is that of a Rydberg state and hence the entire region from very weak ($f \approx 0$) to very strong ($f \approx 1.0$) transitions is observed, as well as many possibilities for the intensity distribution over vibrational modes as caused by geometrical changes upon excitation; the latter behavior depends on the bonding, non-bonding or antibonding characteristics of both MO's involved in the electronic transition. Mixing of Rydberg and valence-shell states is possible and seems to play an important role in photochemical processes (16), at least for smaller systems like NH_3, CH_4, C_2H_6 or butadiene, for example.

Theoretical calculations must be able to account for Rydberg and valence-shell states in an equivalent manner, which objective in ab initio treatments simply requires an extension of the AO basis set compared to conventional ground state treatments to include long-range functions of proper symmetry so as to allow for representation of the expanded charge density of the upper orbital in a Rydberg excited species; the corresponding CI treatments are not formally different for the two types of states, but rather simply involve different reference configurations. Similarly the calculations must be capable of describing ionizations from outer or inner-shell valence species as well as inner-shell excitations on an equal footing, but this requirement does not necessitate a change in the basic MRD-CI procedure, but rather at most in the nature of the corresponding AO basis set to be employed. Inner-shell phenomena are also becoming increasingly popular in chemistry at present because of the information they can give concerning charge distributions (chemical shifts) in a given system, and examples for this type of calculation will hence also be considered in what follows.

B. Spectra of Saturated Systems

The first excited states in simple saturated systems such as water, ammonia, hydrogen sulfide, ethane and propane have essentially Rydberg character. The calculated vertical transition energies ΔE_e (determined as the difference between the electronic energy of ground and respective excited state at the ground state equilibrium nuclear geometry) to the first excited states in water and ammonia and the oscillator strengths f for corresponding transitions are given in Tables 1a and b as a typical example (2,17,19); for these molecules detailed experimental data are available which allow for a clean test of the accuracy of the theoretical results in this instance. It is obvious from the tables that discrepancies between calculated ΔE_e values and the experimental peak energies (intensity maxima in the observed spectrum) are in every instance smaller than 0.2 eV, which is a typical error limit experienced in many other MRD-CI studies

Table 1a Calculated and experimental transtition energies (in eV) and oscillator strength in H_2O.

State	Characterization	MRD-CI ΔE_e	f(r)	Experiment[a] ΔE	f
$^1A_1$	ground state	0.0	-	0.0 X	-
$^3B_1$ $^1B_1$	$b_1 \rightarrow 3s$	6.90 / 7.30	- / 0.059	7.0 / 7.4 A	- / 0.044, 0.052 0.060±0.006
$^3A_2$ $^1A_2$	$b_1 \rightarrow 3py$	9.04 / 9.20	- / -	8.9 / 9.1	- / -
$^3A_1$ $^1A_1$	$3a_1 \rightarrow 3s$	9.01 / 9.80	- / 0.069	9.3 / 9.7 B	- / 0.05
$^3A_1$ $^1A_1$	$b_1 \rightarrow 3px$	9.65 / 10.32	- / 0.013	9.81 / 10.16 D	- / D/C=1.2
$^3B_1$ $^1B_1$	$b_1 \rightarrow 3pz$	9.84 / 9.90	- / 0.012	9.98 / 10.01 C	- / D/C=1.2
$^3B_2$ $^1B_2$	$3a_1 \rightarrow 3py$	10.99 / 11.21	- / 0.003	11.1 / (11.46)	- / ?
$^3B_1$ $^1B_1$	$3a_1 \rightarrow 3px$	11.68 / 11.72	- / 0.0002	- / (11.77)	- / ?

[a] For ΔE values see Ref. (18), for discussion of f values see the original reference (17).

Table 1b Calculated and experimental transition energies (eV) and oscillator strengths for NH_3.

State	Characterization	MRD-CI ΔE_e	f(r)	Experiment[a] ΔE	f
$^1A_1$	ground state	0.0	-	0.0 X	-
$^1A_1$	$3a_1 \rightarrow 3s$	6.27	0.089	6.39 A	0.13, 0.079, 0.088, 0.0696
^1E	$3a_1 \rightarrow 3px, 3py$	7.84	0.002	7.91 B	comparable to C ← X
$^1A_1$	$3a_1 \rightarrow 3pz$	8.21	0.002	≈ 8.14 (7.92-8.6)	30 times smaller than A ← X

[a]For discussion see the original reference (19).

of molecular transition energies employing AO basis sets of similar quality (double-zeta including some polarization functions; 24 contracted gaussians for water). It is also worth mentioning at the time that the results of the calculations were published the experimental literature value for $^3B_1$ was 7.2 eV and only the more recent electron impact measurements (18) bring the previously predicted theoretical value of 6.9 eV in good accord with experiment. The calculated oscillator strengths for the various transitions to the ground state are also in very good agreement with the experimental information available. Since absolute intensities (i.e. f values or alternatively radiative life-times) are difficult to measure, such comparisons are very seldom possible for larger systems. Finally, it should also be mentioned that although the characterization of the upper states of H_2O and NH_3 is made in terms of (the dominant) Rydberg MO's, the wavefunction expansions show (especially for the 3s) some hydrogen admixture, which fact is reflected in the relatively large singlet-triplet splittings for the lower states of water compared to related differences in many other systems in which pure Rydberg character is present.

Other examples of molecules in this category for which theoretical calculations are available are ethane and propane (20,21). In each case the experimental resolution of the spectra is relatively low and for this reason no definitive assignment of the peaks contained therein was possible on this basis alone. The ethane spectrum, for example, is marked by one single strong band (22) with some vibronic structure, whereby the measured total oscillator strength of approximately f= 0.3 indicates an overlapping of bands, the polarization of which was a source of controversy (22,23). The propane spectrum by comparison shows two strong peaks around 8.85 and 9.65 eV respectively with no clear vibrational features and a steady increase of absorption cross-section up to approximately 15 eV. In both of these cases the calculations are able to specify the origin of the broad spectral appearance noted above and are able to unambiguously classify the states, a knowledge which is ultimately important for the understanding of the photochemical behavior of these systems.

In ethane the two highest-lying occupied MO's $3a_{1g}$ and $1eg$ are almost iso-energetic and hence the various Rydberg series with the same upper MO origina-ting from these two valence-shell lower species also lie very close in energy, as is clearly seen from Table 2.

Indeed there are a total of five allowed transitions, two with parallel (to the CC axis) and three with perpendicular polarization which fall in an energy interval of 0.15 eV, and which possess a total oscillator strength of f= 0.280 which is very close to the estimated value of f= 0.3 for the entire band. Furthermore the situation in propane is closely analogous (Fig. 2): in the latter molecule there are three (highest occupied) orbitals which are of nearly equal stability (namely $6a_1$, $4b_2$ and $2b_1$) and hence its low-energy spectrum is calculated to consist of a variety of closely overlapping Rydberg transitions originating from these three orbitals (21). There is one difference, however, namely that while in ethane the first Rydberg members to which transitions are allowed by the dipole selection rules are of 3p symmetry, excitations to the lower-lying 3s Rydberg MO in propane are dipole-allowed for all three series of the larger alkane system.

Table 2   Calculated vertical excitation energies (in eV) and oscillator strengths for the low-lying singlet excited states in ethane.

State	Excitation	f(r)	polarization	ΔE_e
$^1A_{1g}$	ground state	-	-	0.00
^1E_g	$1e_g \rightarrow 3s$	-	$[x^2-y^2, xy]$	9.16
$^1A_{1g}$	$3a_{1g} \rightarrow 3s$	-	$[x^2+y^2, z^2]$	9.21
^1E_u	$1e_g \rightarrow 3p\sigma$	0.056	(x,y)	9.91
$^1A_{2u}$	$3a_{1g} \rightarrow 3p\sigma$	0.144	z	9.86
^1E_u	$1e_g \rightarrow 3p\pi$	0.002	(x,y)	9.99
$^1A_{2u}$	"	0.020	z	9.99
$^1A_{1u}$	"	-	-	10.04
^1E_u	$3a_{1g} \rightarrow 3p\pi$	0.058	(x,y)	10.00

C. Molecules with Low-lying Rydberg and Valence States

The vertical electronic spectrum of a large number of molecules which posses relatively low-lying unoccupied valence-shell MO's has been studies via the MRD-CI and similar computational techniques. Typical examples include formaldehyde, formamide, acetone, thioacetone, ethylene, butadiene, pyrrole, benzene, and

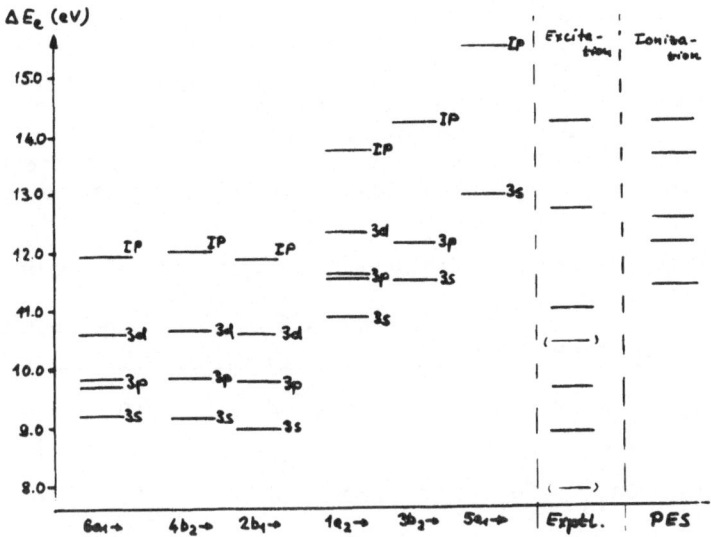

Fig. 2 Calculated Rydberg transitions in propane occuring from excitation of the $6a_1$, $4b_2$, $1a_2$, $3b_2$ and $5a_1$ MO and comparison with corresponding experimental lines and ionization potentials. Values in parentheses are somewhat uncertain.

a large number of triatomic species of AB_2, ABC or HAB type as well as several diatomic species. In the present context only very few examples for polyatomic molecules will be cited to illustrate basic trends in the calculated results.

A number of the simple free radical species which play a role in atmospheric chemistry such as the systems HO_2, HSO, HS_2 and HOCl have been studied in a parallel manner by various experimetal and theoretical techniques. Their low energy transitions are purely intravalence in nature, being found at 0.93 eV for HO_2 (0.88 eV experimental T_o), 1.56 eV for HSO (exptl. 1.77 eV), 0.65 eV for SOH (not yet found experimentally, although calculations predicted it to be more stable than HSO) and 0.85 eV for HS_2, which is currently under experimental investigations. Radiative life-times have been predicted for all these species in their first excited states but these results have not yet been confirmed experimentally. Nearly perfect agreement on details of the spectrum of HOCl has been obtained by two independent theoretical groups (24) but even though the dominant measured peaks can be explained very well on the basis of such calculations, the low-energy broad feature does not fit in at all well with such theoretical results.

A typical example for the occurrence of valence-shell states of various characteristics relatively separated from the corresponding Rydberg states is the ozone molecule, for which the calculated data on singlet states are collected in Table 3 (the equivalent information for the triplets can be found in the original reference (25)). Only the energy range between 2 and 5 eV is well characterized experimentally, being assigned in terms of the weak Chappuis and Huggins bands and the very strong Hartley system, whereas a number of features are known at higher energies (26) which all remain unassigned. The MRD-CI calculations divide the ozone transitions into essentially three groups: first, transitions in the 1 to 5 eV area are found to arise predominantly from excitations out of the energetically neighboring $1a_2$, $4b_2$ and $6a_1$ MO's into the unoccupied $2b_1$ (π^* type) species; secondly, transitions in the energy range following (up to roughly 8 eV) are seen to result from

Table 3 Calculated and experimental vertical excitation energies (in eV) for ozone.

State	Characterization	MRD-CI ΔE$_e$	MRD-CI f	Experimental ΔE	Experimental f
$^1A_1$	ground state	0.0	-	0.0	-
$^1A_2$	$4b_2 \to 2b_1$	1.72	0.0	-	0.0
$^1B_1$	$6a_1 \to 2b_1$	1.95	10^{-4}	2.1 Chappuis	weak
$^1A_1$	$4b_2^2 \to 2b_1^2$ $6a_1^2 \to 2b_1^2$	3.60	10^{-4}	3.5-4.2 Huggins	weak
$^1B_2$	$1a_2 \to 2b_1$	4.97	0.18	4.86 Hartley	strong
$^1A_2$	$6a_1,1a_2 \to 2b_1^2$	6.37	0.0	-	0.0
$^1B_2$	$4b_2,6a_1 \to 2b_1^2$	6.87	2×10^{-4}		
$^1B_1$	$4b_2,1a_2 \to 2b_1^2$	7.26	1×10^{-3}		
	$4b_2^2 \to 2b_1^2$			7.18	broad maximum
$^1A_1$	$6a_1^2 \to 2b_1^2$	7.34	5×10^{-6}		
$^1A_1$	$1a_2^2 \to 2b_1^2$ $1b_1^2 \to 2b_1$	7.60	not calc.		
$^1A_1$	$6a_1 \to 7a_1(\sigma^*)$	9.29	2×10^{-3}	9.32	strong
$^1B_2$	$4b_2 \to 7a_1(\sigma^*)$	10.05	0.1	10.2	
$^1B_2$	$4b_2 \to 3s$	9.21	3×10^{-2}	9.24	
$^1A_1$	$6a_1 \to 3s$	(9.3)	10^{-2}	9.40	

further Rydberg p series

double-excitations involving the same four MO's; and finally, higher transitions are calculated to involve both Rydberg upper states as well as excitations into an antibonding σ^*-type orbital. Again, agreement between measured absorption maxima and the vertical ΔE_e values is very good (Table 3), and it should be especially pointed out that the energy extrapolation technique in the CI procedure is quite important for the treatment of the Hartley band system; the corresponding ΔE_e value would be obtained as 6.7 eV, for example, if selection of configurations (without extrapolation) would be undertaken for a threshold as low as T= 30 µh. The broad maximum at 7.18 eV is predicted to arise from various double-excitation states whereas the strong feature around 9.3 should also be a combination of various states as indicated in the Table.

The systems acetone, thioformaldehyde and butadiene are of more interest to organic chemists than possibly any of the molecules previously discussed and calculations have also been carried out for their electronic spectra which illustrate the applicability of the theoretical method. The peaks labelled B, C and D in the experimental electron loss spectrum (27) sketched in Fig. 3 can be assigned as (n,3s), overlapping (n,3p σ) and (n, 3pn) and finally, (n, 3d) plus (n,4s) Rydberg bands respectively according to the calculations, whereby the very weak shoulder at the low-energy side is seen to arise (similar to the situation in formaldehyde) from the forbidden (n, π^*) transition. The vibrational structure observed in the electron loss spectrum was not further investigated but rather only vertical energy differences were calculated in this instance. For these calculations the number of configurations for each of the acetone states treated was between 50 000 (ground state) and 210 000 $^3(\pi, \pi^*)$, where-

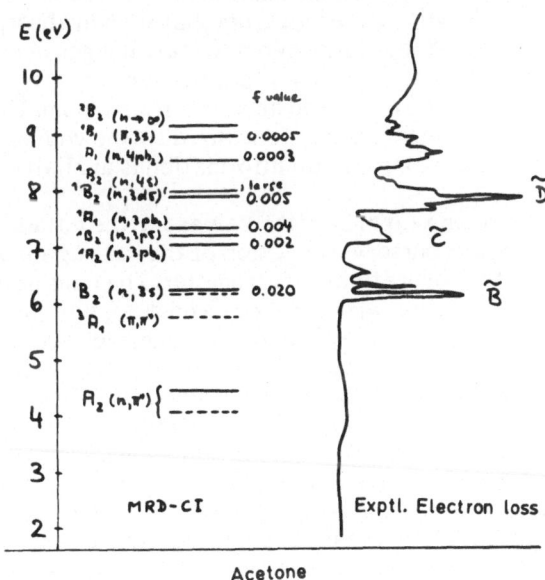

Fig. 3 Comparison of the theoretical spectrum of acetone obtained from MRD-CI calculations with the corresponding experimental data (sketched after the data of Ref. 27).

Table 4 Comparison of theoretically predicted vertical transition energies in
H_2CS with recently observed data (values in eV).

State	Excitation	Predicted ΔE_e Ref. (28) (1974)	observed peak intensity
$^1A_1$	ground state	0.00	0.00
$^3A_1$	$n \to \pi^*$	1.84	1.80 } Ref. (29)
$^1A_1$	$n \to \pi^*$	2.17	2.03 } (1975)
$^1B_2$	$n \to 4s$	5.83	5.83 } Ref. (30)
$^1A_1$	$n \to 4p_y$	6.62	6.59 } (1978)
$^1A_1$	$\pi \to \pi^*$	7.92	5.72 broad peak

as the order of the secular equations actually solved was below 5000 in each
case; this example again points out the practical advantage of the MRD-CI
method combined with selection-extrapolation features. The thioformaldehyde
spectrum (Table 4) is a particularly nice example for the usefullness of the theoretical approach; it is seen that the data obtained by CI calculations (28) afforded
an excellent prediction for the actual H_2CS transition energies measured somewhat later (29,30). The only discrepancy in this comparison is for the $^1(\pi, \pi^*)$
state, but it is interesting in this connection to note that apparently the emission
spectrum for this band is quite different from that seen in absorption, a finding
which is a strong indication that large geometrical changes occur upon $^1(\pi, \pi^*)$
excitation. In turn this fact also suggests that the vertical ΔE_e value does not
necessarily coincide with the measured absorption maximum. The electronic
states of the related compound thioacetone are also known from the calculations
(31) but no corresponding experimental information is available to date.

Finally, calculated and experimental values for the butadiene spectrum
are collected in Table 5. A cursory inspection of the data again points out the
good performance of the theoretical spectrometer. The calculated triplet valence states agree very closely with the measured data, but only after the experimental transition energy of the second state had been reevaluated (a value
of 3.9 eV was thought to be correct until electron-impact measurements around
1969 and various other methods later on determined the value quoted in Table
5); the measured Rydberg peaks (32) can also be correlated in a very straight-for-
ward manner with the calculated data for these species. There is again a single
important exception to this pattern of agreement, however, namely the theoretical and measured location of the strong ^1B_u (π, π^*) transition differ considerably. Since the calculations (33) indicate that geometrical changes in the nuclear
framework as well as interaction with a neighboring (in the perturbed structure)
Rydberg state play an important role in the description of the upper ^1B_u state
it is perhaps not surprising, however, that mere calculation of the vertical ΔE_e
value is not sufficient to describe the major features of this particular transition.

Table 5 Calculated vertical transition energies (in eV) to various low-lying states of trans-butadiene and comparison with existing experimental data.

State	Characterization	ΔE_e	f	Herzberg Notation	ΔE (max. intensity)
$^1A_1$	ground state	0.0	-	X	0.0
^3B_u	$X_2 \to X_3$	3.31	-	a	3.2, 3.2, 3.3[a]
^3A_g	$X_1 \to X_3, X_2 \to X_4$	4.92	-	b	4.8, 4.93, 4.9[a]
^1B_g	$X_2 \to 3s$	6.20	-	B	6.22[b]
^1A_u	$X_2 \to 3p\sigma$	6.53	0.0002		
^1B_u	$X_2 \to 3p\pi$	6.67	0.07	C	6.657 (series 3, n= 3)[b]
2^1A_u	$X_2 \to 3p\sigma$	6.72	0.05		
2^1A_g	valence mixture	7.02	-	E or F	7.06 (series 2, n=3)
2^1B_g	$X_2 \to 3d\sigma$	7.29	-		7.328 (series 4, n= 3)
3^1A_g	$X_2 \to 3d\pi$	7.53	-		7.481 (series 1, n= 3)
2^1B_u	$X_2 \to 4p\pi$	7.96	0.09	G	7.857 (series 3, n= 4) or 8.002 (series 2, n= 4)
n^1B_u	$X_2 \to X_3$	7.67	0.9	A	5.71 - 6.29

a) Data from various experiments; for details see original reference (33)
b) values for the Rydberg states and corresponding assignment from Ref. (32)

D. Ionization energies

To some extent ionized states can be considered as a special class of excited species, and since the CI method is completely general it must also be able to predict the values of ionization potentials (and electron affinities). This expectation is generally borne out in practice, although it has been found that a balanced description of the neutral system (with n electrons) and the positive ion (wtih n-1 electrons) requires a somewhat larger AO basis set than is found to be necessary for equivalent accuracy in the majority of excited state treatments. Similar experience has been noted in the description of negative ions relative to the corresponding neutral species.

A comparison between various ionization potentials obtained via the present MRD-CI method, the MBPT (Green's function) approach (34) and from experiments is presented in Table 6 for ethylene; the same AO basis is thereby employed in both theoretical treatments (35). For the CI calculations the contribution of the full CI space is also estimated in this nature. It is seen that for all practical purposes both theoretical methods (MBPT and CI) perform equally well, as they eventually must when both are carried to the theoretical limit. Furthermore comparison with the experimental data shows clearly that the theoretical determination of the various C_2H_4 ionization potentials is quite good. Additional comparisons of ionization energies obtained from MBPT, CI and experiments are in the literature for B_2H_6 (8) and H_2O, N_2, C_2H_2 and HCN (36), for example.

Table 6 Comparison of ionization potentials (in eV) for various states of ethylene obtained from the MRD-CI method, the MBPT (Green's function) approach and experiment.

Ion	MRD-CI	Full-CI	MBPT (34)	Expt.
$^2B_{3u}$	10.33	10.46	10.45	10.51
$^2B_{3g}$	12.97	13.01	13.04	12.85
^2A_g	14.58	14.66	14.77	14.66
$^2B_{2u}$	15.85	15.88	16.10	15.87
$^2B_{1u}$	19.17	19.04	19.45	19.23 $\}$ 19.10

E. Inner-shell Phenomena

From a theoretical point of view there is no difference between an excitation (or ionization) from a valence-shell or an inner-shell even though the relative energetics, which naturally plays a role in the choice of experimental equipment, is drastically different in the two cases. Various inner-shell excited and ionized states have been calculated using the present MRD-CI method for N_2, C_2H_2 and C_2H_4 for example and Table 7 shows some representative results for C_2H_2. Details of the treatment can be found elsewhere (37); perhaps it should be noted, however, that somewhat larger expansion lengths (configuration spaces of 300 000 - 500 000) for the wavefunction are generally necessary than in more routine valence-shell or Rydberg excited state calculations.

Inspection of Table 6 which lists results for core-ionized, core-excited and shake-up species reveals a similar consistency between theoretical and experimental data as has been observed so far, although the overall accuracy appears to be somewhat lower for such high-energy states. The fact that the optical absorption experiments did not observe the $1s \rightarrow 3s$ and $1s \rightarrow 3p\sigma$ states is not surprising according to the calculations since they predict relatively low intensity (f values) for the corresponding transition.

Table 7 Calculated vertical excitation energies (in eV) and selected oscillator strengths for various core-excited and shake up states relative to the 1s ionization potential in C_2H_2 and comparison with data obtained from two experimental studies.

State	ΔE MRD-CI	ΔE full CI estimate	f	electron impact [a]	optical [b]
$^2\Sigma_g^+(1s^{-1})$	291.2	290.9	-	291.11	
$^1\Pi_u(1s\to\pi_g)$	4.61	(5.21)[c]	0.17	5.29	5.6
$^1\Sigma_u^+(1s\to3s)$	3.16	3.31	0.0010	$\begin{cases}3.0\\3.23\end{cases}$	-
$^1\Sigma_u^+(1s\to3p\sigma)$	2.59	2.70	0.0014	2.94	-
$^1\Pi_u(1s\to3p\pi)$	2.17	2.30	0.0082	$\begin{cases}2.14\\2.3\end{cases}$	2.5
$^1\Sigma_u^+(1s\to3d\sigma)$	1.49	1.50	-	?	?
$^1\Sigma_u^+(1s\to4s)$	1.29	1.18	-	1.1	1.2
$^2\Sigma_g^+(1s^{-1},\pi_u\to\pi_g)$ -8.2		(-7.6)[c]	-		-7.2[d]
$^2\Sigma_g^+(1s^{-1},\pi_u\to3p\pi)$-15.20		-15.00	-		-15.1
$^2\Sigma_g^+(1s^{-1},\sigma\to\sigma^*)$-18.13		-17.50	-		-16.9
$^2\Sigma_g^+(1s^{-1},2\sigma_u\to3s)$-22.54		-21.98	-		-22.1

[a] electron impact values from Ref. 38,39

[b] optical data from Ref. 40

[c] somewhat lower accuracy

[d] PES data from Ref. 41

4. VIBRATIONAL STRUCTURE OF THE SPECTRA

A. General Aspects

In order to match the entire fine structure between an observed and a theoretically predicted band system it is necessary to account for the individual vibrational energy levels of both electronic states involved in the transition, as
schematically indicated in Fig. 4, and not only for their (vertical) ΔE_e value.
In the general treatment it is assumed that the total wavefunction can be represented as a product of electronic and vibrational functions; this seperation
leads to a Schrödinger equation for the vibrational wavefunctions in which the
electronic energy plays the role of the potential in which the nuclear motion
occurs. The electronic potential energy surfaces are obtained pointwise (i.e. by
calculation of the electronic energy for a given nuclear arrangement) and judged
from many comparisons with experimental data (i.e. on the basis of dissociation
energies and spectroscopic constants which describe the shape of the potential
surface near equilibrium) the MRD-CI calculations are able to give a quite reliable description of the entire energy surface all the way from equilibrium to
dissociation. If various bonds are broken it is found that the extrapolation procedure can be quite important (11). Once the potential energy surface is known,
the vibrational energy levels and corresponding wavefunctions can be obtained
quite easily by expansion techniques, for example (38,39). The O_2 molecule presents a convenient test case for this type of calculation, and it is seen that the
vibrational energy levels obtained from the calculated ground state potential
energy curve differ by less than 100 cm^{-1} from the observed levels up to vibrational quantum numbers as high as v''= 22; no comparison with even higher levels converging to the O_2 dissociation limit is possible, however, since no higher
levels have been directly determined experimentally. The zero-point energy is
thereby calculated to be 784.6 cm^{-1}, in quite close agreement with the experimental 787.4 cm^{-1}.

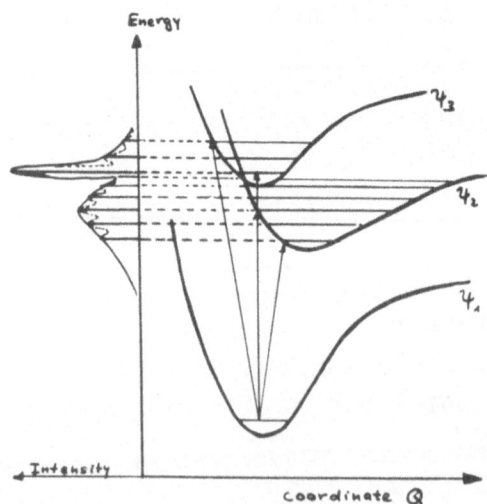

Fig. 4   Schematic diagram illustrating the origin of the vibrational structure of
an electronic transition.

The intensity for the transition between two vibrational energy levels of ground and excited state is governed by two factors, the electronic transition moment $R_{e' \, e''}$ (which depends solely on the character of the electronic states involved in the intercombination) and the overlap between the vibrational wavefunctions of upper and lower levels (the squares thereof are known as Franck-Condon factors). If the electronic transition moment is fairly constant over the range of nuclear displacements which occur in the vibration, the relative intensity of the various vibrational peaks can be described to a good accuracy by the Franck-Condon factors alone.

B. Examples for Calculated Intensity Distributions

A typical example for a calculated and a measured intensity distribution is the $^1A''(^1\Sigma^-) - ^1A'(^1\Sigma^+)$ band of HCN. The molecule is linear in its $^1\Sigma^+$ ground state and bent by approximately 125° in its first excited $^1A''(^1\Sigma^-)$ state. Vertical transitions from the linear ground state to this species are not allowed by the dipole-selection rules since the electronic transition moment for $\Sigma^- - \Sigma^+$ combination is zero. Hence measurable intensity is only found away from the vertical transition in regions where the electronic transition moment is not zero and at the same time the overlaps between ground and excited state vibrational wavefunctions are still sizeable; the most probable transition is predicted to arise for the v'=6 vibrational state in the bending mode (Fig. 5), in very good accord with what is actually measured.

Another problem of long-standing interest in this connection is the singlet-triplet splitting between the two lowest electronic states $X\,^3B_1$ and $^1A_1$ in methylene (42). For some time the question appeared to have become settled as photochemical measurements gradually converged to a value of 9-10 Kcal for this quantity, while ab initio calculations started out at much higher values before 1970 but in more recent years came out rather uniformly with a result between 9 and 13 Kcals. Considerable doubt was brought into the discussion in 1976, however, when an interpretation of the electron detachment measurements (43) for the CH_2^- ion in terms of a vibrational progression in the CH_2 bending mode led to the conclusion that the T_0 value for the $^3B_1 - ^1A_1$ splitting is actually 19.5 Kcal/mole, roughly 10 Kcal higher than previously thought. The PES data on CH_2^- suggested thereby that a very strong peak at 1.05 eV corresponds to $^1A_1 \leftarrow {}^2B_1$ ionization and further that a long progression in the spectrum was caused by the $^3B_1 \leftarrow {}^2B_1$ electron detachment process, whereby extrapolation to the zeroth vibrational level for the $^3B_1$ state on this basis led to the high $^1A_1 - {}^3B_1$ T_0 value of 19.5 Kcal. Ab initio calculations for the CH_2^- ion and the neutral CH_2 species (44) confirm the strong peak (predicted at 1.01 eV after making a theoretical correction for an expected 0.20 eV error, see original reference (44)) with a large Franck-Condon factor (Fig. 6), but suggest that at most four or five of the low-intensity peaks in the electron detachment spectrum actually result from $^3B_1 - {}^2B_1$ transition; instead the remaining two or three lines should be caused by another process (hot bands from $^2B_1$, or transitions involving the second CH_2^- (resonance) state). This new interpretation of the electron detachment data is based on the pattern of calculated energy levels as well as on the computed intensity distribution. Similar calculations have been carried out independently by Harding and Goddard (46) who arrived at essentially the same results, except that their electron affinity results were in much less satisfactory agreement with experiment.

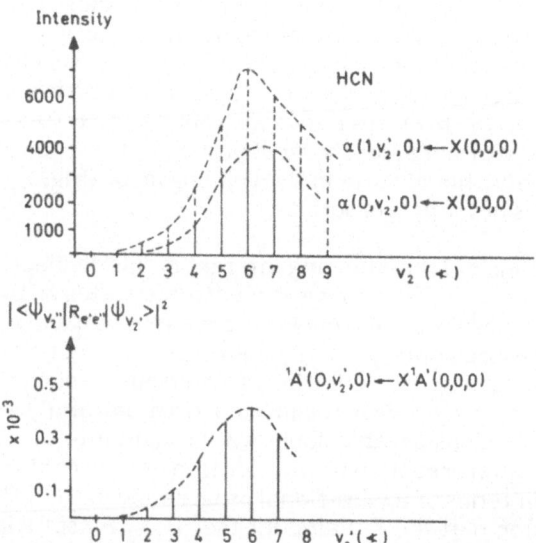

Fig. 5 Comparison of the experimental intensity distribution in the transition to the first excited singlet state of HCN (referred to as the α state) with the corresponding theoretical data, whereby both, the change in electronic transtition moment $R_{e'e''}$ and the vibrational overlap is taken into account.

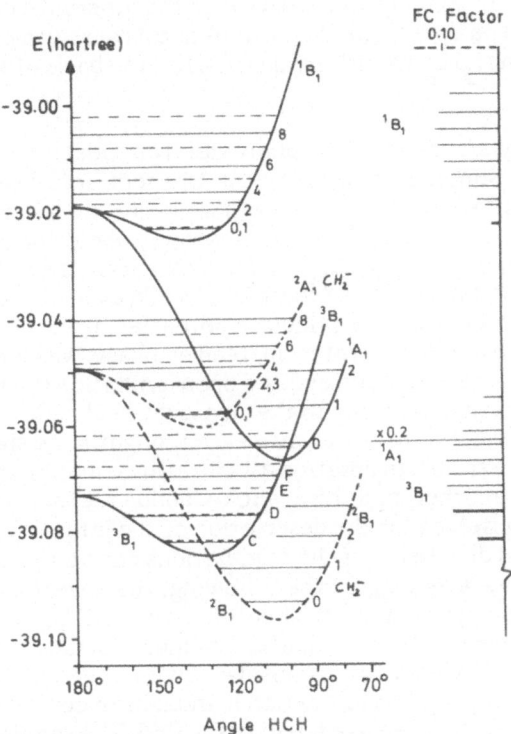

Fig. 6 Calculated potential energy curves for various states of methylene CH_2 and its negative ion and corresponding vibrational levels. Calculated Franck-Condon factors for the $^1A_1 - {}^2B_1$ and $^3B_1 - {}^2B_1$ states are also given; the correlation with the experimental peaks C to F is indicated.

Many more examples of a quite sucessful prediction of vibrational features in electronic band spectra can be found in the literature. Of such studies the work on NH_2 and PH_2 in which rotations are also taken into account to some extent (47) deserves special mention, as well as the calculations of the mixed valence-Rydberg states (48,49) in O_2 which allowed the assignment of well-known but until then unidentified peaks at 9.96 and 10.28 eV, and the detailed study of all possible vibrational species in N_2H_2 which led to an assignment of the inducing modes (antisymmetric bending and torsion) for the otherwise dipole-forbidden $^1(n, \pi)$ transition. The detection of new lines and a reassignment of various spectral features in N_2H_2 was undertaken (50) on the basis of the theoretical prediction.

The vibrational structure in the photoelectron spectrum of ethane is chosen as the last example to be discussed in this section. Since the two upper-most occupied MO's $1e_g$ and $3a_{1g}$ lie very close in energy, as already pointed out in the previous section, there has been controversy about the order of IP's in ethane (ionization out of $1e_g$ or $3a_{1g}$) and over the identification of the observed band structure in the PES. Because there are 18 vibrational coordinates in this instance no attempt was made to calculate potential curves for all such species in the ground and various ionic states; instead emphasis was placed on those vibrations which are expected to be excited upon ionization (and possibly fall in-to the energy range corresponding to the measured peaks), i.e. on those modes for which the equilibrium structure values are thought to be significantly different before and after ionization. In addition all vibrational modes are assumed to be independent of one another, and the Franck-Condon factors are considered to be sufficient in themselves for the description of the intensity distribution in this case. A detailed discussion of the calculations can be found in the original reference (51). The results suggest the following: the lowest energy is obtained for $C_2H_6^+$ in the $^2A_{1g}$ state at a considerable larger CC bond length than in the neutral ground state; structural relaxation includes various deformations in this state and since the CC frequency is considerably smaller than the nearly 1200 cm^{-1} progression observed this ionization is thought to contribute only to the underlying (structureless) intensity found in the PES. The regular structure with a 0-0 transition at 11.56 eV is predicted to be caused by the $^2E_g - ^2B_2$ ioniza-tion, with progressions in both the CC stretching (ν_3) and HCH bending (ν_{11}) vibrations coinciding very closely in the spectrum. If a certain portion of the measured intensity is substracted from the total spectrum as being caused by $^2A_{1g}$ ionization the calculated intensity distribution and the measured va-lues are found to match very well with one another, as is obvious from Fig. 7. It should also be mentioned, however, that the other ^2E_g component has the same symmetry in a distorted C_{2h} nuclear framework as does its $^2A_{1g}$ counterpart, a fact which complicates the appearance of the spectrum beyond 12.6 eV; the strong indication is that the usual separation of electronic and nuclear motion (Born-Oppenheimer approximation) is no longer acceptable, thereby offering at least a qualitative explanation for the existence of some quite irregular struc-ture in terms of sharply avoided crossings of potential surfaces which are expec-ted to occur in this energy region.

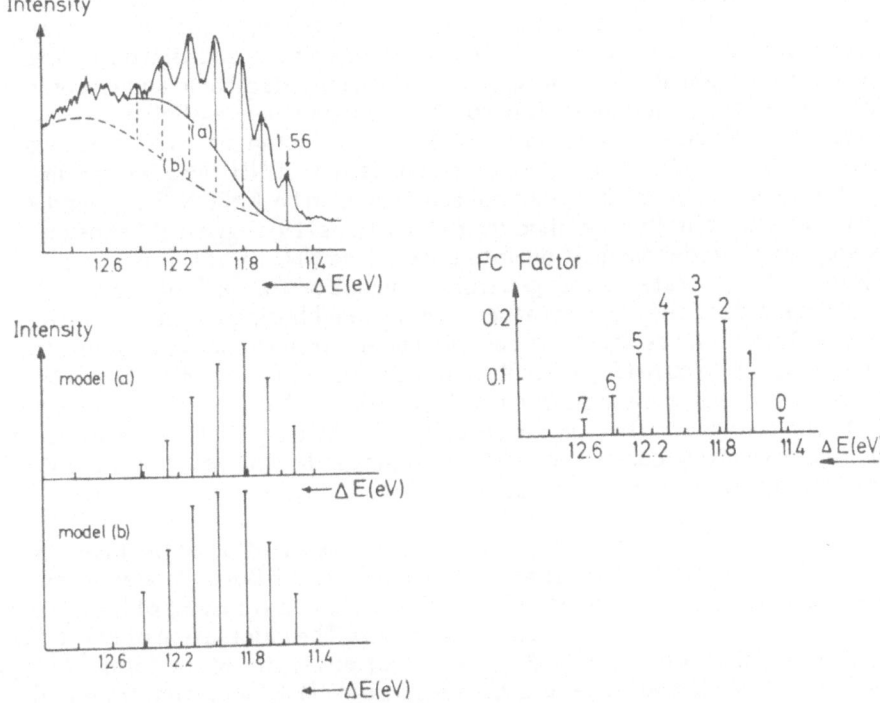

Fig. 7   Comparison of the intensity distribution in the C_2H_6 photoelectron
spectrum (after substraction of underlying intensity according to two
models) with the calculated values.

5. SUMMARY AND OUTLOOK

The present contribution has attempted to demonstrate that present-day
CI calculations are a quite effective tool for predicting details of molecular
spectra in small polyatomic molecules including vibrational features and absolute intensities of transitions between states of various characteristics.
This fact in turn strongly indicates that electronically excited and ionized states
can be described very satisfactorily by such theoretical   methods, giving information about the relative spacing of the various electronic and vibrational
states, their energy behavior with respect to nuclear distortions as well as other
inherent properties. With the rapid development in CI methods which is taking
place to date, both with respect to theoretical as well as technical aspects,
coupled with advances in computational resources, it seems obvious that the extension of such calculations to larger systems than treated thus far is a natural
and straight forward direction in which to proceed.

The understanding and prediction of photochemical behavior of molecules requires the knowledge of large portions of potential energy surfaces, sometimes very far away from the equilibrium structure, and in addition quite often detailed information about the interaction of various electronic and vibronic states is necessary to achieve this purpose. Mixing of valence-shell and Rydberg states seems to be an important factor in some instances, as has been explicitly demonstrated (19) in the photodecomposition of NH_3 into the fragments $NH_2 + H$ (^2S), whereby the NH_2 radical is observed in both the X^2B_1 ground and $^2A_1$ excited states. In this instance the calculations (19) suggest that the primary step for photodecomposition into the excited NH_2 radical is not excitation into the $NH_3 B$ state, as was generally assumed on the basis of the magnitude of the incident energy, but rather into the neighboring Rydberg C state, (see Table 1b) which has only been observed in electron impact but not in optical spectroscopy to date (53). Calculations confirm that population of the $NH_3 A$ Rydberg state leads to dissociation into hydrogen and NH_2 in its X^2B_1 ground state. In various other cases similar experience indicating that the Born-Oppenheimer approximation can be successfully employed for the description of various excited state interactions (48,49) has been forthcoming.

In many other cases, however, a quantitative description of the interaction of such states requires treatments going beyond the Born-Opppenheimer approximation and/or spin-independent Hamiltonians. Extensions of both types are under active investigation in our laboratories. The calculation of non-adiabatic coupling matrix elements (53) between various states of the form $\partial/\partial Q$ and $\partial^2/\partial Q^2$ has become quite feasible for large MRD-CI wavefunctions (54) in past years, and a variational-perturbation method to include them for the description of physical effects has been formulated in the framework of a general CI method (55). Initial results of this nature have been used to explain the quenching mechanism for H_2 flourescence (B $^1\Sigma_u^+$ state) in the presence of rare gas atoms, in particular helium (56), and of the vibronic coupling in various small molecules such as N_2, NH_2 (Renner-Teller effect) and PH_2. Inclusion of a complete treatment of one- and two-electron spin-orbit and spin-spin effects in the MRD-CI package has also been accomplished. Under these circumstances there is a realistic hope to use the theoretical methods based on MRD-CI procedures in the relatively near future in order to study not only spectroscopic features of molecules but also details of photochemical pathways which cannot be dealt with by purely experimental methods.

ACKNOWLEDGEMENT

We would like to thank all our associates, in paricular Drs. Bruna, Perić, Shih and Dipl.Chem. Hess, Hirsch and Runau for considerable help in this research pertinent to the excited states of molecules, and we are grateful to the Deutsche Forschungsgemeinschaft for continued financial support given to our work. The services of the University of Bonn computation center are also gratefully acknowledged.

References

(1)   C.Sandorfy, P.J. Ausloos, M.B. Robin, "Chemical Spectroscopy and Photo-
      chemistry in the Vacuum-Ultraviolet", NASI-Series Vol. 8, D. Reidel Publ.,
      Dordrecht, Holland (1974)

(2)   R.J. Buenker and S.D. Peyerimhoff in "Excited States in Quantum Chemis-
      try", eds. C.A. Nicolaides and D.R. Beck, NASI-Series Vol. 46, D. Reidel
      Publ. Co., Dordrecht, Holland (1978) p. 45-103, 403-416

(3)   I. Shavitt, in "Modern Theoretical Chemistry", Vol. 3 (Methods of Electronic
      Structure Theory) eds. H.F. Schaefer III, Plenum Press New York (1977) p.
      189

(4)   E.R. Davidson, in "TheWorld of Quantum Chemistry", eds. R. Daudel and
      B. Pullmann, D. Reidel Pull. Co. Holland, 1973, p. 17

(5)   I. Shavitt, C.F. Bender, A. Pipano and R.P. Hosteney, J. Comput. Phys. 11
      (1973) 90;
      E.R. Davidson, J. Comput. Phys. 17 (1975) 87;
      W. Butscher and W.E. Kammer, J. Comput. Phys. 20 (1976) 313

(6)   P.J. Bruna, Gaz. Chim. Ital. 108 (1978) 395

(7)   R.J. Buenker and S.D. Peyerimhoff, Theor. Chim. Acta 35 (1974) 33

(8)   R.J. Buenker and S.D. Peyerimhoff and W. Butscher, Mol. Phys. 35 (1978) 771

(9)   J.L. Whitten and M. Hackmeyer, J. Chem. Phys. 51 (1969) 5584

(10) R.J. Buenker and S.D. Peyerimhoff, Theor. Chim. Acta 39 (1975) 217

(11) S.D. Peyerimhoff and R.J. Buenker in "The World of Quantum Chemistry",
      eds. B. Pullmann and R. Parr, D. Reidel Publ. Co. Holland, 1976, p. 213

(12) R.J. Buenker and S.D. Peyerimhoff, Chem. Phys. Letters 29 (1974) 253

(13) M. Zeitz, S.D. Peyerimhoff and R.J. Buenker, Chem. Phys. Letters 58 (1978)
      487

(14) R.J. Buenker, S.D. Peyerimhoff and S.K. Shih, J. Chem. Phys. 69 (1978) 3882

(15) See for example: M.B. Robin, in "Higher Excited States of Polyatomic Mole-
      cules", Vol. 1, Academic Press, New York 1974

(16) C. Sandorfy in "Applications of MO Theory in Organic Chemistry" (Progress
      in Theoret. Organ. Chemistry Vol. 2) Elsevier Publ. Co. Amsterdam (1977)
      p. 384

(17) R.J. Buenker and S.D. Peyerimhoff, Chem. Phys. Letters 29 (1974) 253

(18) A. Chutijan, R.I. Hall and S. Trajmar, J. Chem. Phys. 63 (1975) 892

(19) R. Runau, S.D. Peyerimhoff and R.J. Buenker, J. Mol. Spectry 68 (1977) 253

(20) R.J. Buenker, S.D. Peyerimhoff, Chem. Phys. 8 (1975) 56

(21) A. Richartz, R.J. Buenker and S.D. Peyerimhoff, Chem. Phys. 31 (1978) 187

(22) C. Sandorfy, in "Chemical Spectroscopy and Photochemistry in the Vacuum
      Ultraviolet", eds. C. Sandorfy, P.J. Ausloos and M.B. Robin (Reidel, Dord-
      recht, 1974)

(23) E.M. Custer and W.T. Simpson, J. Chem. Phys. 60 (1974) 2012

(24) R.L. Jaffe and S.R. Langhoff, J. Chem. Phys. 68 (1978) 1638; P.J. Bruna, G.
      Hirsch, S.D. Peyerimhoff and R.J. Buenker, Chem. Phys. Letters 52 (1977)
      442; Can. J. Chem. 57 (1979) 1839

(25) K.H. Thunemann, S.D. Peyerimhoff and R.J. Buenker; J. Mol. Spectry 70
      (1978) 432

(26) R.J. Celotta, S.R. Milczarek and C.E. Kuyatt, Chem. Phys. Letters 24 (1974)
      428; Y. Tanaka, E.B.Y. Inn and K. Watanabe, J. Chem. Phys. 21 (1953) 1651

(27) R.H. Huebner, R.J. Celotta, S.R. Mielczarek and E.C. Kyatt, J. Chem. Phys.
      59 (1973) 5434

(28) P.J. Bruna, S.D. Peyerimhoff, R.J. Buenker and P. Rosmus, Chem. Phys. 3
      (1974) 35

(29) R.H. Judge and G.W. King, Can. J. Phys. 53 (1975) 1927

(30) R.H. Judge, C.R. Drury-Lessard and D.C. Moule, Chem. Phys. Letters 53
      (1978) 82

(31) P.J. Bruna, R.J. Buenker and S.D. Peyerimhoff, Chem. Phys. 22 (1977) 375;
      P.J. Bruna in "Progress in Theoretical Organic Chemistry", Vol. 2, ed. I,
      Csizmadia (Elsevier, Amsterdam, 1977)

(32) R. McDiarmid, J. Chem. Phys. 64 (1976) 514; Chem. Phys. Letters 34 (1975)
      130

(33) R.J. Buenker, S.K. Shih and S.D. Peyerimhoff, Chem. Phys. Letters 44 (1976)
      385

(34) W. v. Niessen, G.H.F. Diercksen, L.S. Cederbaum and W. Domcke, Chem.
      Phys. 18 (1976) 469

(35) K.H. Thunemann, R.J. Buenker, S.D. Peyerimhoff and S.K. Shih, Chem. Phys. 35 (1978) 35

(36) D.P. Chong and S.R. Langhoff, Chem. Phys. Letters 59 (1978) 397

(37) A. Barth, Diplomarbeit Bonn 1978;
      A. Barth, R.J. Buenker, S.D. Peyerimhoff and W. Butscher, Chem. Phys. (1979), in press

(38) M. Tronc, G.C. King and F.H. Read, J. Phys. B 12 (1979) 137

(39) A.P. Hitchcock and C.E. Brion, J. Electr. Spectr. 10 (1977) 317

(40) W. Eberhard, R.P. Haelbich, M. Iwan, E.E. Koch and C. Kunz, Chem. Phys. Letters 40 (1976) 180

(41) R.G. Cavell and D.A. Allison, J. Chem. Phys. 69 (1978) 159

(42) J.F. Harrison, Acc. Chem. Res. 7 (1974) 378

(43) P.F. Zittel, G.B. Ellison, S.V. O'Neil, E. Herbst, W.C. Lineberger and W.P. Reinhardt, J. Am. Chem. Soc. 98 (1976) 3731

(44) S.K. Shih, S.D. Peyerimhoff, R.J. Buenker and M. Perić, Chem. Phys. Letters 55 (1978) 206

(46) L.B. Harding and W.A. Goddard III, Chem. Phys. Letters 55 (1978) 217

(47) M. Perić, R.J. Buenker and S.D. Peyerimhoff, Can. J. Chem. (1979), in press

(48) M. Yoshimine, K. Tanaka, H. Tatawaki, S. Ohara, F. Sasaki and K. Ohno, J. Chem. Phys. 64 (1976) 2254

(49) R.J. Buenker, S.D. Peyerimhoff and M. Perić, Chem. Phys. Letters 42 (1976) 383

(50) R.A. Back, C. Willis and D.A. Ramsay, Can. J. Chem. 56 (1978) 1575

(51) A. Richartz, R.J. Buenker, P.J. Bruna and S.D. Peyerimhoff, Mol. Phys. 33 (1977) 1345

(52) W.R. Harshberger, J. Chem. Phys. 54 (1971) 2504

(53) M. Desouter-Lecompte, J.C. Leclerc and J.C. Lorquet, Chem. Phys. 9 (1975) 147; M. Desouter-Lecompte and J.C. Lorquet, J. Chem. Phys. 66 (1977) 4006; C. Galloy and J.C. Lorquet, J. Chem. Phys. 67 (1977) 4672

(54) G. Hirsch, P.J. Bruna, R.J. Buenker and S.D. Peyerimhoff, Chem. Phys. (1979) in press

(55)   R.J. Buenker, Gaz. Chim. Ital. 108 (1978) 245

(56)   J. Römelt, S.D. Peyerimhoff and R.J. Buenker, Chem. Phys. 34 (1978) 403;
       41 (1979) 133

THE APPLICATION OF AB INITIO QUANTUM CHEMISTRY TO PROBLEMS OF

CURRENT INTEREST RAISED BY EXPERIMENTALISTS

P. S. Bagus
B. Liu
A. D. McLean
M. Yoshimine

IBM Research Laboratory
5600 Cottle Road
San Jose, California   95193

ABSTRACT: Rigorous methods of quantum chemistry can now be used to address problems raised by experimentalists working in a variety of areas from free radical spectroscopy to surface chemistry. The focus of this paper will be on the kinds of results that can be obtained rather than on details of the computations. This will be achieved by presenting the results of a few representative studies performed recently at the San Jose Laboratory. We shall consider first examples that belong to the area traditionally associated with quantum chemistry, the spectroscopy of small molecules. The van der Waals interactions in the dimers He_2, Be_2, and Mg_2 will be discussed. It will be shown that with a suitable model the weak bonding in these systems can be treated with rather high accuracy. The behavior of the dipole moment curves (surfaces) for HCN and CO will be discussed. These curves are difficult to determine with infrared spectroscopy particularly for the portions relevant for highly excited vibrational levels. For CO, the calculations have helped to determine its density in the solar atmosphere. Although these first two examples deal with traditional areas of quantum chemistry; they are particularly important as cases in which definitive results are very likely to be obtained more easily from theoretical than from experimental data. The study of the electronic excitations in the peroxyl radicals, HO_2 and CH_3O_2, considered next show the kind of interplay now possible between theory and experiment. For HO_2, the theory was able to confirm the assignments made for the near infrared absorption spectrum. For CH_3O_2, careful theory, albeit at a simple level of approximation, made <u>possible</u> the interpretation of the

anomolous behavior of satellite bands of the main vibronic
transitions. The kinds of properties discussed in these examples
can also be obtained for larger systems; the Wolff rearrangement
discussed in another paper is an excellent case of this. The
final example will discuss an application to what is, in principle,
an even larger system, the interaction of an atom with a solid
surface. The interactions of H, F, and Cl with the (111) surface
of Si, modeled by molecular clusters of up to ten Si atoms, will
be discussed and shown to lead to an explanation for the greater
reactivity of F than Cl with a Si surface.

INTRODUCTION

    Following the development of quantum mechanics in the 1920's
work began to apply this formalism to obtain information about
the electronic structure of atoms and molecules. This work continues
to the present and it forms a discipline which is now called
quantum chemistry. There were some remarkably accurate results
obtained in the early days of quantum chemistry for 2-electron
systems: for example, the work of Hylleras[1] in 1930 for the
ionization potentials of the He iso-electronic sequence and that
of James and Coolidge[2] in 1933-1935 for the dissociation energy
and potential energy curve of H_2. However, for more complex
systems the calculations were rather crude and involved severe
approximations. With the availability and steady increase in the
power of digital computers, more accurate calculations of
electronic wavefunctions became possible. Along with the increase
in computing power, there was substantial work done to develop
and implement mathematical and numerical methods to solve
Schrödinger's equation "accurately" for many-electron
(significantly greater than 2 electron) systems.[3]

    These two parallel developments gave great impetus to ab initio
quantum chemistry and, for example, the compendium of Richards et al.
shows that ab initio wavefunctions have been obtained for a large
number of molecules. However, the goal of quantum chemistry is
not one merely of comparing computed properties with experimental
observations. It's proper goals are to help us understand the
significance and systematics of these observations; to provide us
with reliable information for cases in which experiment is either
impossible or very difficult; and to provide us with an independent
source of information which will complement and help to assess
the accuracy of experiment. In effect, quantum chemistry must
take its place as another of the spectroscopies of modern chemistry
albeit one very different from all the others. Yet this very
difference, which centers about the ability to directly construct
and evaluate theoretical models, is what makes quantum chemistry
unique and particularly valuable. Studies which meet these goals
are now appearing in the literature[5] and we fully expect them to
become more common in the coming years.

One essential requirement for meeting these goals is that quantum chemical calculations be accurate or, at the very least, that they have known reliabilities and uncertainties. Ab initio calculations have always had, in principle, the ability to satisfy this requirement and, as we shall demonstrate, can, in practice, be applied to an ever increasing number of problems. A second set of important requirements center about the interplay of theory and experiment. It is desirable, of course, that calculations aimed at resolving a problem posed by experimentalists become available in a reasonable time after the problem is known. It is also desirable that quantum chemists understand clearly the nature of the problem so that they can design their calculations to best resolve the problem. Finally, it is desirable to have quantum chemistry make real predictions; that is to have calculated values for observables to suggest and to stimulate new experimental work. These requirements mean that the calculations must be timely and this has traditionally been a problem for ab initio work. However, with the twin developments in computational power and methodology mentioned above, it is now possible for ab initio quantum chemistry to achieve this timeliness. We plan to demonstrate that this is indeed the case in this paper.

Specifically, we will consider four examples drawn from our own recent work at the IBM San Jose Research Laboratory. Our intent is to use these examples to demonstrate the current status of ab initio quantum chemistry in achieving the goals described above. We plan also to use these examples to demonstrate the range of problems that can be treated and the kinds of models and accuracy that are possible (or needed) for different sorts of problems. These examples are based on calculations performed with the program system ALCHEMY. This system was developed primarily by us although we are indebted to several collegues for their substantial assistance with certain segments of the programs.[6] ALCHEMY is an extensive set of programs which include: (1) the capability to perform self-consistent field, SCF, calculations for general open shell configurations; (2) the capability to perform multi-configuration SCF, MCSCF, calculations with a fairly large number of configurations; (3) the capability to perform large scale configuration interaction, CI, calculations with an unrestricted - a so called "open-ended" - way to select the classes of configurations used; and (4) the capability to directly calculate many observables from the molecular wavefunctions. We will not present very much information about the details of the calculations on which our examples are based. These details will be given elsewhere. We will, however, make some general remarks concerning our approach to ab initio quantum chemistry.

In the next section, Models in Ab Initio Quantum Chemistry, we will discuss our approach - essentially our philosophy - for designing computational models. We will point out that ab initio calculations are far from being free of parameters and we will consider criteria and procedures for testing whether the models

and parameters chosen will lead to meaningful results. In each
of the succeeding sections, we will present the four representative
examples that we have chosen. The first two deal with small
systems – diatomic or linear triatomic molecules. The focus here
will be on the accuracy that can be obtained and on the fact that
the computational models used must reflect the chemistry involved
in the problem. In fact, these models should be different for
different chemical problems. The other two examples deal with
much larger systems and here the concern is not so much with high
accuracy as with being able to obtain a qualitative understanding
of the mechanisms which lead to the observed phenomena. These
examples involve answering questions and explaining problems which
arose from experimental work carried out at our laboratory.

MODELS IN AB INITIO QUANTUM CHEMISTRY

In principle, the central problem of quantum chemistry is to
find solutions of Schrödinger's equation,

$$H\Psi = E\Psi ,$$ (1)

for ground and excited states of molecules. (We limit ourselves
to discrete, square-integrable wavefunctions Ψ and we consider
only the nonrelativistic clamped nucleus Born-Oppenheimer form of
the Hamiltonian, H.) This problem has been solved in closed form
only for the H atom. For all other systems, we must consider
approximate solutions to Eq. (1). This is true even for two
electron systems although very accurate wavefunctions have been
obtained for the He iso-electronic series[7] and for H_2.[8] Our
ab initio approach for finding approximate solutions to Eq. (1) is
based on the use of molecular orbitals, MO's, and the configuration
interaction method.[9] This is the most commonly used approach in
ab initio work and is used – in somewhat different forms – for the
work reported in two other papers in this volume; those by
S. Peyerimhoff and by W. Meyer and P. Rosmus.

Ab initio quantum chemistry is usually believed to be different
from semi-empirical and empirical work in that it is free of
adjustable parameters. However, this is not all true as can be
seen from the following. The CI wave function is expanded in a
basis of n particle terms, Φ, which are denoted configuration
state functions,[9] CSF,

$$\Psi = \Sigma_k c_k \Phi_k .$$ (2)

The CSF's are linear combinations of Slater determinants which
are anti-symmetrized products of a one-particle basis, the MO's,
φ. The MO's themselves are expanded in an elementary basis set,
χ;

$$\varphi = \Sigma c_p \chi_p \; . \tag{3}$$

The summations of Eqs. (2) and (3) are always truncated from the
infinite sums required for an exact solution to Eq. (1). Even
though the linear coefficients, C_k and c_p are determined by
application of the variational principle, the choice of the sets
of functions, Φ, φ, and χ, represents a parameterization of the
solution of Eq. (1). A sound choice will lead to meaningful results
and a poor device to meaningless results.

However, the parameterization of ab initio calculations is
quite different from that of semi-empirical or empirical
calculations. First, it is often possible in ab initio calculations
to examine the convergence of various properties with respect to
the size of the basis sets, both one- and n-particle. While these
convergence tests need not and should not be done always, they
must be done for important representative cases in order to obtain
reliable estimates of the truncation errors. Such convergence
studies are not possible, essentially by definition, for empirical
methods. This is especially clear if we recall that, in these
methods, the parameters are normally adjusted to fit experimental
data. As the number of parameters is increased, it is easier to
fit more data and more difficult to assign any meaning to the
parameters. The second way in which ab initio parameterization is
different follows precisely from the fact that parameters are not
chosen to fit experiment. They are selected according to general
criteria. For the one particle bases, φ and χ, the principle
criterion is the convergence with respect to basis set size
discussed above. This application of this criterion is normally
straightforward although care must be employed and thought given
to the problem to be solved. For example, if states may have
Rydberg character then diffuse functions must be included. For
the n particle basis, the criteria lead to the selection of classes
of configurations. The principles upon which these classes are
selected must reflect the essential features of the chemistry of
the problem. We refer to a selection of a set of classes of CSF's
as a CI model. We shall, in the first two applications described
below, give specific examples of CI models. The establishment of
a suitable CI model is not at all straightforward and often
requires a qualitative understanding of the problem at hand. The
model may well be refined during the course of a series of
calculations. It is also likely that a particular model will be
suitable for a specific class of problems; different models may
be necessary for different problems.

It is important to comment on the role that obtaining a low
total energy for a molecule, $E = \langle \Psi | H | \Psi \rangle / \langle \Psi | \Psi \rangle$, plays - or actually
does not play - in selecting a suitable CI model. All observables
involve measurements of energy differences; differences between
the energies of different electronic states or differences of the
energies at different nuclear geometries. There is virtually no
interest in the accurate calculation of the total energy. Thus,

it is clear that the essential requirement for a good CI model is
that it provides wavefunctions and energies that are equally good
for all states and/or geometries of interest. It is necessary to
find models or approximations such that the errors in the
calculated wavefunctions cancel. A model which leads to
wavefunctions which have a lower total energy may well also lead
to poorer results than a model which produces wavefunctions with
higher energies but where there is a more complete cancellation
of errors. Our first two applications provide excellent evidence
for the fact that a low total energy is not a generally valuable
criterion for a good CI model.

VAN DER WAALS BONDED DIMERS

    Two of us, B.L. and A.D.M., have made extensive studies of
the weak van der Waals interaction in He_2, Be_2 and Mg_2 dimers.[10,11]
The accurate theoretical determination of these interactions poses
special and rather difficult problems. Why this is so may be seen
by considering He_2. Here, the best estimate of the attractive
well depth, from the scattering data of Burgmans et al.,[12] is
$10.6°K \approx 7$ cm$^{-1} \approx 8 \times 10^{-4}$ eV. The correlation error of the SCF treatment
of two He atoms is ~2 eV.[13] Thus, the interaction energy, E_{INT}, at
the well minimum is a factor of 4×10^{-4} smaller than the absolute
energy lowering which must be obtained in an accurate CI
calculation of the total energy of He_2. Even with rather large
Slater type, STO, basis sets, it is very difficult to obtain an
absolute accuracy in a CI calculation of the He atom better than
~0.1% or ~2×10^{-3} eV for 2 He atoms.[14] The absolute error in a He_2
CI calculation derives principally from limitations in the one
particle basis. Liu and McLean[10] have shown that a He_2 calculation
which aims to obtain low total dimer energies will lead to
artificially deep wells because of a systematic "basis set
superposition error."

    Clearly, it is necessary to develop a CI model which can lead
to an accurate $E_{INT}(R)$ despite large errors in the total energy
of the dimer. [We define $E_{INT}(R) = E(R,dimer) - 2E(atom)$.] Liu and
McLean[10,11] have developed such a model and call it the interacting
correlated fragment, ICF, model. In the ICF model, the occupied
MO's are localized so that interatomic correlation can be clearly
distinguished from intra atomic correlation. In this way, intra
atomic correlation can be added in a systematic fashion and the
convergence of $E_{INT}(R)$ can be studied. The method is most easily
described by considering the simplest level, denoted ISCF, where
the zeroth order dimer wavefunction, Ψ, is the molecular SCF
wavefunction. For He, $\Psi_o = \{\sigma_A^2 \sigma_B^2\}$; where $\sigma_A = (1\sigma_g + 1\sigma_u)/\sqrt{2}$ and
$\sigma_B = (1\sigma_g - 1\sigma_u)/\sqrt{2}$ are the localized SCF MO's. The ISCF CI includes
all CSF's of the forms $\{\sigma_A^2 \sigma_B u\}$, $\{\sigma_A \sigma_B^2 u\}$, and $\{\sigma_A \sigma_B uv\}$ where u and
v are virtual MO's. The double excitation matrix element

$$< \{\sigma_A \sigma_B uv\} | H | \{\sigma_A^2 \sigma_B^2\} > \to 0$$

for $R \to \infty$. The matrix element between the singly excited CSF's and Ψ_o is zero at all R by Brillouin's theorem. Thus the contribution of the excited CSF's to the wavefunction vanishes at large R. In effect, Ψ(ISCF) includes only inter-atomic correlation. At higher levels of approximation, $\Psi_o = \{\Phi_A \Phi_B\}$, where Φ represents a partially correlated atomic wavefunction in the dimer. The ICF wavefunction includes all those singly and doubly excited CSF's - with respect to all the CSF's which form Ψ_o - for which the interaction with Ψ_o vanishes at large R. In other words, only inter-atomic correlation is added to the intra-atomic correlation already in Ψ_o. For the ICF2 level results presented below, the forms of the correlated fragment functions, Φ, are:

$$
\begin{aligned}
&\text{He:} \quad A~1s^2 + B~2s^2 + C~2p^2 \\
&\text{Be:} \quad A~2s^2 + B~2p^2 \\
&\text{Mg:} \quad A~3s^2 + B~3p^2~.
\end{aligned}
\tag{4}
$$

The localized orbitals, ns and np, and the CI coefficients, A, B, and C, are determined from multi-configurational molecular calculations. These calculations, at both ISCF and ICF2 levels, have been performed using large STO basis sets and $E_{INT}(R)$ is essentially converged with respect to the one particle bases. The results for the well depth, D_e, and the position of the well minimum, R_e are given in Table 1.

Table 1. D_e and R_e for He_2, Be_2, and Mg_2

	R$_e$ (Å)			D$_e$		
	ISCF	ICF2	EXPT[a]	ISCF	ICF2	EXPT[a]
He$_2$	2.94	3.03	2.97	12.1°K	10.7°K	10.57°K
Be$_2$	2.46	2.51	----	2900 cm^{-1}	810 cm^{-1}	----
Mg$_2$	3.52	3.89	3.88	1248 cm^{-1}	436 cm^{-1}	430 cm^{-1}

[a]For He$_2$, see Ref. 12 and for Mg$_2$, Ref. 15.

The ICF2 results for He$_2$ are in very good agreement with the experimentally derived values. The ICF2 D_e is significantly better than the ISCF value and is within 0.1°K of experiment. This is all the more remarkable since the correlated fragment function, Φ, for He, Eq. (4), includes less than 90% of the intra-atomic correlation.[13] Thus, the absolute error in the total energy of Ψ(ICF2) is ~2000°K; four orders of magnitude larger than the 0.1°K error in D_e! In Fig. 1, theoretical, ICF2, and experimental $E_{INT}(R)$ are compared over a large range of R. The agreement is particularly impressive in view of the fact that the energy scale

is logarithmic. The excellent agreement shown is over three orders
of magnitude of E_{INT}!

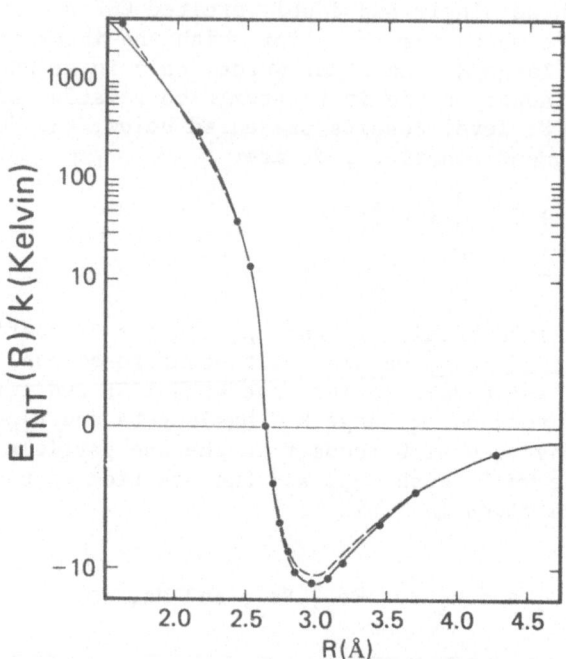

Fig. 1.   He$_2$ interaction potentials in degrees Kelvin. The solid
          curve is from the scattering data of Ref. 16; the dashed
          curve from the bulk data of Ref. 17. The circles are the
          theoretical ICF2 values from Ref. 11.

The results for Mg$_2$ in Table 1 and the $E_{INT}(R)$ curves in Fig. 2
present a new and, at first sight, puzzling feature. The ICF2
results are in rather good agreement with experiment.[15] However,
the ISCF results are very poor and give a well which is much too
deep at a distance which is too short. The ICF model provides a
way to understand this puzzle. There is a "near degeneracy"
between the ns^2 and np^2 configurations of the alkali earth
atoms.[13,18] For Mg, the two configuration MCSCF wavefunction[18] is
$\Phi = 0.96 \ 3s^2 + 0.28 \ 3p^2$. It is essential to take account of this large
$3p^2$ atomic contribution to Φ before meaningful molecular results
can be obtained. Thus, it is to be expected that the ICF2 results
for Mg$_2$ will be as or, actually, slightly more accurate than the

He$_2$ ISCF results since the near degeneracy atomic correlation is not present[13] for He. In Table 2, we give vibrational transition energies, $\Delta G_{v \rightarrow v+1}$. The close agreement with experimental values further demonstrates the accuracy of $E_{INT}(R)$ obtained with the ICF2 model.

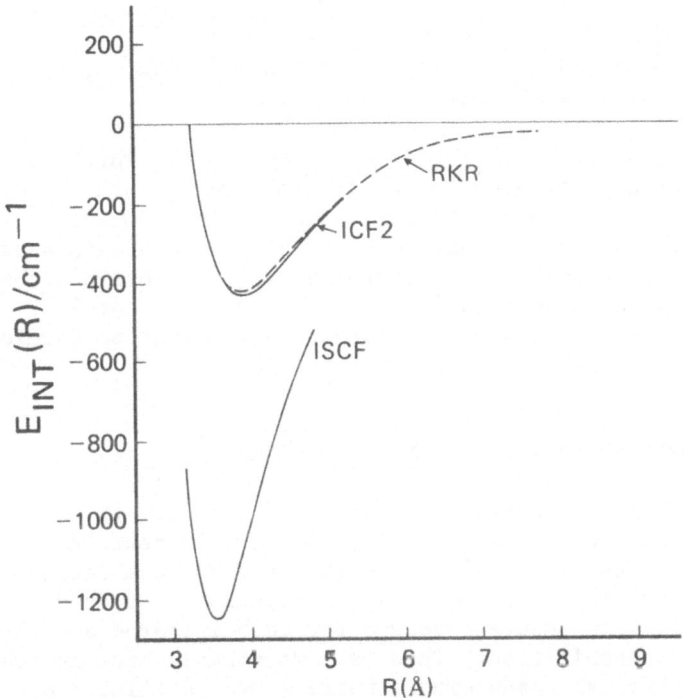

Fig. 2.   Mg$_2$ interactive potentials. The solid RKR curve is from the spectroscopic measurements of Ref. 15. Theoretical, dashed, curves are for the ISCF and ICF2 models.

Also included in Tables 1 and 2 are predictions for Be$_2$ where experimental results are not, as yet, available. The well depth for Be$_2$ is predicted to be about twice as large as for Mg$_2$. The excellent results for both He$_2$ and Mg$_2$ are compelling evidence for the correctness of the Be dimer predictions.

This work has shown that a suitable model, ICF, leads to interaction energies of weakly bonded dimers with small errors; errors which are, in fact, vastly smaller than the absolute errors

Table 2. Vibrational transition energies, $\Delta G_{v \to v+1}$. The
experimental values for Mg_2 are from Ref. 15.

	Mg_2		Be_2
	ICF2	EXPT	ICF2
$\Delta G(cm^{-1})$			
$0 \to 1$	49.9	47.88	202
$1 \to 2$	46.5	44.72	152
$2 \to 3$	43.0	41.63	---

of the total energies of the dimers themselves. Furthermore,
models based on obtaining low total dimer energies have been
shown[10,19] to lead to artificially deep wells. However, we
certainly do not expect this model to be applicable to systems
where a strong chemical bond is formed and the interplay between
intra- and inter-atomic correlation is much more complex. In the
following section, we show that other models must be developed in
this case.

SPECTROSCOPIC PROPERTIES OF SMALL MOLECULES

     The spectroscopic properties of molecules covers a very large
number of phenomena, this section, however, is restricted to just
two topics. The first relates to infrared and the second to
optical (electronic transitions) spectroscopy. It is quite
important to know accurate values for both infrared and electronic
transition probabilities. This is particularly true in connection
with astrophysical measurements where known oscillator strengths
can be combined with observed emission intensities to determine, for
example, the density of matter in various objects. Yet, oscillator
strengths, especially for the species and states observed in
stellar and interstellar media, are often very difficult to measure
experimentally.

     In the first subsection, we shall consider force constants,
dipole moments and dipole moment derivatives, and vibrational
energy levels. (Some of these properties are also discussed in
the paper of Meyer and Rosmus in this volume although for different
molecules and obtained by a somewhat different method.) We shall
compare the results of several different CI models for these
properties. The object of the comparison is to obtain a better
understanding of the kinds of correlation needed to make accurate
predictions. In the second subsection, we shall present results
for electronic transition probabilities for some diatomic
molecules. (Electronic transition probabilities, again for
different molecules, are also discussed in the paper by Peyerimhoff

in this volume.) Here, we will not discuss the CI model used but
will content ourselves with demonstrating that calculated results
can be quite accurate. In fact, for both vibrational and
electronic transition probabilities, the results of properly
designed calculations may often be better than the available
experimental data.

A. Infrared Related Properties.

The results presented here are based on the work of
Kirby-Docken and Liu[20] and of Liu et al.[21] for the ground states
of CO and HCN, respectively. The CI models to be compared here
will be introduced first. The models will be illustrated by
reference to CO and HCN.

The first model is one which includes the SCF ground state
CSF, $3\sigma^2 4\sigma^2 5\sigma^2 1\pi^4$ for both CO and HCN, and all symmetry allowed
CSF's which may be formed by exciting either one or two electrons
from the valence orbitals into virtual MO's. (It is usual to
neglect the correlation contribution involving core electrons
since, to a very good approximation, it is constant for all
internuclear geometries of interest.[22]) The occupied orbitals are
taken as the SCF MO's. The model is referred to as a singles and
doubles CI, SDCI. It is straightforward and methods and programs
have been developed to apply it in a routine fashion in many
cases.[23]

The second model includes the CSF's formed by distributing
the valence electrons in all possible ways among the valence
orbitals. For CO and HCN, there are 10 valence electrons. For
CO, the valence orbitals are 3σ-6σ and 1π and 2π; the 6σ and 2π
are unoccupied in the SCF CSF. For HCN, there is, in addition a
7σ valence MO which correlates, at large internuclear separation,
to H(1s). The valence MO's are obtained from an MCSCF calculation
which often uses a subset of these CSF's. The subset being chosen
so that the MCSCF wavefunction dissociates to a product of atomic
SCF wavefunctions.[20,21] This model is referred to as a valence
CI, VCI.

The last model considered is an extension of the VCI and is
called a first order CI, FOCI. It includes: (1) the VCI CSF's;
and (2) CSF's which have one electron in the VCI virtual orbital
space. The full set of the second class is generated by
distributing n-1 electrons in the valence orbitals and one in the
virtual orbitals. Constraints are normally imposed to reduce the
number of CSF's in this set,[20,21] however, the selection with
constraints is still by class and energy is not used as a criterion
for selecting any particular CSF.

The physical effects included in these models may be summarized
as follows: The VCI wavefunction includes the CSF's necessary
for proper dissociation and to take account of internal,

near-degeneracy, correlation.[24] The FOCI wavefunction includes, in addition, the CSF's necessary to take account of semi-internal correlation[24] and to describe effects related to relaxation and polarization of the valence electrons. The SDCI does not dissociate properly but it does include CSF's to take account of external correlation[24] and, to a much greater extent than the FOCI, of intra-atomic correlation.

The CO calculations of Ref. 20, based on a modestly large STO elementary basis set, included 196, 2196, and 5244 CSF's for the VCI, FOCI, and SDCI calculations, respectively. The total energies (in hartrees) at R=2.2 bohr, near the equilibrium bond distance, R_e, are:

VCI	-112.9129
FOCI	-112.9470
SDCI	-113.0967 .

It is not surprising that the SDCI gives the lowest energy since it contains the largest number of CSF's and includes the largest amount of external correlation. Other results, presented in Tables 3 and 4, show that the SDCI results are often poorer than the others.

Table 3. Vibrational energy differences, $\Delta G_{v+1/2}=G(v+1)-G(v)$, in cm^{-1}, for the $X^1\Sigma^+$ state of CO. The error of each calculation is given in parenthesis.

v	VCI	FOCI	SDCI	EXPT[a]
0	2128.3 (-15.0)	2140.3 (-3.0)	2235.9 (+92.6)	2143.3 ---
1	2102.4 (-14.4)	2114.7 (-2.1)	2213.9 (+96.1)	2116.8 ---
2	2052.1 (-11.9)	2064.0 (0.0)	2168.3 (+104.3)	2064.0 ---

[a]See Ref. 25.

The vibrational energy levels obtained from the calculated potential curves, see Table 3, show that the poorest results are from the SDCI while the best are from the FOCI where the error is ~0.1%. Similar results are found for the dipole moment curve, $\mu(R)$. The result of a least square fit of the calculated $\mu(R)$ to the Taylor series,

$$\mu(R) = M_0 + M_1(R-R_e) + M_2(R-R_e)^2 + M_3(R-R_e)^3 , \qquad (5)$$

is given in Table 4. Although the SDCI does reasonably well for
$\mu(R_e) = M_0$, it describes the dependence of μ on R (M_1, M_2, and M_3)
least well. The FOCI values of these latter coefficients are as
reliable as the empirical values.[20] So far we have really just
normalized on CI models against known data. The results in Fig. 3
show that Kirby-Ducken and Liu were able to go beyond reproducing
know quantities. In this figure the FOCI $\mu(r)$ is compared with
the empirical dipole function of Ref. 27. In the region around
R_e, the theoretical and empirical curves closely parallel each
other. However, for large R, the theoretical curve has the correct
behavior and goes smoothly toward zero corresponding to nonbonded
neutral C and O atoms. In contrast, the empirical function goes
to large negative values corresponding to substantial C^-O^+
polarity. The incorrect behavior of the empirical function at
large R is not surprising since it is based on data only for
transitions among the first four vibrational levels.[27] However,
$\mu(R)$ must be known accurately for large R in order to obtain
transition intensities involving levels with high vibrational or
rotational quantum number; i.e., just those levels of particular
interest in astrophysics.

Table 4. Calculated and empirical dipole moment functions
for $CO(X^1\Sigma^+)$. The expansion coefficients, M_i, are defined in
Eq. (5) and are for R in bohrs and μ in Debye. The error of
each calculation is given in parenthesis.

	VCI	FOCI	SDCI	EMPIRICAL[a]
M_0	-0.260 (-0.138)	-0.319 (-0.197)	-0.082 (+0.040)	-0.122 ---
M_1	1.726 (+0.081)	1.578 (-0.067)	-1.870 (+0.225)	-1.645 ---
M_2	-0.001 (+0.041)	-0.042 (0.000)	+0.092 (+0.134)	-0.042 ---
M_3	-0.508 (-0.158)	-0.376 (-0.026)	+0.479 (+0.829)	-0.350 ---

[a]M_0 from Ref. 26; all other values from Ref. 27.

Similar results, shown in Tables 5 and 6, have been obtained
by Liu et al.[21] for $HCN(X^1\Sigma^+)$. In Table 5, the normal mode force
constants for stretching obtained from SCF and VCI wavefunctions
are compared with the values obtained from SDCI wavefunctions[28]
and with experimentally derived values.[29] Although, the SDCI leads
to a significant improvement over SCF, the computationally simpler
VCI leads to the best results.

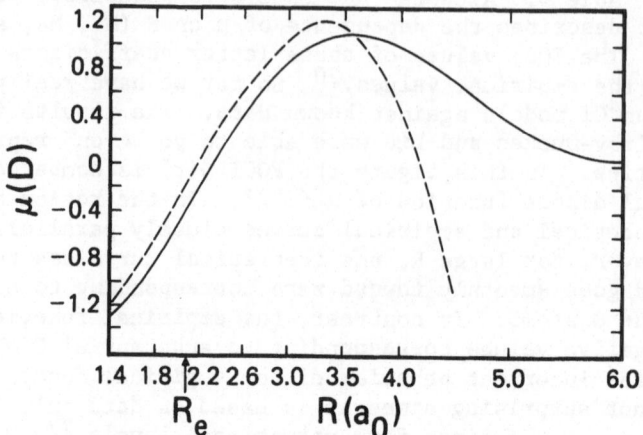

Fig. 3. Theoretical (solid curve) FOCI and empirical (dashed curve) dipole moment functions, in Debye, for $CO(X^1\Sigma^+)$. The equilibrium bond distance, R_e, is marked.

Table 5. Force constants for HCN in mdyne/Å. The CH stretch is denoted k_{11}, the CN stretch is k_{22}, and the coupling term is k_{12}. The error of each calculation is given in parenthesis.

	SCF[a]	VCI	SDCI[a]	EXPT[b]
k_{11}	3.43 (+0.31)	3.00 (−0.12)	3.27 (+0.15)	3.12 ± 0.01 ---
k_{22}	12.14 (+2.75)	9.02 (−0.37)	10.33 (+0.94)	9.39 ± 0.09 ---
k_{12}	−0.18 (+0.03)	−0.22 (−0.01)	−0.15 (+0.06)	−0.21 ± 0.08 ---

[a] See Ref. 28.
[b] See Ref. 29

The most striking improvement of VCI over SDCI is for the CN triple bond stretch. Clearly, the higher order excitations included in the VCI model are required to properly describe this stretch. The dipole moment at equilibrium geometry and its derivatives with respect to CH and CN bond lengths, μ'_{CH} and μ'_{CN}, are given in Table 6. It is interesting to note that only the relative signs of μ'_{CH} and μ'_{CN} are known from experiment.[21] The accuracy of the magnitudes of the calculated VCI values establishes the absolute signs as $\mu'_{CH}>0$ and $\mu'_{CN}<0$.

Table 6.   Calculated and experimental values for the equilibrium dipole moment and its first derivatives. Errors of the calculations are in parenthesis.

	SCF	VCI	EXPT[a]
μ (D)	3.21 (+0.22)	2.94 (-0.05)	2.99 ----
μ'_{CH} (D/Å)	1.34 (+0.28)	1.02 (-0.04)	1.06 ----
μ'_{CN} (D/Å)	0.55 (+0.87)	-0.34 (-0.02)	-0.32 ----

[a]For an analysis of the experimental data see Ref. 21.

The comparisons presented above, for both CO and HCN, show two important things. First, that it is possible to obtain quantitatively accurate results for $\mu(R)$, for vibrational energy levels and for related quantities. Second, they provide us with important information about the kinds of correlation effects that must be treated. It is necessary to include CSF's which describe the stretching of bonds to form separate atoms and which describe internal correlation.[24] These CSF's are included in the VCI model. The semi-internal correlation[24] included in the FOCI model leads to substantial improvements over VCI. Both VCI and FOCI lead, however, to results which are generally superior to the computationally more demanding SDCI.

B. Electronic Transition Probabilities.

The principle quantity of concern in this subsection is the band oscillator strength, $f^{v'}_{v''}$, given by[30]

$$f^{v'}_{v''} = 2/3 \ g\Delta E \left[\int \Psi^*_{v'}(R) \underset{\sim}{M}(R) \Psi_{v''}(R) dR \right]^2 \qquad (6)$$

where $\Psi_{v'}$ and $\Psi_{v''}$ are vibrational wavefunctions for the upper and lower electronic states, respectively, ΔE is the transition energy, and g is the degeneracy of the final electronic state. The electronic transition moment, $\underset{\sim}{M}(R)$, is

$$\underset{\sim}{M}(R) = \int \psi^*_{\alpha'}(\underset{\sim}{r},R) \; \Sigma_{\underset{\sim}{r}_i} \; \Psi_{\alpha''}(\underset{\sim}{r},R) d\underset{\sim}{r} \; ; \tag{7}$$

where $\Psi_{\alpha'}$ and $\Psi_{\alpha''}$ are the electronic wavefunctions for the two states. Other measures of the transition probability, e.g., radiative lifetimes, can be obtained easily[30] from $f^v_{v''}$. For the examples we will consider, the Ψ_α used in Eq. (7) for the evaluation of $\underset{\sim}{M}(R)$ are rather extended and accurate CI wavefunctions. The details of these wavefunctions will not be discussed here; for two out of the three examples considered, these details have been published and will be referenced. The vibrational wavefunctions, $\Psi_{v'}$ and $\Psi_{v''}$, required for Eq. (6) have been obtained using experimental RKR potential curves rather than the calculated CI curves. We conservatively estimate that this procedure leads to theoretical values for $f^v_{v''}$ accurate to ±10%.

Table 7. Band absorption oscillator strengths for the $X^3\Sigma^+_g$-$B^3\Sigma^-_u$ transition in O_2.

v'	v''=0		v''=1	
	CALC	EXPT[a]	CALC	EXPT[a]
0	3.06×10^{-10}	3.45×10^{-10}	8.55×10^{-9}	----
1	3.65×10^{-9}	3.90×10^{-9}	9.26×10^{-8}	----
2	2.23×10^{-8}	2.38×10^{-8}	5.14×10^{-7}	5.35×10^{-7}
4	3.02×10^{-7}	3.21×10^{-7}	5.78×10^{-6}	6.15×10^{-6}
6	1.84×10^{-6}	1.91×10^{-6}	2.95×10^{-5}	3.15×10^{-5}
8	6.57×10^{-6}	6.68×10^{-6}	8.96×10^{-5}	9.40×10^{-5}
10	1.55×10^{-6}	1.57×10^{-6}	1.83×10^{-4}	1.91×10^{-4}
12	2.54×10^{-5}	2.53×10^{-5}	2.66×10^{-4}	2.73×10^{-4}
14	3.02×10^{-5}	3.03×10^{-5}	2.85×10^{-4}	2.95×10^{-4}

[a]See Ref. 32.

We present below results for the transition probabilities for transitions in three different molecules. We have chosen these transitions because they represent cases where the status of the agreement between theory and experiment is rather different.

The first is the $X^3\Sigma_g^+$-$B^3\Sigma_u^-$ transition in O_2 studied by
Yoshimine.[31] Here, there is extensive and reliable experimental
data, summarized by Krupenie,[32] available for comparison with
Yoshimine's theoretical results. A selected comparison is given
in Table 7. The range of v' and of values of $f_{v''}^v$ is rather wide.
It is clear from Table 7 that the agreement between the theoretical
and experimental $f_{v''}^v$ is indeed within 10% for values of f over a
range of six orders of magnitude.

The second transition considered is the $X^1\Sigma^+$-$A^1\Pi$ of CH^+ studied
by Yoshimine et al.[33] A comparison of theory and various
experimental values for the absorption f_0^0 is given in Fig. 4. The
values of f are plotted against the year in which the work was
reported.

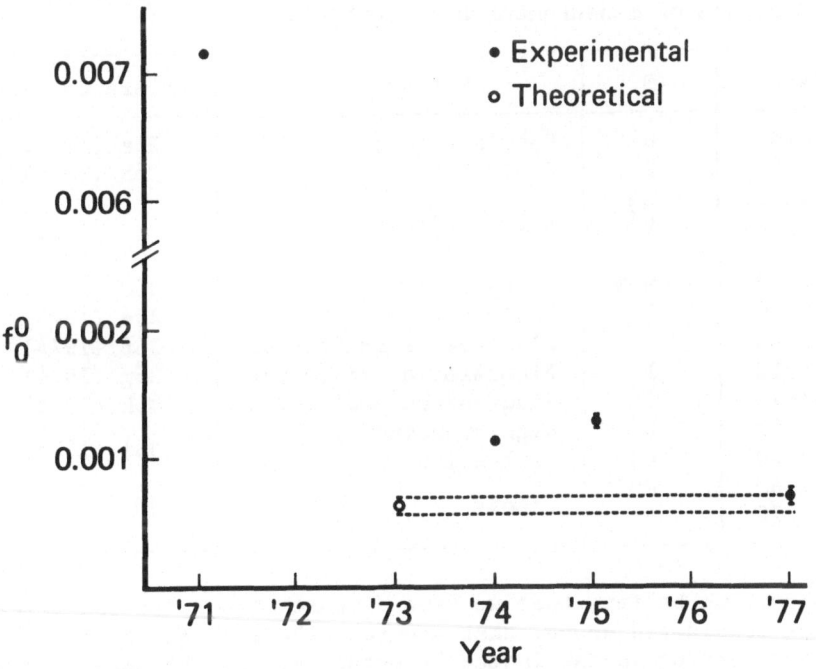

Fig. 4. Experimental (X) and theoretical (0) values of the band
absorption oscillation strengths for CH^+ $X^1\Sigma^+$-$A^1\Pi$. Error
estimates are given for the theoretical, Ref. 33, and the
most recent experimental values. The experimental f_0^0
are reviewed in Ref. 34.

It can be seen that the latest measurement,[34] using a high
frequency deflection technique, is in agreement with the results
of calculations reported four years earlier. Previous measurements
differ from the calculated value; one, the 1971 value, by almost
a factor of 10. In other words, the accuracy of the measurements
have now caught up to that of the calculation.

The last transition considered is the $X^2\Pi$-$A^2\Sigma^+$ of OH studied
by Liu et al.[35] In Table 8, we compare various experimental values
for the lifetime of the v'=0 level of the A state in specific
rotational levels, N, with the theoretical values. Although some
of the measured values fall within the theoretical error estimates,
the most recent values of Brzozowski et al.[36i] are clearly outside
of the estimated accuracy of the theory. It will be interesting
to see if further experimental work will support the theory or
not.

Table 8. Experimental and Theoretical values for the lifetime
of the $A^2\Sigma^+$ state of OH in the v'=0 vibrational level. The
rotational level is listed under N. The year of publication
and method of measurement are identified.

τ(nsec)	N	Method	Reference (Year)
503 ± 50	0	Hook's Method	36a (1967)
620 ± 8	2	Hanle Effect	36b (1971)
803 ± 80	1		
697 ± 70	2	Phase Shift	36c (1972)
783 ± 75	3		
755 ± 50	low N	Delayed coincidence	36d (1973)
580 ± 50	2	Hanle Effect	36e (1973)
820 ± 40	2	Fluoresence Excitation	36f (1974)
780 ± 13	1	Fluoresence Excitation	36g (1974)
689 ± 7	1	Fluoresence Excitation	36h (1975)
758 ± 20	0	High Frequency	
755 ± 20	4	Deflection	36i (1978)
636 ± 65	0		
656 ± 70	4	Theory	35

In sum, we believe that we have demonstrated, with the three
examples considered above, that accurate theoretical values of
oscillator strengths can indeed be obtained. In the first example,
the theory is in agreement with extensive accurate experimental
data; in the second, the most recent measurements are in agreement
with theory; and in the last, the A-X system of OH, they are close
to theory but still significantly outside of the estimated
theoretical error. We have sufficient confidence in the
theoretical method to believe that the theoretical values for OH
are more accurate and we expect that further experimental work
will demonstrate this.

ELECTRONIC ABSORPTION SPECTRA OF PEROXYL RADICALS

The previous examples have dealt with cases where it was
necessary to use theoretical methods which would lead to results
of high accuracy. In this example, we consider a case where the
SCF approximation - with its limited accuracy - has been used
successfully. An aspect of special interest and value in this
example is that it involved a timely interaction between
theoreticians and experimentalists; both groups working in our
laboratory. The theoretical work was able to lead to the
interpretation of an unforseen, unusual, and puzzling feature of
the absorption spectra of the methyl peroxyl radical.

The experimental work involved is that of Hunziker and
Wendt.[37,38] They have taken high resolution absorption spectra
for the electronic transition between the ground, $^2A''$, and excited,
$^2A'$, states of several peroxyl radicals. A sketch of the structure
of the simplest of these radicals, HO_2, is:

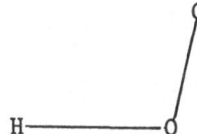

The absorption is in the near infrared at $\sim 1.4\mu \approx 0.9$ eV. A spectrum
for HO_2 is shown in Fig. 5a. The observed bands are assigned[37] to
a progression in the O-O stretch: where the vibrational
transitions are between the ground state $v_s''=0$ level to the excited
state levels v_s' shown in the figure. (The dashed portion of the
curve is an overtone of the OH vibration.) The converging series
of satellite bands following each vibrational transition are due
to the rotation of H about O_2. The analysis of the satellite
structure indicates that there is little change of the HOO angle
between the two electronic states of the transition.[37]

McLean and Yoshimine[39] have carried out SCF and CI calculations
on both the $^2A''$ and $^2A'$ states of HO_2. The calculations used a
modest sized contracted Gaussian basis set, CGTO, denoted double
zeta plus polarization.[3] Their results for equilibrium geometry
and term energy are given in Table 9. The wavefunctions indicate
that the open shell electron in the $^2A''$ ground state can be roughly
described as an oxygen $2pz(\pi)$ electron and that the open shell
electron in the $^2A'$ state is an O $2pxy(\sigma)$ electron. Thus, the $^2A''$
to $^2A'$ excitation can be characterized as being "localized" on O_2
which is consistent with the observed O-O stretching progression.[37]
In this case, it would be expected that the largest change in
geometry between the two states would be for R_{OO} and that the
changes for R_{OH} and $<HOO$ would be much smaller. This is indeed
the case as may be seen from Table 9. The calculated CI geometries
and term energy are reasonably close to the available

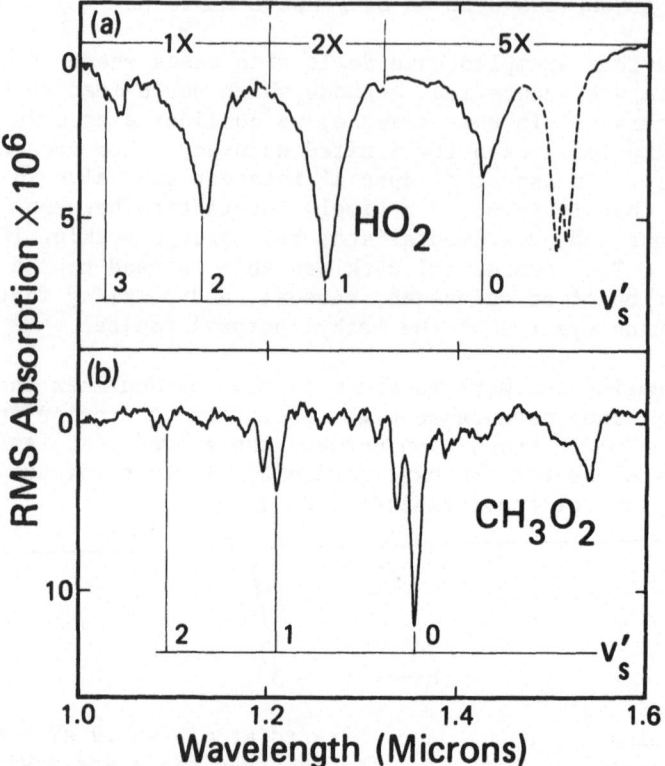

Fig. 5. Absorption spectra, from Ref. 38, for HO_2 and CH_3O_2. The $^2A'$ vibrational levels, v'_8, are marked. The absorption scale should be multiplied by the factors shown for HO_2.

experimentally derived values. Thus, these results support and confirm the assignments[37,38] of the spectra of HO_2, Fig. 5a. They are also rather similar to the results of previous theoretical studies[41] of HO_2. The essential point that we wish to call attention to comes from a comparison of the SCF and CI results. Although the SCF term energy is significantly different from the CI value (25% smaller), the SCF and CI equilibrium geometries are quite close to each other. It is, thus, reasonable to expect the SCF method to lead to similarly accurate geometries in related molecules.

Table 9. Calculated equilibrium geometries and term energies for the $^2A''$ ground and $^2A'$ excited states of HO_2 and CH_3O_2.

		HO_2			
		R_{HO} (Å)	R_{OO} (Å)	<HOO	Te(eV)
$^2A''$	SCF	0.95	1.30	105.7°	---
	CI	0.96	1.33	105.6°	---
	EXPT	0.98[b]	1.34 ± 0.02[a]	104.1°[b]	---
$^2A'$	SCF	0.95	1.36	103.5	0.57
	CI	0.96	1.395	102.7	0.735
	EXPT	---	1.41 ± 0.03[a]	---	0.88[a]

		CH_3O_2			
		R_{CO} (Å)	R_{OO} (Å)	<COO	<HCO
$^2A''$	SCF	1.46	1.38	113°	106°
$^2A'$	SCF	1.46	1.43	111°	106°

[a]See Ref. 37.

[b]See Ref. 40.

Based on this expectation, McLean and Yoshimine undertook SCF calculations, again with a double zeta plus polarization CGTO basis set, for the ground and excited states of CH_3O_2. The structure of CH_3O_2 is obtained by substituting CH_3 for the H atom in HO_2 and the electronic structure of the two molecules is expected to be rather similar. All geometrical parameters were optimized for both $^2A''$ and $^2A'$ states and particular attention was given to the staggered and eclipsed conformations of the CH_3 group. The equilibrium conformation for both states is staggered as shown in the sketch below:

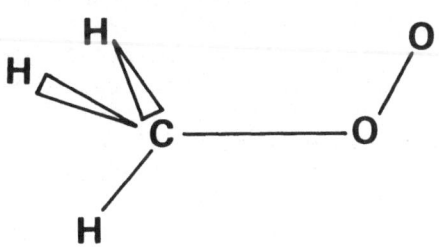

The eclipsed conformation is the conformation of maximum energy for the rotation of CH_3. Some relevant SCF geometrical parameters for the equilibrium, staggered conformation of the $^2A''$ and $^2A'$ states are given in Table 9. As for HO_2, the major difference between the geometries of the two states is that R_{OO} is longer in the excited, $^2A'$, than the ground, $^2A''$, state. Thus, the electronic absorption spectra of CH_3O_2 would be expected to be similar to that of HO_2.

This spectra is shown in Fig. 5b. The same progression in the 0-0 vibrations, v_8', of the $^2A'$ state are observed as for HO_2. In CH_3O_2, the progression is shifted by ~ 300 cm^{-1}=0.04 eV to higher energy. There is, however, an unusual and unexpected behavior of the satellite bands on the v_8' transitions. The bands are quite pronounced and they form a diverging progression. For HO_2, the satellites were converging and were assigned to the rotation of H about O_2. The analogous rotational structure cannot be observed for CH_3O_2 since the larger mass of CH_3 will lead to a very small separation of the rotational levels. Clearly there must be a different origin for the CH_3O_2 satellite bands.

We may consider the role of rotational (torsional) motion of the CH_3 group.[38] The quantum numbers for this motion are denoted $v_{t,i}''$ and $v_{t,i}'$ for the ground and excited states, respectively, where $i=0,1,2,\ldots$ indicates the torsional level. The assumption that the electronic transition moment is independent of the normal coordinate for torsional motion is reasonable since the $^2A''$ to $^2A'$ excitation is localized on O_2. Then the intensity of $v_{t,i}''-v_{t,k}'$ transitions will depend only on the Franck-Condon factor, $\int \chi_{t,i}''(q)\chi_{t,k}'(q)dq$. Assuming that the torsional potential is harmonic, transitions i to i+1 (actually i to i+k, k odd) are forbidden by symmetry since the two states have different parity. Further the Franck-Condon factors can be expected to be considerably larger for i to i than for i to i+2 or higher transitions. Thus, the $v_{t,i}''-v_{t,i}'$ torsional transitions should be the only ones to contribute significantly to the absorption spectra.

McLean and Yoshimine[39] have calculated the barrier between the staggered and eclipsed conformations of CH_3. They find that the barrier to rotation is 2.0 kcal/mole for the $^2A''$ states and is much larger, 3.5 kcal/mole, for the $^2A'$ excited state. This means that the curvature of the torsional potential curve will be larger for the $^2A'$ than for the $^2A''$ state. The separation of the excited state torsional levels, $\Delta v_t' = v_{t,i+1}' - v_{t,i}'$, will be larger than $\Delta v_t''$ for the ground state. Hence, the transitions $v_{t,i}''-v_{t,i}'$ will form a divergent series in i. This is shown schematically in Fig. 6. This divergent series is exactly what is required to explain the origin of the CH_3O_2 satellite bands. They are asigned[38] as hot bands of the torsional motion of CH_3 about the C-O bond.

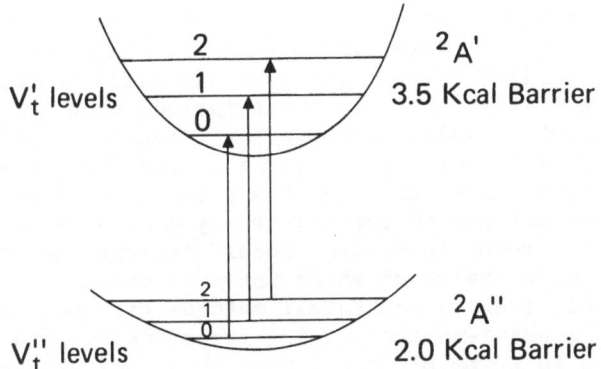

Fig. 6.   Schematic representation of the potential curves and
          energy levels for torsional motion of the CH_3 group around
          the C-O bond for the $^2A''$ and $^2A'$ electronic states of
          CH_3O_2. The torsional hot bands, $v''_{t,i} - v'_{t,i}$, are shown to
          form a divergent series.

     There are three points to stress about this example. First
it was the calculated result that the rotational barriers were
different for the ground and excited states that lead to the
assignment of the satellite bands in the CH_3O_2 spectra. The
calculations did not merely confirm an assignment; they made the
assignment possible! Second, the calculations were undertaken and
completed in a relatively short time after the problem was raised
by the experimental work. The time was sufficiently short that
the assignment could be included in the paper which reported the
observations.[38] Third, SCF wavefunctions, based on a relatively
simple and computationally easy level of approximation, were
perfectly adequate for this purpose. However, these wavefunctions
could be used with confidence only because their accuracy was
known to be sufficient for the problem under consideration. This
is an excellent example of how theoretical and experimental work,
performed essentially in parallel, can together lead to a better
understanding of spectroscopic problems.

MOLECULAR ORBITAL CLUSTER MODELS FOR CHEMISORPTION

     In this section we shall be concerned with an application of
quantum chemistry quite different from those discussed above.
This application is to the interaction of atoms with a solid
surface, specifically to chemisorption and other problems related

to electronic structure at a surface. The theoretical methods
that are used to study these problems may be placed into two
general groups. First, those that take explicit account of the
extended nature of the surface. These include band structure
formulations which normally assume a two dimensional periodicity
for the surface (substrate) and for the overlayer of adsorbed
atoms. And second, approaches which are referred to as cluster
models where the surface is represented by only a small number of
atoms. Here, the focus is on the "local" interaction of an
adsorbate atom with the atoms which form the surface cluster. In
these approaches, quantum mechanical methods are used which treat
the cluster as a quasi-molecule. The properties of the "molecule"
are interpreted in terms of observables at a real surface.
Semi-empirical as well as ab-initio methods have been used to
determine the electronic structure of the molecular clusters. A
brief discussion and references for these two groups of methods
is given in Ref. 42.

The obvious disadvantage of the cluster model approach is its
neglect of the "lateral" interactions among the adsorbed
atoms[43] that occur on a real extended surface. The advantage of
the approach is that it may be able to provide an accurate and
detailed description of the nature of the local bonding and
interaction of an adsorbate with the surface. In this context,
quantities of particular interest are those which are related to,
or may be obtained from, the interaction potential energy curves
or surfaces. These include, for example, equilibrium geometries,
surface vibrational frequencies, and chemisorption binding energies
at different sites.[42,44] These are considerably more difficult to
determine experimentally at a surface than for a molecule.[45] Yet
it is for just such quantities that the use of the methods of
ab initio quantum chemistry has a special advantage. As has been
shown in the previous sections, these methods are able to lead to
accurate descriptions of these quantities for real molecules.
They may be expected to do so for surfaces as well provided that
a cluster model is a satisfactory representation of the surface.
However, with few exceptions,[46] for the other methods used in
either the extended or cluster model approach, the surface geometry
is assumed rather than determined. The results to be described
below are based on SCF wavefunctions calculated using CGTO basis
sets of modest size. This type of calculation, applied to real
molecules, gives results of modest accuracy for equilibrium
geometries and force constants and gives qualitative guides for
relative energies at different points on a potential surface.[3] In
fact, until more is learned about the convergence of various
properties with respect to cluster size,[44] it is not clear that
it is generally useful to go beyond SCF calculations for these
cluster models.

The principal subject of this section is a study by Seel and
Bagus[47] of the interaction of F and Cl with a Si surface. The
objective of the study is to help to understand and explain the

different reactivities of these radicals with this surface. However, we shall first review the results of a study by Hermann and Bagus[48] of the adsorption of H on the (111) face of a Si crystal, denoted H/Si(111). The purpose of the review is to explain the choice of the surface cluster and other aspects of the theoretical approach. It is also to compare the theoretical results with experimental data in order to demonstrate the accuracy and suitability of the cluster model approach. (There is rather little data available for F and Cl on Si which is directly comparable with the calculated properties.)

As is the case for many surfaces, Si(111) relaxes and reconstructs; i.e., the positions of the surface atoms in the first few layers of the crystal change from their positions in the bulk.[49] In the cluster models[47,48] used for H/Si(111) and for F and Cl/Si(111), these changes are neglected and the Si atoms are chosen to have their bulk geometry.[50] (There is evidence that the displacements of the surface Si atoms on Si(111) are reasonably small.[51]) On this ideal Si(111) surface, each surface Si atom has three nearest neighbors in the second layer, as shown in Fig. 7a, rather than the four nearest neighbors for a tetrahedrally coordinated bulk atom. Assuming sp^3 hybridization, as in the bulk, each surface atom has one unpaired electron, called a "dangling bond," which has essentially sp_z character. For H/Si(111), the largest cluster used contained four Si atoms:[48] one on the surface (first layer) and its three nearest neighbors in the second layer, see Fig. 7b. In order to simulate the remainder of the crystal, each edge Si atom was given three H atom neighbors placed at the Si-H distance in SiH_4. This ensures that every Si atom in the cluster has its proper atomic coordination: three for the surface Si and four, one Si atom and three H atoms, for the second layer Si. The H atoms force the second layer Si atoms to have the bulk sp^3 hybridization and effectively saturate the edges of the cluster: this is called "embedding".

It is generally agreed[48] that H atoms adsorb directly over the surface Si atoms on Si(111). Thus, the chemisorption of H was studied by adding an H atom, denoted H_a to distinguish it from the nine embedding H atoms, above the surface Si as shown in Fig. 7b. In this directly overhead site, H_a forms a two-electron single bond with the Si "dangling bond". SCF calculations were performed with a double zeta plus polarization CGTO basis set and a potential energy curve was obtained for $(Si_4H_9)H_a$ as a function of the distance of H_a from the surface Si atom. The potential curve was used to determine the equilibrium Si-H_a distance, r_e, the binding energy of H_9 to Si_4H_9, D_e, and the harmonic vibrational energy, ω_e.

The calculated values of these properties are given in Table 10 where they are compared with experimentally derived values for D_e and ω_e. Considering the correlation errors for the D_e for a single two electron bond, we would expect the SCF D_e to be too small by

(a)

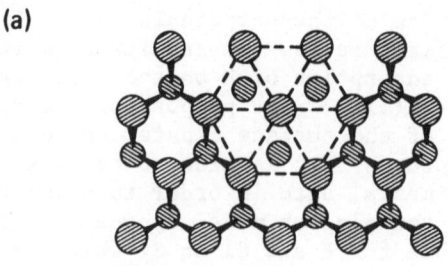

(b) $(Si_4H_9)H_a$

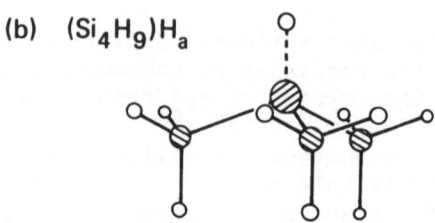

⊘ Si atom First Layer
⊖ Si atom Second Layer
○ Hydrogen

Fig. 7.   (a) Schematic of the Si(111) surface showing first and
          second layer Si atoms. The two kinds of three-fold sites,
          open and eclipsed are indicated by dashed lines.   (b) The
          $(Si_4H_9)H_a$ cluster used for H/Si(111) in the directly
          overhead site.

~1 eV[3]. However, the calculated value is quite close, within 0.2 eV,
to the experimental value.[52] The convergence of D_e for various
cluster sizes[48] suggests that the value will not change by more
than ~0.2 eV for larger cluster representations of the surface.
Thus, we believe that the experimental D_e is likely to be in error
and that the actual binding energy of H/Si(111) should be ~3.5 eV.
The SCF ω_e is 10% larger than the experimental value. This error
is characteristic for SCF vibrational frequencies for real
molecules.[3] There is no experimental information for r_e; however,
the SCF value is a reasonable one. It is close to both the SCF
and experimental Si-H distance in silane and it is close to the
value of r_e obtained by Appelbaum and Hamann from a band structure
type calculation.[46] In general, it seems that the shape of the

interaction potential curve converges more rapidly with respect
to cluster size than does the absolute value of the depth of the
minimum.[44]

Table 10. Calculated and experimental values for H/Si(111).
The H-Si bond distance, r_e, the chemisorption binding energy,
D_e, and the vibrational energy, ω_e, are given for the directly
overhead adsorption site.

	SCF	EXPT[a]
r_e (Å)	1.49	----
D_e (eV)	3.0	3.1-3.2
ω_e (cm^{-1})	2274	2073

[a]See Ref. 52 for D_e and Ref. 53 for ω_e.

There are two major conclusions from this study of H/Si(111)
that we wish to call attention to here. First, the use of a
relatively simple but suitably chosen cluster leads to quite
reasonable results for the interaction potential curve. Second,
the accuracy of the theoretical results provides strong evidence
for the localized character of the bonding. It also provides a
sound justification for the use of a cluster model to represent
the surface-adsorbate interaction.

We turn now to consider the interaction of F and Cl with Si.[47]
There is considerable interest in this interaction because of the
role of halogen containing radicals in the plasma etching of Si
and Si compound surfaces.[54-56] Plasma etching is playing an
increasingly important role in the manufacture of semiconductor
micro-circuits and other devices. The particular problem addressed
here concerns the different reactivity of F and Cl radicals with
an Si surface, either a polycrystalline film[55,57] or a single
crystal.[58,59] Fluorine radicals, arising at the surface, for
example, from the dissociative adsorption of XeF$_2$, react
spontaneously to form volatile SiF$_4$ at a rapid rate.[57,58] However,
Cl radicals, arising, for example, at the surface from the
dissociative adsorption of Cl$_2$, do not react spontaneously at any
appreciable rate.[55] Instead, on Si(111), they form an ordered
overlayer being adsorbed, almost certainly, at directly overhead
sites.[59] Yet both SiF$_4$ and SiCl$_4$ are volatile species and the
reactions to form these molecules at a Si surface are very
exothermic; that to form SiF$_4$ by 397 kcal/mole=17.2 eV[60,61] and that
to form SiCl$_4$ by 157 kcal/mole=6.8 eV.[60]

Specifically, Seel and Bagus[47] have investigated the
possibility of penetration of a Si surface by halogen atoms.

Penetration is a reasonable step toward the formation of SiX_4 (X=F,Cl) since it is likely to lead to surface species with a relatively high coordination of halogen atoms. These species are likely to be reaction intermediates in the formation and desorption of SiX_4.[62] Consider the three-fold, equilateral triangle, sites on Si(111) shown by dashed lines in Fig. 7a. Each surface atom is surrounded by six of these sites. For three of the sites, there is a second layer Si atom in the center of the triangle. The second layer atom is reasonably close, 0.78Å below, to the surface; these sites are called eclipsed sites. For the other three sites, the first Si atom below is in the fourth layer and is considerably, 3.9Å, below, the surface; these sites are called open sites. They have rather open structures and, on Si(111), are the most likely sites at which an atom could penetrate the surface. Suppose that halogen atoms are able to penetrate the lattice at three adjacent open sites forming an SiX_3 complex. It would then be easy to form SiX_4 by adsorption of another halogen atom directly over the central Si. Because of the strong internal bonds in the very stable SiX_4,[60] it is likely to be only weakly bound to the rest of the surface and should easily desorb into the gas phase. This hypothetical reaction path on Si(111) depends on surface penetration, the first step, being possible. For a polycrystalline Si surface, it is not possible, of course to propose a reaction in so much detail. However, the Si(111) open site can still serve as a model for sites which are likely to be favorable for surface penetration.

The interaction of F and Cl with Si(111) has been studied at the directly overhead, the eclipsed, and the open sites.[47] Here, we consider only the results for the open site at which the adatom approaches normal to the center of the equilateral triangle. The surface cluster chosen to model this site contains 10 Si atoms and 13 embedding H atoms, $Si_{10}H_{13}$. The Si atoms represent four layers of the surface; three each in the first, second, and third layers and one in the fourth layer. The fourth layer atom is in the center of the triangles formed by the three atoms in each of the three other layers. As for H/Si(111), the embedding H atoms are used to form a tetrahedral environment for the Si atoms and to effectively force sp^3 hybridization. The first, second, and fourth layer Si atoms each have one H neighbor; the third layer Si atoms have two H neighbors. The $Si_{10}H_{13}$ cluster is shown in Fig. 8.

Because the $Si_{10}H_{13}$ cluster is considerably larger than the Si_4H_9 cluster used to study H adsorption,[48] smaller basis sets have been used for calculations on this cluster. The Si CGTO basis set is minimal for the 1s, 2s, and 2p core and double zeta for the valence shells; it does not contain polarization functions. Similar basis sets, valence double zeta without polarization functions, have also been used for the F and Cl atoms. The point group symmetry for $Si_{10}H_{13}$ and $Si_{10}H_{13}X$ (X=F,Cl) is C_{3v} with irreducible representations a_1, e(doubly degenerate), and a_2. The cluster has three dangling bonds, one at each surface Si atom.

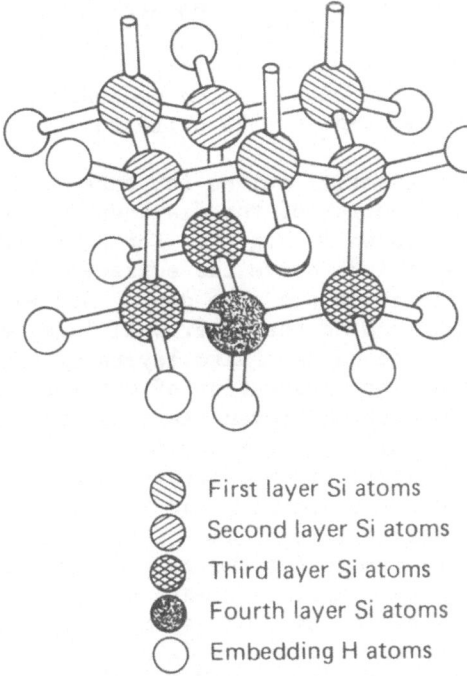

⬡ First layer Si atoms
◕ Second layer Si atoms
⬣ Third layer Si atoms
● Fourth layer Si atoms
○ Embedding H atoms

Fig. 8.   The four layer $Si_{10}H_{13}$ cluster used to model the open
          site of Si(111). The dangling bonds of the first layer
          Si atoms are indicated.

The ground state of the $Si_{10}H_{13}$ substrate cluster is found to
have the open shell configuration $a_1 e^2$ where the a_1 and e orbitals
are different combinations of the essentially sp_z dangling bonds.
In order to simplify the treatment of the cluster, the energies
and wavefunctions have been obtained for the average of
configurations[63] of the $a_1 e^2$ open shells rather than for a specific
spin and spatial coupling. Several configurations, again using
the average of configurations formalism, were investigated for
$Si_{10}H_{13}F$ and $Si_{10}H_{13}Cl$ in order to determine the one with the
lowest energy. In order to understand the bonding of the halogen
to Si, it is helpful to recall that the atomic ground state p^5
configuration may, in the C_{3v} symmetry of the cluster, be either
$e^4 a_1^1$ with a hole in p_z or $e^3 a_1^2$ with a hole in p_{xy}. The lowest
energy configuration for $Si_{10}H_{13}F$ has an e^2 dangling bond open
shell. The dangling bond of a_1 symmetry has formed a bonding
combination with the singly occupied F $2p_z$ orbital. The lowest

energy configuration for $Si_{10}H_{13}Cl$ has $a_1^1e^1$ dangling bond open shells. Here, the open e^3 shell ($3p_{xy}$) of Cl has formed a bond with a surface Si electron of e symmetry. The difference in bonding between these two cases arises because Cl is larger than F. For example, for F, $<r>_{2p}=0.57$Å while for Cl, $<r>_{3p}=0.97$Å.

The interaction potential curves for these lowest configurations as a function of the distance of the halogen atom from the surface are shown in Fig. 9. For F, there is a well with a minimum of 0.6 eV at 1.3Å above the surface. In addition, there is a second well with about the same depth, 0.5 eV, at 1.4Å below the surface. The minimum of the inner well lies, as shown in Fig. 9a, between the second and third layers of the surface. The height of the barrier for penetration of the surface is ~1 eV. For Cl, the situation is quite different as shown in Fig. 9b. As for F, there is an outer well with a minimum at 1.9Å and a depth of 0.4 eV. However, the barrier to penetration of the surface is much larger, ~13 eV. The higher barrier for Cl than for F is a consequence of the larger size of Cl. The SCF model is known to have reasonably large errors for the calculation of potential energy surfaces.[3] A reasonable estimate of the error of the SCF barrier height is ~1 eV; a generous estimate is ~2 eV. These errors are much smaller than the over 10 eV difference between the calculated penetration barriers for F and Cl.

From these results, it is reasonable to conclude that F atoms are very likely to penetrate a Si surface at suitable sites; i.e., those which are similar to the open site on Si(111). The calculated barrier at this site is likely to be somewhat too high because of the neglect of surface relaxation and reconstruction in the cluster model. The surface Si atoms can be expected to change their positions in such a way as to accomodate the F atom and hence, to lower the barrier and to increase the depth of the inner well. The errors of the SCF calculation are also likely to lead to a barrier which is too high.[64] Thus the actual barrier for F atoms at the open site will be less than the calculated 1 eV. It is also likely that the same effects would lead to a lower barrier for Cl penetration. However, a generous estimate for the lowering of ~2 eV will still leave a penetration barrier which is greater than 10 eV! Thus, without the addition of energy from some external source, Cl atoms will not be able to penetrate the surface. These cluster model results indicate that penetration is a reasonable path for the formation of surface species which are intermediates in the spontaneous reaction to form volatile SiF_4. There are, in fact, some observations which provide evidence for the existence of such intermediates.[62] The fact that the barrier for Cl atoms is too high to allow them to penetrate the surface is consistent with the observation that a spontaneous reaction to a volatile product does not occur.[55] However, when energetic radiation, say ~500 eV Ar^+ ions, is incident in addition to Cl, a reaction is clearly observed.[55]

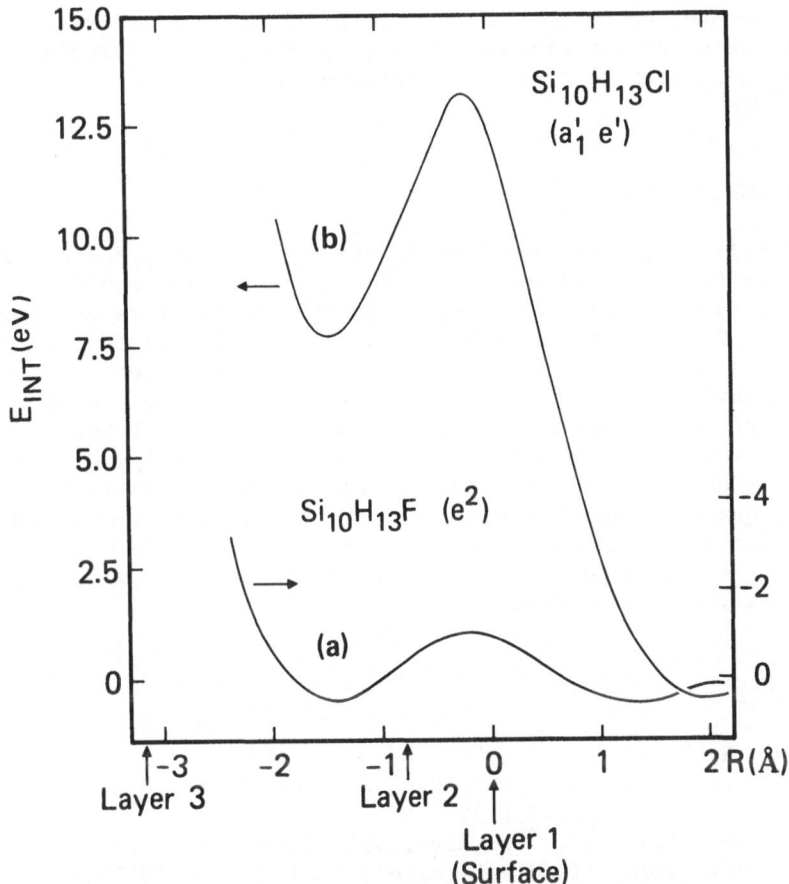

Fig. 9. Interaction potential curves for $Si_{10}H_{13}F$ (a) and
$Si_{10}H_{13}Cl$ (b) as a function of the distance, R, of the
Halogen atom from the surface. The zero of the
interaction energy, E_{INT}, is the sum of the energies of
the substrate cluster and the halogen atom. The positions
of the layers of the Si surface are marked. The right
hand scale for E_{INT} is for (a) and the left for (b).

The work described in this section shows that ab initio quantum chemistry can be applied to study the interaction of atoms with a solid surface. In particular, the work on F and Cl/Si shows that it can be used to provide insight about problems of current technological interest. The methods of ab initio quantum chemistry have also been shown to give results of useful, albeit rather modest, accuracy for reasonably large systems. For example, the $Si_{10}H_{13}Cl$ cluster contains 170 electrons and 11 heavy (i.e., nonhydrogen) atoms.

CONCLUDING REMARKS

In this paper, we have demonstrated that ab initio quantum chemistry is now a very versatile and powerful approach for research in problems that involve electronic structure and chemical bonding. The examples that have been presented have ranged from the weak Van der Waals bonding in He_2 to chemisorption and reaction on a Si surface. In the last two examples - dealing with the spectra of peroxyl radicals and with surface interactions - the theory was shown to be responsive in meeting and helping to resolve current problems posed by experimentalists in our own laboratory. In sum, quantum chemistry can properly be regarded as an important and effective compliment to experiment. It has reached the stage where it is available and, indeed, necessary for the solution of a wide variety of chemical problems.

REFERENCES

1. E. A. Hylleraas, Z. F. Phys. 65, 209 (1930).
2. H. M. James and A. S. Coolidge, J. Chem. Phys. 1, 825 (1933);
        J. Chem. Phys. 3, 129 (1935).
3. H. F. Schaefer III, "The Electronic Structure of Atoms and
        Molecules," (Addison-Wesley, Reading, MA, 1972).
4. W. G. Richards, T. E. H. Walker, and R. K. Hinkley, "A
        Bibliography of Ab Initio Molecular Wave Functions,"
        (Clarendon Press, Oxford, 1971); W. G. Richards,
        T. E. H. Walker, L. Farnell, and P. R. Scott, "Bibliography
        of Ab Initio Molecular Wave Functions. Supplement for
        1970-1973," (Clarendon Press, Oxford, 1974).
5. G. Richards, Nature 278, 507 (1979).
6. We wish, in particular, to acknowledge, Prof. J. Almlöf,
        Dr. M. Dupuis, and Prof. J. Hinze.
7. C. L. Pekeris, Phys. Rev. 112, 1649 (1958); Phys. Rev. 115,
        1216 (1959); Phys. Rev. 126, 1470 (1962).
8. W. Kolos and L. Wolniewicz, J. Chem. Phys. 41, 3663 (1964);
        J. Chem. Phys. 43, 2429 (1965).
9. P. S. Bagus, B. Liu, A. D. McLean, and M. Yoshimine,
        "Application of Wave Mechanics to the Electronic Structure
        of Molecules Through Configuration Interaction," in "Wave

Mechanics: The First Fifty Years," ed. W. C. Price,
S. S. Chissick, and T. Ravensdale (Butterworths, London,
1973) p. 99.

10. B. Liu and A. D. McLean, J. Chem. Phys. 59, 4557 (1974).

11. B. Liu and A. D. McLean, to be published.

12. A. L. J. Burgmans, J. M. Farrar, and Y. T. Lee, J. Chem. Phys.
64, 1345 (1976).

13. N. Sabelli and J. Hinze, J. Chem. Phys. 50, 684 (1969).

14. C. F. Bunge, Theoret. Chim. Acta (Berl.) 16, 126 (1970).

15. W. J. Balfour and A. E. Douglas, Can. J. Phys. 48, 901 (1970);
K. C. Li and W. C. Stwalley, J. Chem. Phys. 59, 4423
(1973); C. R. Vidal and H. Scheingraber, J. Mol. Spectros.
65, 46 (1977).

16. J. M. Farrar and Y. T. Lee, J. Chem. Phys. 56, 5801 (1972).

17. D. E. Beck, Mol. Phys. 14, 311 (1968).

18. Y. K. Kim and P. S. Bagus, Phys. Rev. A8, 1739 (1973).

19. M. R. A. Blomberg and P. E. M. Siegbahn, Int. J. Quantum Chem.
14, 583 (1978).

20. K. Kirby-Docken and B. Liu, J. Chem. Phys. 66, 4309 (1977).

21. B. Liu, K. M. Sando, C. S. North, H. B. Friedrich, and
D. M. Chipman, J. Chem. Phys. 69, 1425 (1978).

22. See, for example, A. Pipano, R. R. Gilman, and I. Shavitt,
Chem. Phys. Lett. 5, 285 (1970).

23. B. O. Roos and P. E. M. Siegbahn, Modern Theoretical Chemistry,
Vol. 3, p. 277, H. F. Schaefer III, Ed. (Plenum, New York,
1977).

24. H. J. Silverstone and O. Sinanoglu, J. Chem. Phys. 44, 1899
(1966) and references therein.

25. A. W. Mantz, J. K. G. Watson, K. Narahan Rao, D. L. Albritton,
A. L. Schmeltekopf, and R. N. Zare, J. Mol. Spectrosc.
39, 180 (1971).

26. C. A. Burrus, J. Chem. Phys. 28, 427 (1958); B. Rosenblum,
A. H. Nethercot, Jr., and C. H. Townes, Phys. Rev. 109,
400 (1958).

27. L. A. Young and W. J. Eachus, J. Chem. Phys. 44, 4195 (1966).

28. U. Wahlgren, J. Pacansky, and P. S. Bagus, J. Chem. Phys. 63,
2874 (1975).

29. I. Suzuki, M. A. Pariseau, and J. Overend, J. Chem. Phys. 44,
3561 (1966).

30. M. Yoshimine, A. D. McLean, and B. Liu, J. Chem. Phys. 58,
4412 (1973).

31. M. Yoshimine, to be published.

32. H. Krupenie, J. Phys. Chem. Ref. Data 1, 423 (1972).

33. M. Yoshimine, S. Green, and P. Thaddeus, Astrophys. J. 183,
899 (1973); the electronic wavefunctions are reported in
S. Green, P. S. Bagus, B. Liu, and A. D. McLean, Phys.
Rev. A5, 1614 (1972).

34. P. Erman, Astrophys. J. 213, L89 (1977).

35. B. Liu, S. Chu, and M. Yoshimine, to be published; the
electronic wavefunctions are reported in S. Chu,
M. Yoshimine, and B. Liu, J. Chem. Phys. 61, 5389 (1974).

36. a. J. Anketell and A. Pery-Thorne, Proc. Roy. Soc. A301, 343
       (1967).
    b. R. L. deZafra, A. Marshall, and H. Metcalf, Phys. Rev. A3,
       1557 (1971).
    c. B. G. Elmergreen and W. H. Smith, Astrophys. J. 178, 557
       (1972).
    d. R. A. Sutherland and R. A. Anderson, J. Chem. Phys. 58,
       1226 (1973).
    e. K. R. German, T. H. Bergeman, E. M. Wernstork, and R. N. Zare,
       J. Chem. Phys. 58, 4304 (1973).
    f. K. H. Becker, D. Haaks, and T. Tatarczyk, Chem. Phys.
       Lett. 25, 564 (1974).
    g. J. H. Brophy, J. A. Silver, and J. L. Kinsey, Chem. Phys.
       Lett. 28, 418 (1974).
    h. K. R. German, J. Chem. Phys. 62, 2584 (1975); ibid. 63,
       5252 (1975).
    i. J. Brzozowski, P. Erman, and M. Lyra, Physica Script 17,
       507 (1978).
37. H. E. Hunziker and H. R. Wendt, J. Chem. Phys. 60, 4622 (1974).
38. H. E. Hunziker and H. R. Wendt, J. Chem. Phys. 64, 3488 (1976).
39. A. D. McLean and M. Yoshimine, to be published.
40. Y. Beers and C. J. Howard, J. Chem. Phys. 64, 1541 (1976).
41. J. C. Gole and E. F. Hayes, J. Chem. Phys. 57, 360 (1972);
       R. J. Blint and M. D. Newton, J. Chem. Phys. 59, 6220
       (1973); W. A. Goddard III, "Lecture Notes, School on the
       Fundamental Chemical Basis of Reactions in the Polluted
       Atmosphere," C. W. Kern, Ed. (Battelle Research Center,
       Seattle, WA, 1973), p. 254.
42. K. Hermann and P. S. Bagus, Phys. Rev. B17, 4082 (1978).
43. See, e.g., I. P. Batra and S. Ciraci, Phys. Rev. Lett. 39,
       774 (1977); I. P. Batra, K. Hermann, A. M. Bradshaw, and
       K. Horn, Phys. Rev. B20, 801 (1979).
44. C. W. Bauschlicher, P. S. Bagus, and H. F. Schaefer, IBM J.
       Res. and Dev. 22, 213 (1978); P. S. Bagus, H. F. Schaefer,
       and C. W. Bauschlicher, to be published.
45. See, for example, the papers in the sections on "Atomic and
       Molecular Scattering from Surfaces" and "Aspects of
       Surface Chemical Bonding," in "Topics in Surface
       Chemistry," E. Kay and P. S. Bagus, Eds. (Plenum, New
       York, 1978).
46. A. significant exception is the work by J. A. Applebaum and
       D. R. Hamann, Phys. Rev. Lett. 34, 806 (1975).
47. M. Seel and P. S. Bagus, to be published.
48. K. Hermann and P. S. Bagus, Phys. Rev. B20, 1603 (1979).
49. See, e.g., K. C. Pandey, IBM J. Res. and Dev. 22, 250 (1978).
50. R. W. G. Wyckoff, "Crystal Structures," (Interscience, New
       York, 1964) Vol. II.
51. J. A. Appelbaum and D. R. Hamann, Phys. Rev. Lett. 31, 106
       (1973); A. Redondo, W. A. Goddard III, T. C. McGill, and
       G. T. Surratt, Solid State Comm. 20, 733 (1976);
       P. S. Bagus, unpublished results.

52. M. Henzler and G. Schulze, unpublished.
53. H. Froitzheim, H. Ibach, and S. Lehwald, Phys. Lett. A55, 247
     (1975).
54. H. F. Winters, J. W. Coburn, and E. Kay, J. Appl. Phys. 48,
     4973 (1977); J. W. Coburn and H. F. Winters, J. Vac. Sci.
     Technol. 16, 391 (1979).
55. J. W. Coburn and H. F. Winters, J. Appl. Phys. 50, 3189 (1979).
56. H. F. Winters, J. Appl. Phys. 49, 5165 (1978).
57. H. F. Winters and J. W. Coburn, Appl. Phys. Lett. 34, 70
     (1979).
58. Y. Y. Tu and H. F. Winters, to be published.
59. P. K. Larson, N. V. Smith, M. Schlüter, H. H. Farrell, K. M. Ho,
     and M. L. Cohen, Phys. Rev. B17, 2612 (1978).
60. JANAF Thermochemical Tables, Nat. Stand. Ref. Data Ser., Nat.
     Bur. Stand., No. 37 (1971).
61. V. I. Pepkin, Y. A. Lebedev, and A. Y. Apin, Zh. Fiz. Khim.
     43, 1564 (1963).
62. T. J. Chuang, J. Appl. Phys., in press.
63. J. C. Slater, "Quantum Theory of Atomic Structure,"
     (McGraw-Hill, New York, 1960) Vol. I.
64. P. S. Bagus and M. Seel, to be published, have shown that a
     two configuration MCSCF calculation leads to a lower
     barrier than an SCF calculation for the penetration of
     H on Si.

COMPUTER CHEMISTRY STUDIES OF ORGANIC REACTIONS:

THE WOLFF REARRANGEMENT

Joachim Bargon[*]
Kiyoshi Tanaka
Megumu Yoshimine

IBM Research Laboratory
5600 Cottle Road
San Jose, California   95193

ABSTRACT

The Wolff Rearrangement (WR) describes the isomerization of α-carbonyl carbenes into ketenes. Originally observed during the irradiation or thermal decomposition of α-diazoketones, such carbene intermediates also occur in other areas of organic chemistry, for example during the reaction of oxygen with acetylene.

We have studied the WR by ab initio quantum chemistry, mapped out the potential energy surface of this reaction, and compared the relative stabilities of the four possible reaction products (hydroxyacetylene, oxirene, formylmethylene and ketene). Furthermore, we have computed the activation barrier to rearrangement of the α-carbonyl carbene to ketene both on the electronic singlet and triplet potential surface. The results are shown to depend on the level of approximation, in particular on the quality of the basis set and on the consideration of electron correlation energy corrections. The data obtained for the highest level of sophistication explain many experimental findings such as the spectroscopists' inability to observe formylmethylene or oxirene. Hydroxyacetylene is predicted to be rather stable but inaccessible via the WR.

[*] IBM World Trade Postdoctoral Fellow. Present Address: Department of Chemistry, Faculty of Science, Hokkaido University, Sapporo, Japan.

I. INTRODUCTION

Isomers with the composition C_2H_2O occur as products or intermediates in a variety of chemical reactions. The most significant pathways to these isomers and to their higher homologues are the thermolysis or photolysis of α-diazoketones, the reaction of oxygen (^1D) with acetylene, the photolysis of ketenes and the reactions of carbenes with CO as outlined in Fig. 1a. The most important isomers of the C_2H_2O family are depicted in Fig. 1b. Of them, ketene (Ia) is the most stable and the only one known. The other three, namely formylmethylene (IIa), hydroxyacetylene (IIIa), and oxirene (IVa) are elusive, even though they are significant as either likely or speculative reaction intermediates.

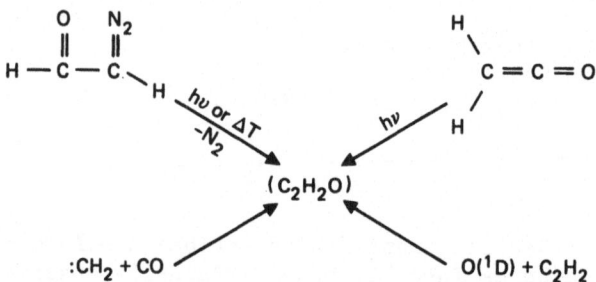

Fig. 1a. Reactions leading to C_2H_2O

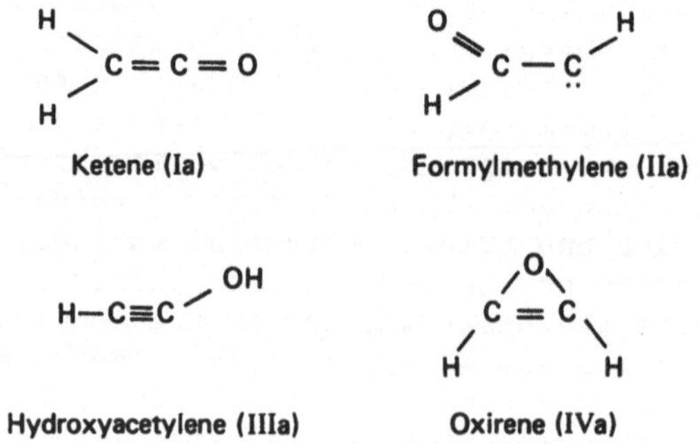

Ketene (Ia)

Formylmethylene (IIa)

Hydroxyacetylene (IIIa)

Oxirene (IVa)

Fig. 1b. The C_2H_2O isomers

Even though it is possible, at least in principle, to enter the potential energy surface containing the four C_2H_2O isomers from many entry points, the by far most common approach is via nitrogen elimination from α-diazoketones (V) which yields a carbonyl carbene (II) and nitrogen according to Scheme 1.

Scheme 1.

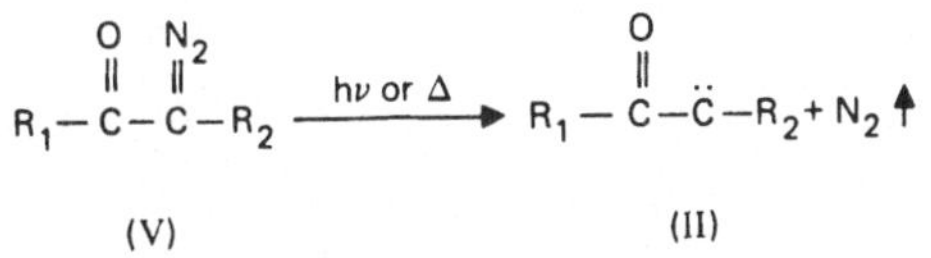

$$R_1-\overset{\overset{O}{\|}}{C}-\overset{\overset{N_2}{\|}}{C}-R_2 \quad \xrightarrow{h\nu \text{ or } \Delta} \quad R_1-\overset{\overset{O}{\|}}{C}-\ddot{C}-R_2 + N_2 \uparrow$$

(V)   (II)

In the presence of water the reaction proceeds most likely via the sequence

Scheme 2.

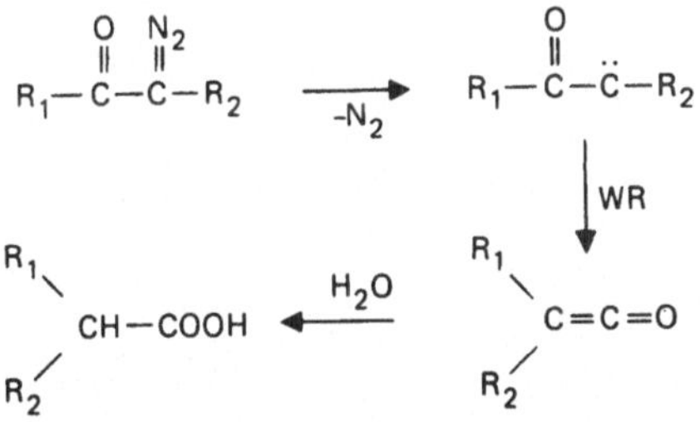

$$R_1-\overset{\overset{O}{\|}}{C}-\overset{\overset{N_2}{\|}}{C}-R_2 \quad \xrightarrow{-N_2} \quad R_1-\overset{\overset{O}{\|}}{C}-\ddot{C}-R_2$$

$$\Big\downarrow \text{WR}$$

$$\overset{R_1\searrow}{\underset{R_2\nearrow}{}}CH-COOH \quad \xleftarrow{H_2O} \quad \overset{R_1\searrow}{\underset{R_2\nearrow}{}}C=C=O$$

V→II→I→ acid, as outlined in Scheme 2. This conversion was discovered by Wolff[1] and independently by Schroeder.[2] The isomerization of the α-carbonylcarbene (II) into ketene (I) has become known as the Wolff Rearrangement (WR). Whether other intermediates such as for example oxirenes (IV) play a role in the WR has been a controversial subject matter,[3] as will be outlined in the following. For this purpose both experimental evidence and previous theoretical predictions as to the involvement of intermediates in the WR will be reviewed. The results of a new state-of-the-art ab initio calculation will be presented, in the light of which the experimental findings will then be discussed.

THE REACTION MECHANISM OF THE WOLFF REARRANGEMENT

General Considerations and Experimental Evidence

Even though the WR is known since the turn of this century and,[4] inspite of the fact that it is both of great synthetic utility and of technological significance,[5] the details of the reaction mechanism have remained the subject of an ongoing discussion.[3]

Thus Eistert[6] was first in pointing out that oxirenes (IV) could
be intermediates of the WR. The discoverers already had discussed
the migration of substituent R_1 from C_1 to C_2 in V according to
Path A in Scheme 3. The postulation of an oxirene as an intermediate
by Eistert made Path B a more likely one. To date all attempts to
provide direct experimental evidence for oxirene intermediates
have failed, but there is accumulating indirect evidence for their
occurrence.

Scheme 3.

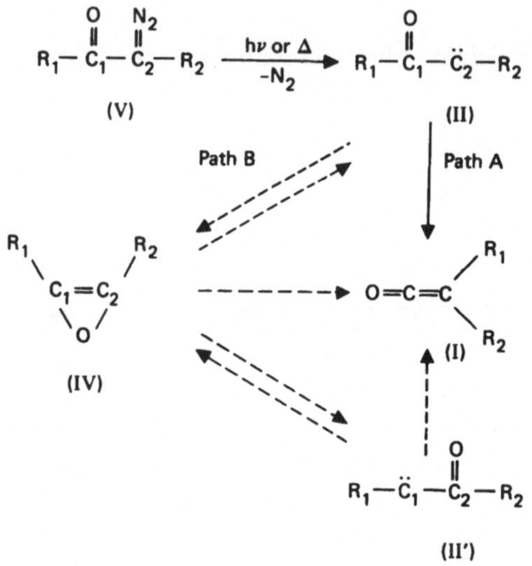

Thus Krantz[7] isolated diazoacetaldehyde (Va) in argon and in an
ethylene matrix, respectively, and induced its decomposition and
WR by photolysis. Even when conducting the photolysis at 8K he
obtained exclusively ketene with no evidence for the intermediacy
of IVa whatsoever.

Very recently Pacansky and Coufal[5b] investigated the
electron-beam induced WR of a more complex α-diazo ketone in bulk
as a thin film at 8K, and they found no evidence for an oxirene
intermediate either. They observed the formation of the ketene,
however.

Efforts to obtain indirect evidence for the intermediacy of
IV in the WR have initially yielded rather controversial results.
Thus Franzen[8] used ^{14}C-labeled azibenzil (Vb) and after either
photolysis or thermolysis in the presence of water, he recovered
all the ^{14}C-activity in the carbonyl carbon of diphenylacetic acid
(VIb) obtained according to Scheme 4. His results led him to the
postulation that oxirenes should not occur as intermediates of
the WR.

Scheme 4.

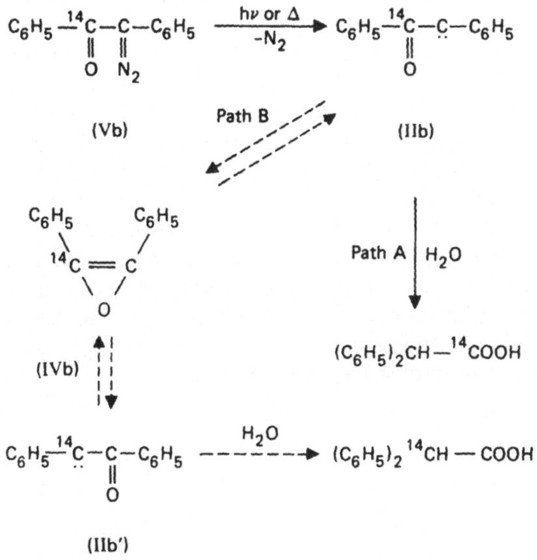

Carbon-13 labeling techniques had been applied as early as 1942 to study the WR of $[1-^{13}C]$-diazoacetophenone, but according to Hugget et al.,[9] only $[2-^{13}C]$-phenylacetic acid was found and thus no evidence for oxirene participation. Strausz and coworkers,[10] however, came to very different conclusions using ^{13}C isotopes to probe Scheme 4. They used mass spectroscopy as an analytical tool to determine the whereabouts of the ^{13}C label in the product VIb. According to their findings about 50% of Vb decompose via IIb to VIb directly, whereas the remaining 50% rearrange via Path B, i.e., via IIb, IVb, IIb'→VIb'.

Subsequently, Zeller and coworkers[11] followed the decomposition of ^{13}C-labeled Vb with both ^{13}C-NMR and mass spectroscopy. They confirmed the findings of Strausz et al.[10] for the photolysis initiated decomposition, but estimated the degree of oxirene participation (Path B in Scheme 4) to be between 23-31%. They also studied the thermolysis and electron impact induced degradation of Vb, however, in these two composition modes they found no ^{13}C scrambling and thus no evidence for oxirene (IVb) participation.

In a related effort Zeller[12] investigated the photolysis of the parent compound diazoacetaldehyde (Va) containing a ^{13}C label at the carbonyl position, again using both ^{13}C-NMR as well as mass spectroscopic techniques. He found that after photolysis of Va 8% of the carbonyl label had migrated, thus invoking an oxirene (IVa) participation of about 13-16%. He attributed the apparent difference in the levels of oxirene participation during the photodecompositions of Va and Vb, respectively, to the greater ease of hydrogen migration than that of phenyl,[3] thus indirectly suggesting a higher production rate and/or lifetime of IVb over IVa.

A basically different approach to obtaining evidence for
oxirene intermediates during the WR of diazoketones was taken by
Martin and Sammes.[13] These authors did not use isotopic labeling
techniques, but concentrated on investigating the rearrangement
of assymmetrically dialkyl substituted open chain diazoketones.
According to Scheme 5, the intermediate carbonyl carbene (II) can
give rise to 2 different α,β-unsaturated ketones, if an oxirene (V)
occurs as an intermediate. The ratio of the two isomeric alkenyl
ketones can and was used by Martin and Sammes to estimate the
degree of oxirene participation. This method, though elegant and
without requirements for custom synthesis or specifically labeled
precursors, is of course restricted to higher members of the
α-diazoketone family, and not suitable for investigations of the
parent compounds, for example, Va or Vb.

Martin and Sammes[13] found that both photochemically as well
as thermally induced decomposition of α-diazoketones involved
oxirene participation leading to isomeric α,β-alkenyl ketones,
whereas the metal (copper or silver) catalyzed decomposition
occurred without oxirene intermediates. The same authors also
investigated the effect of temperature on the amount of oxirene
participation during both the thermal and photochemical
decomposition of various α-diazoketones. They found that the
degree of oxirene participation increases with temperature in
either mode of decomposition.[13] They explained the disrepancies
between their own as well as the similar results of Strausz et al.[10]
to earlier ones in the literature[8,9] in the following way: Huggett
and his coworkers[9] had searched for oxirene intermediates under
the conditions of the Arndt-Eistert synthesis.[4] This approach
requires silver oxide as a catalyst, which according to Martin
and Sammes[13] does not involve oxirene participation. Franzen,[8]
on the other hand, had carried out his investigations of the

photolysis using a long wavelength light source ("Sunlamp"),
whereas Strausz et al.[10] as well as Martin and Sammes[13] had used
higher energy medium pressure mercury lamps. Thus the latter
authors concluded that the WR rearrangement of Vb as studied by
Franzen[8] was probably proceeding by a "low-energy reaction
profile," thereby by-passing oxirene intermediates.

By inspection of the accumulated experimental data it seems
that the level of oxirene participation is dependent on the:
1) particular substituents (cf. difference in oxirene participation
from Va versus Vb as a precursor), 2) photon energy, and
3) temperature. According to Martin and Sammes[13] the "matrix" in
which the WR is investigated is also of importance. Furthermore
in solution, the level of oxirene participation seems to increase
with increasing polarity of the solvent.

Another study of the degree of oxirene participation in the
WR was launched by Russell and Rowland,[14] who studied the
photolysis of ^{14}C-carbonyl labeled ketene. Their investigations
were further complemented by those of Montague and Rowland,[15] who
found evidence for extensive oxirene participation from ^{14}C
scrambling data obtained during the reaction of singlet ^{14}CH$_2$
(generated from the photolysis of H$_2$^{14}C=CO) with CO.

Table 1 lists the estimated levels of oxirene participation
during the decomposition of a variety of α-diazoaldehydes and
α-diazoketones.

Table 1. Amount of Estimated Oxirene Participation During
          the Decomposition of Various α-diazocarbonyl
          Compounds.

| Substituents | | Mode of | | % | | |
R$_1$	R$_2$	Decomposition	Method	Oxirene	Year	Ref.
C$_6$H$_5$	H	metal catalyzed	^{13}C	"none"	1942	9
C$_6$H$_5$	C$_6$H$_5$	photolysis	^{14}C	"none"	1958	8
C$_6$H$_5$	C$_6$H$_5$	"	^{13}C	50%	1968 1970	10
C$_6$H$_5$	C$_6$H$_5$	"	^{13}C	23–31%	1972	11
alkyl	alkyl	"	isomerization	f(T)	1972	13
H	H	"	matrix isolation	"none"	1973	7
H	H	"	^{13}C	13–16%	1977	12

Oxirenes as Intermediates in Miscellaneous Reaction Mechanisms

Apart from their role in the WR, oxirenes have been considered elsewhere in organic reaction mechanisms, and evidence in their favor has occasionally been claimed. Typically, however, subsequent reinvestigations showed the earlier claims to be erroneous. Thus the first report of a (supposedly) successful synthesis of an oxirene dates back to 1870, when Berthelot[16] described what he thought to be methyl oxirene. Then in 1931 Madelung and Oberwegner[17] claimed that the reaction of desyl chloride yielded IVb, which was subsequently disproved by Dauben et al.,[18] however.

Just like the WR, the reaction mechanism of the oxidation of acetylenes has been a controversial subject matter, and here as well oxirenes have been claimed and disputed as hypothetical intermediates.[19,20] Thus Schlubach and Richaud[21] had attributed an oxirene structure to intermediates obtained during oxidations of acetylenes with peracids,[19] but subsequently Franzen[19f] showed this assignment to be wrong.

Other Approaches to Oxirenes

In part because of a considerable theoretical interest in oxirenes (due to their potential 4π antiaromatic character[22]), numerous efforts have been undertaken to get direct evidence and spectroscopic data. The approaches discussed so far have typically attempted to form the three-membered heterocyclic ring from an acyclic precursor. Alternate efforts to be discussed in the following have tried to attempt the formation of oxirenes from precursors with the heterocyclic ring already preformed but still lacking the desired carbon-carbon double bond. The approaches can be classified according to Scheme 6.

Scheme 6.

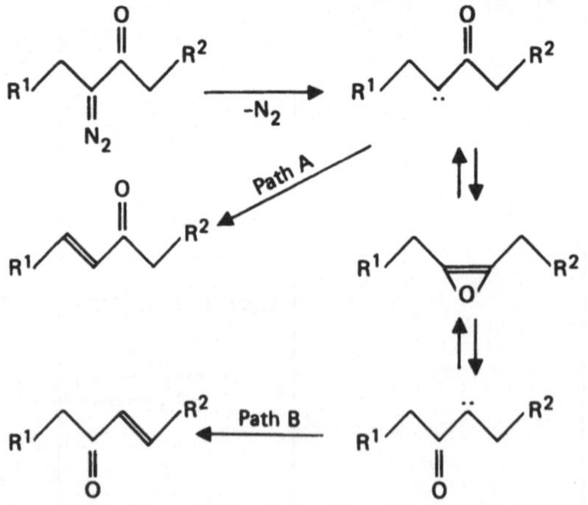

They have in common that they start from an oxirane, either VII
or VIII, from which molecules xy are eliminated either by α- or
β-elimination. In general, however, results have been
disappointing, or at best nonconclusive. Thus thermally induced
α-elimination of benzene as the molecule xy from a suitable
bicyclic precursor (VIIIa) yielded ketene as expected, but allowed
no conclusion as to an intermediate oxirane–carbene (IX) or its
potential rearrangement product oxirene,[23] according to the
sequence outlined in Scheme 7.

Scheme 7.

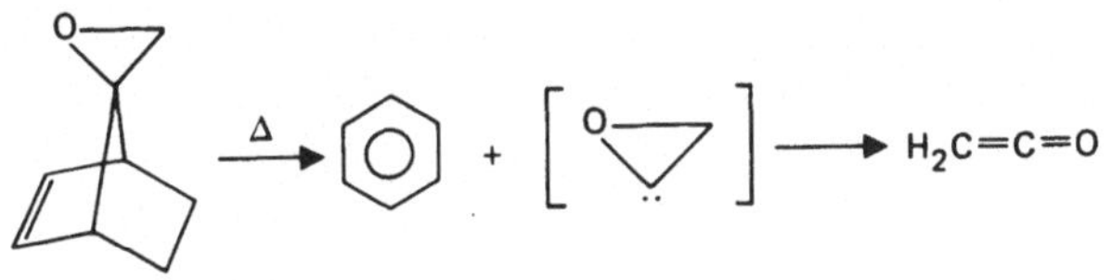

(VIIIa)   (IX)

Precursors for oxirene generation via β-elimination have also been
described,[24] among them VIIa, from which the formation of
phenyloxirene (IVc) has been claimed[25] to occur via a Norrish Type
II photoelimination of acetophenone according to the steps shown
in Scheme 8.

Scheme 8.

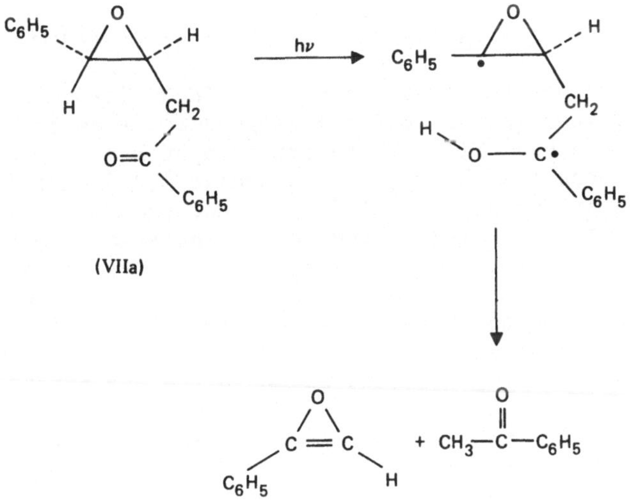

(VIIa)

(IVc)

Thus, even though there has been no success in obtaining direct evidence for the existence of oxirenes, data are accumulating in favor of their intermediacy in organic reaction mechanisms.

Sulfur Analogous Compounds, the C_2H_2S System

The sulfur analogous case of the WR, i.e., the C_2H_2S system, is interesting for comparison, because of some striking similarities and some profound differences: First of all of the tautomeric forms shown in Scheme 9,

Scheme 9.

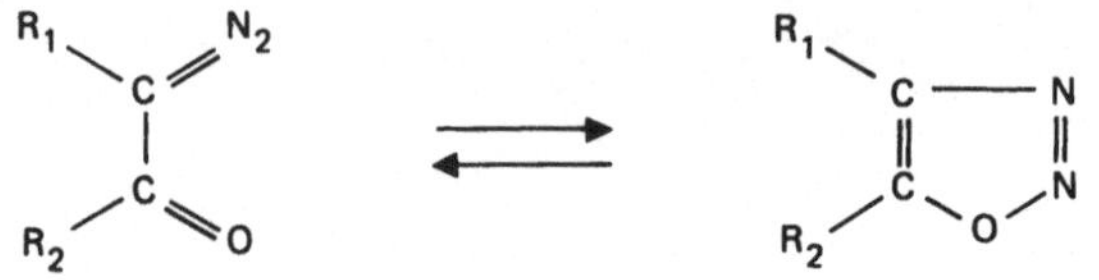

α – Diazoketone (V)   1,2,3, – Oxdiazole (X)

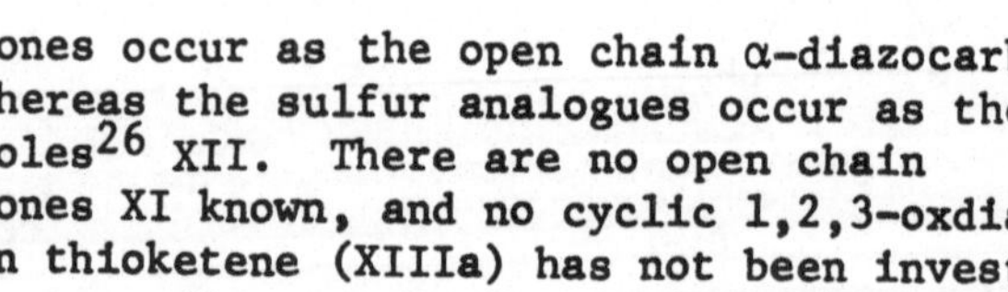

α – Diazothioketone (XI)   1,2,3 – Thiadiazole (XII)

the α-diazoketones occur as the open chain α-diazocarbonyl compounds V, whereas the sulfur analogues occur as the cyclic 1,2,3-thiadiazoles[26] XII. There are no open chain α-diazothioketones XI known, and no cyclic 1,2,3-oxdiazoles X either.[26] Even thioketene (XIIIa) has not been investigated until recently.[27-29] According to the matrix isolation studies of Krantz et al.,[27b-d] the C_2H_2S isomers XIIIa, XVa and XVIa have all been obtained via photolysis of 1,2,3-thiadiazole XIIa. The above C_2H_2S isomers seem to be stable enough to allow their characterization by infrared spectroscopy at 8K in an argon matrix. Photoelectron spectra, however, taken in the gas phase, yield no evidence for the occurrence of any other C_2H_2S isomers but thioketene.[27f-g] The validity of the infrared derived evidence,[27b-d] for the formation of XVI in particular, has been questioned,[28] but independent evidence for its occurrence has been obtained[28] via photolysis of XII. The photolysis of 1,2,3-thiazoles has first been investigated by Kirmse and Horner,[29a] and subsequently extensively by Zeller et al.[29b] who also studied the electron-impact induced fragmentation of these compounds.[29c] In all these cases the intermediate occurrence of thiirene and/or other isomers of the C_2H_2S family has been claimed. The fact that these elusive intermediates can also be obtained via the photolysis of isothiazoles[29d] (XVII) has been taken as additional proof for the validity of the claim[28b-d] that these intermediates are indeed the isomers XIII, XV, and XVI. The photolysis of XVII is believed to follow Scheme 10:[29d]

Scheme 10.

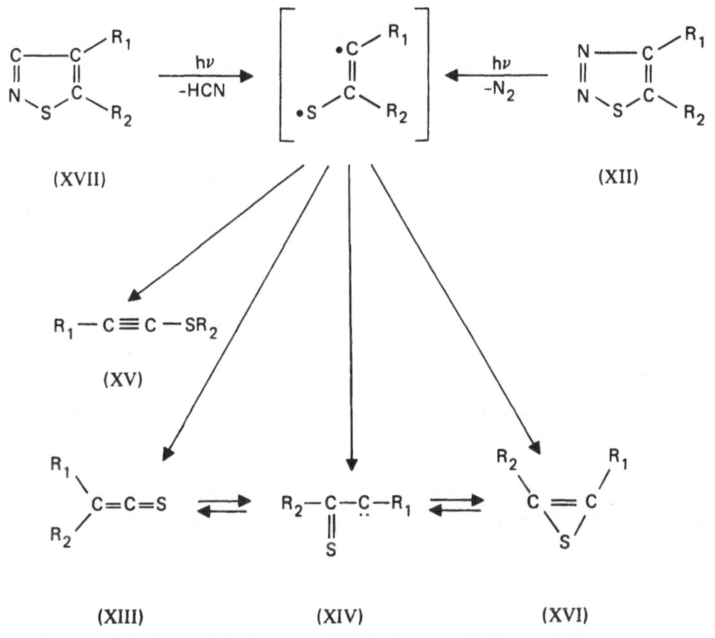

(XVII)   (XII)

$R_1-C\equiv C-SR_2$

(XV)

(XIII)   (XIV)   (XVI)

Scheme 10 shows the close relationship between the photolysis of XVII and XII, respectively.

    Some confusion and controversy remain as to the genuine nature of the C_2H_2S isomers, but evidence seems to accumulate supporting the claim that both thiirene and mercaptoacetylene as well as substituted forms thereof have been made and characterized. Theoretical efforts to explain their stability and their properties will be discussed later (vide infra).

THEORETICAL STUDIES

General Considerations

    The controversial experimental results combined with the challenging nature of oxirene and of the other associated isomers

have rendered the WR an attractive target of previous theoretical
calculations. The elusive oxirene intermediate alone has been
the subject matter of numerous theoretical studies, the results
of which reflect as much controversy over its existence as the
experimental data summarized earlier. In the absence of reliable
experimental data, the role of the theoretician and of his
calculations can be twofold:

a.   to guide the experimentalists in their search

b.   to predict the theoretical properties and the modes of
      formation and decay of unknown species.

With this goal in mind, theoreticians have tried for the past
decade to determine the shape of the potential energy surface of
the WR, using extended Hückel[30] (EH), semiempirical[31] (MINDO/3
and NDDO), as well as ab initio methods of various degrees of
sophistication.[32] Whereas the EH calculations predict oxirene to
be less stable than either formylmethylene and ketene (by 36 and
69 kcal/mol, respectively),[30] the MINDO/3 results predict the
opposite, namely that oxirene should be more stable than
formylmethylene by 18–20 kcal/mol,[31] though less stable than ketene.
This latter result roughly agrees with thermochemical estimates
by Wentrup.[3b] Early ab initio results revealed that formylmethylene
and oxirene should have approximately the same stability, their
heat of formation being about 70 kcal/mol higher than that of
ketene, with oxirene being slightly more stable (by ~0.4 kcal/mol)
than formylmethylene.[32b]

For reliable conclusions none of these theoretical results
proved useful, since they all reflect known shortcomings of the
methods employed. Thus EH methods are known to <u>overestimate</u> ring
strains,[33] and thus one would have expected the <u>EH</u> results to
penalize oxirene severely. MINDO/3, on the other hand, especially
in its original form, is also known to <u>underestimate</u> ring strain.[34]
Thus the ab initio results[32b] might well be close to reality, but
they suffered from a low quality basis set and a lack of
optimization of the geometries of the C_2H_2O isomers. Application
of the above corrections, however, suggests that the difference
in stabilities between formylmethylene and oxirene is most likely
small. Consequently a theoretical treatment of the WR will only
be meaningful if it is conducted on the ab initio level with a high
degree of accuracy and sophistication. At first glance these
latter qualifiers may seem superflous and confusing. They are
very much to the point however, because even though "ab initio"
taken literally implies a calculation by first principles, i.e.,
without approximation, some simplifications and constraints are
typically adopted in practice. They either serve to cut down the
computational effort to a size that can be handled technically or
economically, but more frequently they are intrinsic consequences
of the necessarily imperfect formulations of a many electron
problem.

The essential differences between various strains of common
ab initio procedures typically stem from two main sources, namely:

a.   the size and type of the basis set,

b.   the steps taken to improve the quality of the approximate
      wavefunctions employed, for example via energy variational
      calculations, since the Schrödinger equation of a many
      electron problem is not amenable to direct and thus exact
      solution.

The purpose of this study was to assess the consequences of:
a) using more or less expanded basis sets, and b) consideration of
electron correlation energy corrections, as obtained by
configuration interaction (CI) calculations.

In the following, the typical approach taken in ab initio
calculations[35] are outlined first, and the characteristic
simplifications and modifications are described together with
their computational consequences. Subsequently ab initio
calculations on various levels of sophistication are compared,
and the error bars of the most advanced calculation are assessed.

The Hierarchy of Various Ab Initio Calculations

An ab initio study starts by assuming that the nuclei of a
given molecule are initially fixed in space at some chosen
geometry. With this assumption the Schrödinger equation of the
system can be separated into two equations, one describing the
motion of the electrons within the fixed nuclear framework, and
the other the vibrations and rotations of the nuclei
(BORN-OPPENHEIMER Approximation). The electronic Schrödinger
equation then contains the nuclear configuration as a parameter.
The resulting Hamiltonian is truncated by eliminating insignificant
contributions, such as magnetic interactions (for the purpose
discussed here). The Schrödinger equation for the many electron
problem containing the above Hamiltonian is then translated into
an integral form. Energy-variational principles then yield
sufficiently accurate wavefunctions and energies, at least for
the ground states, and with restrictions, also for the excited
states. Via the so-called orbital approach the real
electron-electron potentials are being replaced by an effective
independent particle potential. Now the solutions can be expressed
as products of spin orbitals or as single Slater determinants.
Further refinements of the crude wavefunction so obtained are
achieved by variation of the spatial parts of the spin orbitals,
using minimization of the total energy as the criterion.

The molecular orbitals are typically synthesized as linear
combinations of atomic orbitals (LCAO expansion). Here, the
different strains of ab initio approaches tend to branch for the

first time, because the choice of the basis becomes crucial. The
accuracy of the eventual results should improve according to the
following hierarchy:

   a.   Minimal Basis Set: one basis function per atomic orbital.

   b.   Extended Basis Set: several basis functions per atomic
        orbital. The angular momentum quantum numbers are
        restricted to be the same as in atomic ground states.

   c.   Extended Basis Set plus Polarization Functions: includes
        basic functions with higher angular momentum or those
        not centered on the nuclei.

The first extension of the minimal basis set (Point b) allows for
less restricted adjustments (relaxation) of the molecular geometry.
The second extension (Point c) allows for a more appropriate
consideration of polar (resonance) structures.

   A second branching of ab initio treatments due to differences
of the basis can result from the analytical form of the basis
functions used. Thus, it is customary to use either exponential
(i.e., hydrogenic or Slater) or Gaussian basis functions. Although
the Slater functions are in general better suited to describe
molecular orbitals than Gaussian functions, the latter are
preferred for molecular systems involving more than two atoms
because of computational difficulties with the former. Since the
number of Gaussian functions needed is larger than that of Slater
functions, contracted Gaussians, i.e., fixed linear combinations
of Gaussian functions,[36] are commonly used to reduce the dimension
of the one-particle space. The optimization of the molecular
orbitals is achieved, starting from a chosen basis set, via the
self-consistent field theory (SCF) according to Roothaan.[37] From
the multitude of SCF solutions the only well-defined solution,
the Hartree-Fock solution, is obtained for a complete and thus
infinitely flexible basis.

   The still remaining error is defined as the correlation energy
error of the system.[38a] It is the consequence of the simplification
in the SCF procedure due to describing the correlated motion of
the electrons by only an average electron repulsion field. To
consider the correlation effect requires more sophisticated
wavefunctions than the single determinant form. Improvements may
be achieved by a variety of steps, among them: introduction of
interelectronic coordinates, by the multiconfigurational SCF
procedure (MCSCF) or by configuration interaction (CI).[38b] The
last two procedures expand the wavefunction into a linear
combination of configurations, which in turn consist out of a
linear combination of determinants. In the MCSCF procedure a
limited number of different types of configurations are considered
but the molecular orbitals are optimized simultaneously.[39] In the
so-called conventional CI many configurations are included which

are commonly obtained via excitation of orbitals derived from the reference configurations. They may either represent SCF or MCSCF solutions to the remaining virtual orbitals.

Theoretical Studies of the Wolff Rearrangement

As discussed earlier, there have been various attempts to elucidate the potential energy surface and thus the energetic details of the Wolff Rearrangement.[3,30-32] However, the results have led to a confusing picture as to the relative stabilities of oxirene and formylmethylene. The more recent ab initio studies using "double-zeta" quality basis sets (SCF(DZ)) have been reviewed by Strausz and coworkers.[32f,g] According to these authors oxirene is 12 kcal/mol less stable than formylmethylene and 86 kcal/mol less stable than ketene on the SCF(DZ) level. These same authors also predicted an activation energy of about 6 kcal/mol for the formylmethylene/ketene interconversion, and an activation energy for the formylmethylene/oxirene rearrangement of about 19 kcal/mol.

Recent SCF and SCEP calculations of Dykstra[32h] have shown, however, that the consideration of polarization functions can affect the relative stabilities at least of the isomers ketene, oxirene, and hydroxyacetylene by several kcal/mol, the magnitude depending on the particular isomer.

In this study we have tried to consider both the effects of polarization functions as well as of electron correlation energy corrections for the relative stabilities of the C_2H_2O isomers Ia-IVa. Furthermore, we have assessed the corrections necessary for the activation energies along the rearrangement paths from one isomer to another. This contribution deals primarily with the results obtained on the singlet ground state potential energy surface of the C_2H_2O system. Results obtained for the excited surfaces (singlet and triplet) are included only to a limited extent, since additional work is still in progress.[40]

Calculational Details

Basis sets were constructed from contracted Gaussian functions (CGF). The first set was the 4-31G basis set, made up of 31 CGF's as proposed by Ditchfield et al.[41] The second set was a "double-zeta" polarization (DZ-P) basis set of 58 CGF's as applied by McLean et al.[42] to the isoelectronic molecular system HCNO. This set is derived from von Duijnevelt's[43] atomic SCF basis set {C(9s5p/4s2p), O(9s5p/4s2p) and H(5s/2s)}, which was further supplemented by d-functions for C and O, and by p-functions for H, using the exponents 0.7327 for carbon, 0.85 for oxygen, and 0.95 for hydrogen, respectively. Further details and contraction coefficients for this DZ+P basis set will be given elsewhere.[40]

For initial exploratory calculations we used the 4-31G basis
set instead of the "double-zeta" (DZ) basis, since it is
computationally more economical. Subsequently, all results
obtained on the 4-31G level were further refined using the DZ+P
basis set described above. Interestingly, similar results for
both optimized geometries and relative stabilities are obtained
from either the DZ or the 4-31G basis set, even though the absolute
energies derived with the 4-31G basis are higher than those
obtained from the DZ basis.

Wavefunctions

CI calculations were carried out at two levels:

a.   at the single configuration SCF level, and

b.   at the correlated wavefunction level.

The number of configuration state functions (CSF's) considered
for each C_2H_2O isomer are listed in Table 2.

Table 2.   Number of CSF's Used in the SDCI Calculations.

Isomer	Symmetry	Number of CSF's	
		(4-31G)	(DZ+P)
Ketene	C_{2u}	3754	16979
Hydroxyacetylene	C_s	7167	32969
Formylmethylene	C_1	13041	62481
Oxirene	C_{2v}	3621	16583

For the correlated wavefunction approach the CI wavefunction was
used which included all singly and doubly excited configurations
with respect to the closed shell SCF configuration. The lowest
three orbitals, corresponding to oxygen 1s and 2 times carbon 1s
are always doubly occupied and are thus not correlated.

In order to be able to assess the contributions of the
polarization functions separately from those due to correlation
energies, four wavefunctions were used, which were labeled
SCF(4-31G), SCF(DZ+P), SDCI(4-31G) and SDCI(DZ+P). Thereby the
symbol 4-31G designates the basis set without polarization
functions, essentially equivalent to DZ, whereas DZ+P designates
the one incorporating polarization functions. SDCI implies
consideration of the correlation energies as derived from the CI
treatment above, which are excluded on the mere SCF level. The
differences between these four sets of calculations allow to

separately determine the contributions of the polarization functions and of the correlation energy, which in turn allows to assess the energy convergence pattern as a function of increased sophistication on the ab initio level. From the convergence behavior we hoped to obtain some feeling for the accuracy of our results.

The geometries of the isomers were individually optimized, first on the 4-31G level followed by a full optimization on the SCF(DZ+P) level. In addition, the geometries of ketene, hydroxyacetylene, and oxirene were further refined and optimized with the SDCI(DZ+P) wavefunction. For formylmethylene this last refinement could not be done, because there is no more activation barrier to the rearrangement formylmethylene→ketene on the SDCI(DZ+P) level, as will be discussed later.

The program libraries used in this calculation included the ALCHEMY program system,[44] the direct CI program,[45] and the GAUSSIAN 70 program system.[46]

Geometries

Table 3 lists the optimized structural parameters for the C_2H_2O isomers obtained in this study together with those from recent calculations of other authors,[32g,h] and where available, experimental data[47] for comparison.

Table 3. Optimized Structures and Total Energies of the Lowest Singlet State of the C_2H_2O Isomers (cf. next page)

ISOMER / APPROXIMATION	BOND DISTANCES (A)				ANGLES (DEGREE)				TOTAL ENERGY (a.u.)
Ketene	C_1-C_2	C_2-O	C_1-H		$\sphericalangle H_1C_1H_2$				
SCF(4-31G)	1.297	1.164	1.070		120.0				-151.49493
SCF(DZ)[32h]	1.311	1.171	1.070		120.1				-155.67244
SCF(DZ+P)	1.309	1.147	1.073		122.0				-151.76225
CI(DZ+P)	1.319	1.167	1.077		122.2				-151.16008
Expt.[47]	1.314	1.161	1.079		122.3				——
	(±0.01)	(±0.01)	(±0.002)		(±0.2)				
Hydroxyacetylene	C_1-C_2	C_2-O	C_1-H	$O-H_2$	$\sphericalangle C_1C_2O$	$\sphericalangle C_2OH_2$			
SCF(4-31G)	1.185	1.317	1.049	0.954	178.3	115.3			-151.44420
SCF(DZ)[32h]	1.193	1.322	1.051	0.953	180.0	115.6			-151.62738
SCF(DZ+P)	1.186	1.302	1.059	0.948	178.4	110.3			-151.70628
CI(DZ+P)	1.204	1.318	1.064	0.960	177.8	108.7			-152.10241
Formylmethylene (Fig. 2)	C_1-C_2	C_2-O	C_2-H_2	C_1-H_1	$\sphericalangle C_1C_2O$	ψ	β	α	
SCF(4-31G)	1.420	1.220	1.098	1.085	129.8	108.6	3.7	61.5	-151.37225
SCF(DZ)[32g]	1.400	1.260	1.100	1.050	130.0	110.0	0	60.0	-151.55386
SCF(DZ+P)	1.429	1.193	1.111	1.089	129.6	107.7	10.2	69.0	-151.64910
Oxirene	C_1-C_2	C_2-O	C_1-H_1		$\sphericalangle C_1C_2O$	$\sphericalangle C_1C_2H$			
SCF(4-31G)	1.248	1.552	1.054		66.3	162.3			-151.35397
SCF(DZ)[32h]	1.261	1.548	1.057		66.1	162.5			-151.53543
SCF(DZ+P)	1.250	1.465	1.067		64.7	162.6			-151.62339
CI(DZ+P)	1.270	1.491	1.071		64.8	162.1			-152.02890

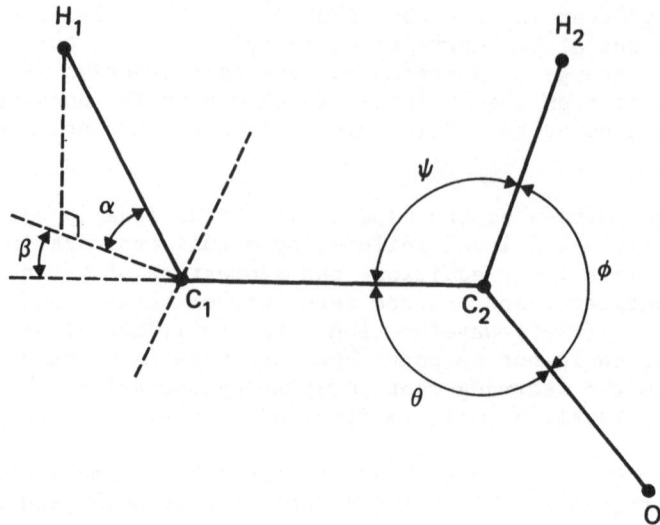

Fig. 2. Formylmethylene. Definition of geometrical parameters,
all atoms except H_1 are in a plane.

For ketene the SDCI(DZ+P) results are in excellent agreement
with the experimental values. Since all other C_2H_2O isomers are
unknown, no further comparison between experimental and theoretical
parameters is possible. Judged from the agreement found for
ketene, however, we assume that the bond lengths predicted for
all the elusive isomers are accurate within ±0.01A on the
SDCI(DZ+P) level, and that the bond angles are accurate within
±1°. The geometry of formylmethylene cannot be optimized on the
SDCI(DZ+P) level, since there is no more energy barrier with
respect to its rearrangement into ketene on this level of
approximation. Therefore, formylmethylene is only defined on the
SCF potential energy surface, and thus its SCF geometry (for this
isomer only) is used in the following.

Effect of Polarization Functions and of Correlation Energy
Corrections upon Geometry

The effect of the polarization functions (d-functions for C
and O; p-functions for H) on the bond lengths is rather small,
except for oxirene. They tend to shorten the bond lengths by
about 0.01 to 0.02A. The C-O bond of oxirene, however, is
shortened by as much as 0.08A. Dykstra[32h] had previously noted
a similar shortening of this C-O bond by comparing his SCF(DZ)
and SCF(DZ+P) results for oxirene.

The correlation energy effect is opposite since it tends to
increase the average bond length again by 0.01 to 0.02A. The two
effects seem to cancel as far as the bond lengths are concerned,
and therefore, previous calculations of small molecules on either
the DZ or (4-31G) level[48] have been remarkably successful.

For the bond angles the effect of the polarization functions
is sometimes as much as ±7 degrees, but the correlation effect is
small. Therefore, the SCF(DZ+P) wavefunctions usually give
reliable bond angles.

The optimized structure of hydroxyacetylene on the SDCI(DZ+P)
level is interesting insofar as the C≡C–O skeleton is predicted
to be nonlinear. The predicted bond angle ⅩCCO of 177.8° is in
striking agreement with the angle ⅩOCN in the isoelectronic
molecule HOCN, for which McLean et al.[42] recently predicted a
theoretical value of 177.7°.

Rotational Constants

Hydroxyacetylene has not yet been observed terrestrially, but
it is believed to occur in dense interstellar clouds, where both
the radicals •OH and •C≡C–H occur. Combination of these might
well yield hydroxyacetylene, not only in space, but perhaps also
in the laboratory, for example, from a suitable precursor in a
matrix isolation experiment (vide infra). For an experimental
search the theoretically predicted physical constants can provide
a useful guidance, and for astronomical purposes the rotational
constants of this elusive molecule are desirable. The latter have
been compiled for all C_2H_2O isomers from the SDCI(DZ+P)
wavefunctions, and the agreement between these theoretical data
and the experimental values[49] for the only known isomer ketene is
within ±0.1 GHz, which we believe is a likely error bar for the
predicted rotational constants for hydroxyacetylene and the other
isomers as well. Details will be given elsewhere.[40]

Relative Energies and Stabilities

The effects of polarization functions and correlation energies
upon the relative energies and stabilities depend upon the specific
isomer as listed in Table 4. For ketene both contributions tend
to lower its energy. For formylmethylene both effects are
substantial (6-8 kcal/mol), but they about cancel one another.
Hydroxyacetylene and oxirene are both stabilized with respect to
ketene by both polarization functions and correlation energy
corrections. For oxirene, the contribution of the polarization
functions is small (1 kcal/mol), whereas that of the correlation
energy correction is more substantial (5 kcal/mol).
Hydroxyacetylene, for example, on the (4-31G) level is 31.8 kcal/mol
less stable than ketene. This difference is reduced by 3 kcal/mol

when polarization functions are invoked, and by an additional
1 kcal/mol because of correlation energy corrections. Additional
corrections of the correlation energy ($\Delta E(Q)$), necessary because
of quadrupole excitations, are estimated according to a formula
suggested by Davidson et al.[50] The resulting corrected energies
are referred to under the label SDQCI. Figure 3 illustrates the
dependence of the relative energies with respect to the most stable
C_2H_2O isomer ketene as a function of the level of approximation
on the ab initio level.

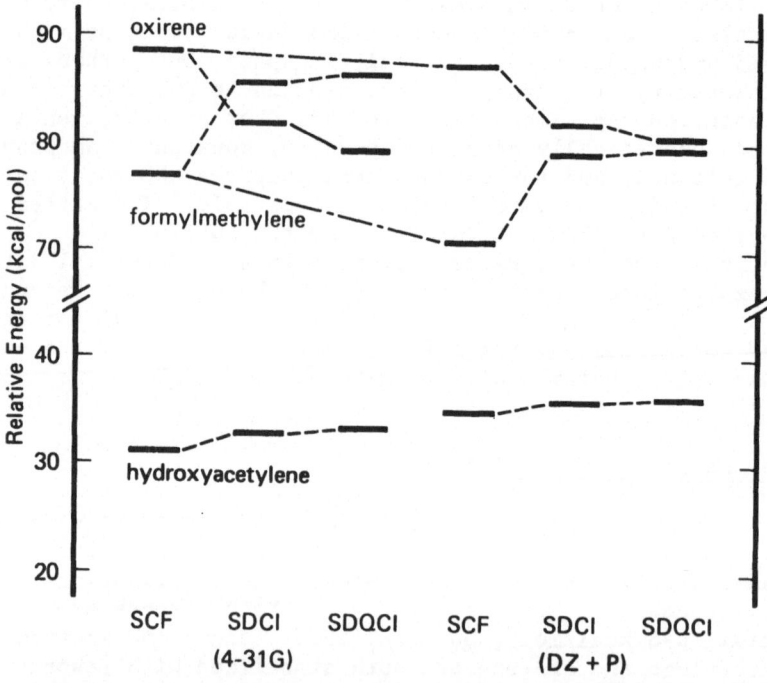

Fig. 3. Effect of polarization functions and correlation energy
on the relative energies of the isomers (dashed lines
indicate the effect of the correlation energy corrections,
whereas dash-dot lines show the effect of polarization
functions).

Table 4a. Total Energies of C_2H_2O Isomers for Different
Geometries and Levels of Approximation.

ISOMER	GEOMETRY	TOTAL ENERGY (HARTREE)		
	Wavefunction for which optimized	APPROXIMATION LEVEL		
		SCF	SDCI	SDQCI
KETENE	SCF(4-31G)	-151.49493	-151.77929	-151.80768
	SCF(DZ+P)	-151.76225	-152.16007	-152.20439
	SDCI(DZ+P)	-151.76121	-152.16104	-152.20621
HYDROXYACETYLENE	SCF(4-31G)	-151.44420	-151.72651	-151.75443
	SCF(DZ+P)	-151.70628	-152.10242	-152.14605
	SDCI(DZ+P)	-151.70514	-152.10359	-152.14820
OXIRENE	SCF(4-31G)	-151.35397	-151.64883	-151.68131
	SCF(DZ+P)	-151.62339	-152.02890	-152.07593
	SDCI(DZ+P)	-151.62207	-152.03016	-152.07839
FORMYLMETHYLENE	SCF(4-31G)	-151.37225	-151.64272	-151.67004
	SCF(DZ+P)	-151.64910	-152.03387	-152.07733
	SDCI(DZ+P)	----	----	----

Table 4b. Relative Energies of C_2H_2O Isomers (Ketene=0) for
Different Geometries and Ab Initio Approximations.

ISOMER	GEOMETRY	RELATIVE ENERGY [kcal/mol]		
	Wavefunction for which optimized	APPROXIMATION LEVEL		
		SCF	SDCI	SDQCI
HYDROXYACETYLENE	SCF(4-31G)	31.8	33.1	33.4
	SCF(DZ+P)	35.1	36.2	36.6
	SDCI(DZ+P)	35.2	36.1	36.4
OXIRENE	SCF(4-31G)	88.5	81.9	79.3
	SCF(DZ+P)	87.2	82.3	80.6
	SDCI(DZ+P)	87.3	82.1	80.2
FORMYLMETHYLENE	SCF(4-31G)	77.0	85.7	86.4
	SCF(DZ+P)	71.0	79.2	79.7
	SDCI(DZ+P)	--	--	--

The stability of the various isomers to dissociation has also
been determined, in particular the dissociation energies (D_e) for
the reactions:

a.   ketene $\longrightarrow CH_2(a^1A_1) + CO(X^1\Sigma^+)$

b.   hydroxyacetylene $\longrightarrow C_2H(X^2\Sigma^+) + OH(X^2\Pi)$

c.   oxirene $\longrightarrow C_2H_2(X^1\Sigma_g^+) + O(^1D)$.

The experimental value for the first of these three reactions is
the only one known[51] (D_e=85.4 kcal/mol). From this experimental
result and from our theoretical values it follows that the relative
energy of oxirene as determined here (82±2 kcal/mol) is below the

dissociation limit of ketene. Further details will be given elsewhere.[40]

REARRANGEMENT PATHS AND ACTIVATION ENERGIES

General Considerations

The search for the rearrangement path on the potential energy surface of the C_2H_2O system was conducted as follows: A suspected path was chosen and an appropriate reaction coordinate was selected, along which points were picked in such a way that they represented the entire path about equally well. Each path was then optimized initially on the SCF(4-31G) level, thereby mapping out a considerable portion of the potential energy surface surrounding the suspected path. This was achieved by relaxing all geometrical parameters other than the reaction coordinate. Subsequently the path was refined on the SCF(DZ+P) level, again with full optimization of the geometries. Correlation energy corrections were then introduced via SDCI(DZ+P) calculations. This hierarchy of increasing sophistication provided for a rather economical approach without loss of accuracy.

Four rearrangement paths were considered, namely:

Path (A): ketene $\rightleftharpoons$ formylmethylene; reaction coordinate $\Psi \equiv \nleq C_1 C_2 H_2$.

Path (B): formylmethylene $\rightleftharpoons$ oxirene; reaction coordinate $\theta \equiv \nleq C_1 C_2 O$.

Path (C): ketene $\rightleftharpoons$ hydroxyacetylene; reaction coordinate $\phi \equiv \nleq O C_2 H$.

Path (D): formylmethylene $\rightleftharpoons$ hydroxyacetylene; reaction coordinate like for (C).

Figure 2 defines the reaction coordinates and all other geometrical parameters. Figure 4 illustrates the energy profiles for the rearrangements along path (A) and (B). Figure 5 shows the energy profiles for those along path (C) and (D). A full account of the optimized geometrical parameters for the selected points and for all the isomers defined along the above rearrangement paths will be given elsewhere[40] for both the SCF(DZ+P) as well as for the SDCI(DZ+P) level.

Figure 4 illustrates the findings referred to earlier, namely that there is no activation barrier to rearrangement from formylmethylene to ketene on the SDCI(DZ+P) level of approximation; therefore, the geometry of formylmethylene is not defined on this level. Furthermore, this result implies that one of the two vicinal hydrogens in formylmethylene can move freely to become a

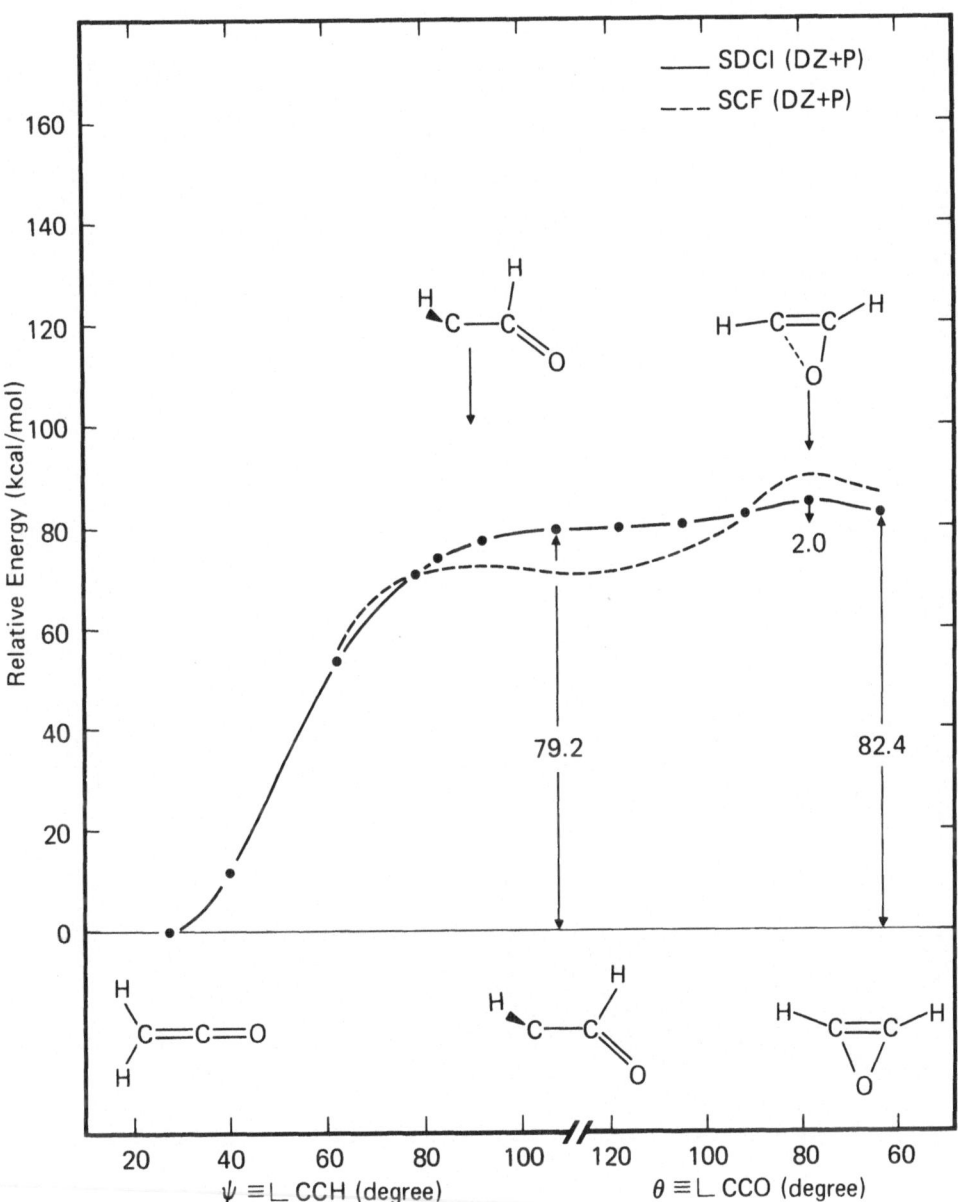

Fig. 4.   Energy profile for rearrangement along paths (A) and (B).

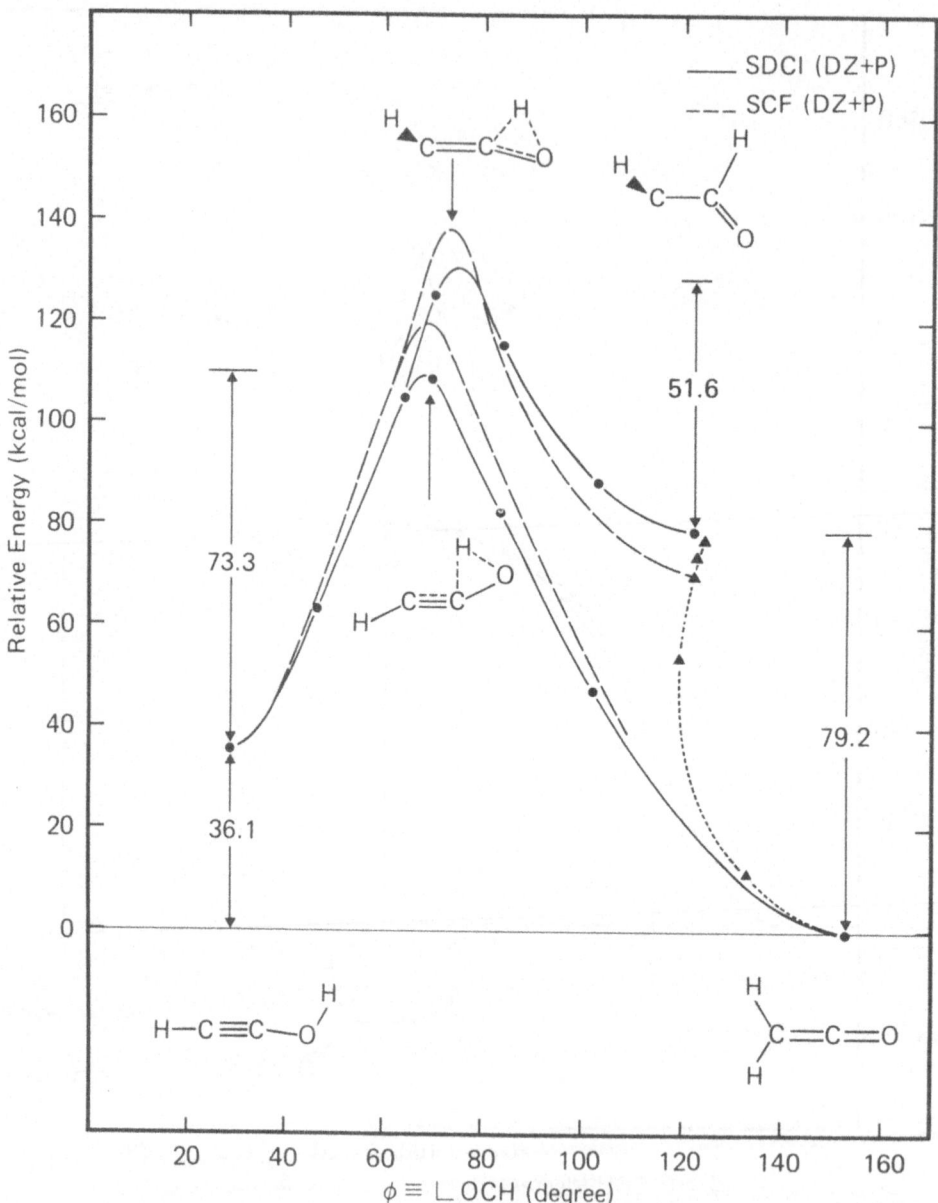

Fig. 5. Energy profile for rearrangement along paths (C) and (D).

geminal hydrogen in ketene (1,2 shift). On the SCF(DZ+P) level,
however, there is a barrier of 0.9 kcal/mol to overcome for this
rearrangement along path (A).

Oxirene, once formed, is predicted to be only marginally stable
with respect to ring opening (path (B)), since the calculated energy
barrier is only 2 kcal/mol. The path for its rearrangement to
ketene via formylmethylene illustrated in Fig. 4 (Path (A) and
Path (B)) is indeed the energetically most favorable for
isomerization on the ground state potential surface.

Paths (C) and (D) reflect some interesting findings pertaining
to the hypothetical stability of hydroxyacetylene (Fig. 5). That
this molecule is still an elusive species seems to be due to the
fact that it is unlikely to be formed during the decomposition of
diazoethanone (Va), because the rearrangement of formylmethylene
into hydroxyacetylene is associated with a theoretical activation
barrier of 51.6 kcal/mol (Path (D)).

Similarly the direct rearrangement of hydroxyacetylene into
ketene (Path (C)) is associated with a high barrier of 73.2
kcal/mol. Therefore, hydroxyacetylene should be quite stable
monomolecularly, in spite of its relative energy of only
36.1 kcal/mol above ketene. The indirect rearrangement of
hydroxyacetylene via formylmethylene into ketene (or oxirene),
i.e., Path (D), is associated with even a higher activation barrier
of 94.6 kcal/mol, rendering isomerization of hydroxyacetylene into
any other C_2H_2O isomer an unlikely event. Table 5 lists the
activation barriers to interconversion of the C_2H_2O isomers as a
function of different ab initio approximations. Previous data
obtained by Strausz et al.[32g] with a double zeta basis set are
listed for comparison. Their data are rather close to our (4-31G)
results.

Table 5. Activation Energies (kcal/mol) for Interconversions
of the Isomers Ia to IVa for Various Levels of
Approximation.

APPROXIMATION ISOMERIZATION	SCF(4-31G)	SCF(DZ)[32g]	SCF(DZ+P)	SCDI(DZ+P)
Ia→IIa	81.2	80.0	71.9	79.2
IIa→Ia	7.3	5.7	0.9	0
IIa→IVa	20.7	19.1	19.0	5.1
IVa→IIa	6.1	7.3	2.8	2.0
Ia→IIIa	126.8	--	119.0	109.4
IIIa→Ia	95.0	--	83.9	73.2
IIIa→IIa	95.2	--	103.9	94.6
IIa→IIIa	50.0	--	67.6	51.6

Ia=ketene, IIa=formylmethylene, IIIa=hydroxyacetylene,
IVa=oxirene

Table 5 shows that the effect of polarization functions, i.e.,
the difference between basis sets (DZ+P) and (4-31G) (or also DZ),
is substantial, sometimes as much as 18 kcal/mol, the details
depending on the isomerization path. The correlation energy
corrections, i.e., the difference between SCF(DZ+P) and SDCI(DZ+P)
results, are also quite large, and sometimes of the same order of
magnitude as the polarization function effects (≤16 kcal/mol).
Tanaka and Yoshimine[40] have been able to show that the effects
due to polarization functions and due to electron correlation
energies are essentially additive in this case studied here.

EXCITED STATES

Triplet Potential Surface

Calculations for the excited states are still in progress,[40]
and therefore, only preliminary findings can be communicated. For
oxirene the triplet energy is found to be about 100 kcal/mol above
the ketene ground state. Formylmethylene in its lowest triplet
state is predicted to have a relative energy of about 60 kcal/mol
above ketene, which is about 20 kcal/mol below the singlet ground
state of formylmethylene.

Excited Singlet States

The lowest excited singlet state of oxirene is predicted to
be about 150 kcal/mol above the ketene ground state. Preliminary
data indicate that the excited singlet potential surface crosses
that of the ground state in the vicinity of the formylmethylene
geometry. The consequences of this finding are still being
evaluated. Excited states of ketene have been calculated by others
previously.[52]

Diazoethanone

In order to complete our understanding of the Wolff
Rearrangement, diazoethanone (Va) was also studied in its ground
state on the SCF(4-31G) level. The results indicate that the
C_2H_2O skeleton of this precursor has a geometry essentially similar
(bond lengths within 0.02A) to that of formylmethylene, the only
difference being that Va is planar. The decomposition of Va into
formylmethylene and nitrogen is predicted to be endoenergetic by
about 29 kcal/mol. Czismadia et al.[30] had determined this energy
difference before with extended Hückel methods to be about
35 kcal/mol.

Figure 6a illustrates the major aspects of the isomerization
of the C_2H_2O isomers in their singlet states. Fig. 6b depicts our
preliminary conclusions for the triplet state.

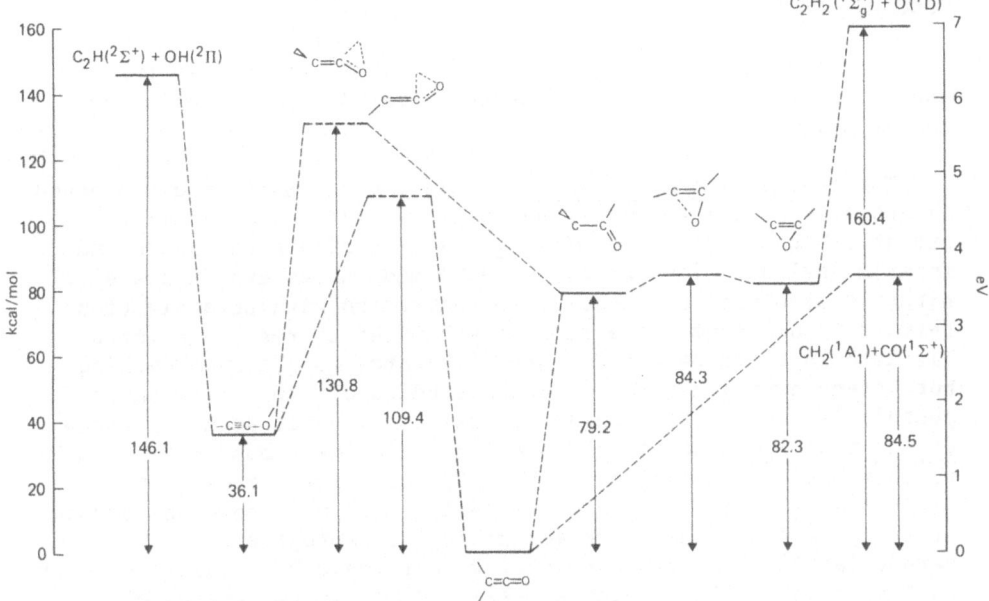

Fig. 6a. Relative energies of C_2H_2O isomers in their singlet ground states.

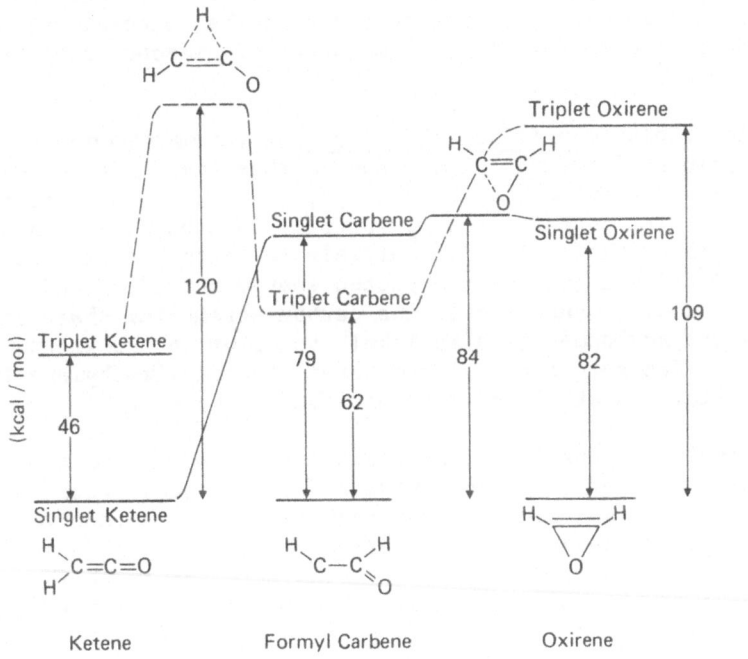

Fig. 6b. Relative energies of C_2H_2O isomers in their singlet and triplet states (preliminary data).

CONCLUSIONS

From the potential surfaces of the C_2H_2O, studied so far in the ground and in the excited states, the following conclusions can be drawn.

The <u>thermal decomposition</u> of diazoethanone most likely proceeds exclusively on the ground state surface. This is especially probable because of the similar geometries of diazoethanone and formylmethylene. The latter, once formed in its ground state by either thermolysis of photolysis, rearranges without activation barrier into ketene. The lack of a barrier to rearrangement of IIa into Ia explains the failure of Krantz[7] to observe anything but ketene during matrix isolation studies of the diazoethanone photolysis at temperatures as low as 8K. Thus it seems that the <u>direct photolysis</u> of diazoethanone must follow ground rules similar to those of the thermolysis, which appears likely because the excited singlet state of oxirene is 150 kcal/mol above the ground state of ketene. Therefore scrambling of isotopically carbon-labeled diazoethanone during photolysis[10,11] seems to occur on the ground state surface as well. (For excited states of diazoethanone vide infra.) Because of the energy barrier in the oxirene⇌formylmethylene path, the amount of carbon scrambling and thus of oxirene participation (Table 1) should reflect the excess vibrational energy, and thus depend upon the temperature, as has been demonstrated for higher homologues of diazoethanone by Martin and Sammes.[13]

The <u>triplet sensitized photolysis</u> of diazoethanone will instead yield triplet formylmethylene, and is thus the starting point for a rather different chemical behavior,[3] due to the different shape of the triplet potential energy profile for the C_2H_2O system. The <u>photolysis of ketene</u> is also unlikely to reach the excited singlet state of oxirene (unless the photon energy should exceed 150 kcal/mol). Consequently the carbon scrambling observed by Russel and Rowland[14] is also likely to occur on the ground state surface. The path from excited ketene to formylmethylene in its ground state is still under investigation, however.[40]

Formylmethylene is not a stable isomer, which explains the lack of direct experimental evidence for its occurrence during the Wolff Rearrangement.[3,7] Oxirene has so low an activation barrier to rearrangement into ketene that it will be most difficult to trap this elusive molecule. Furthermore, it is unlike to be a long-lived intermediate during the WR. Kinetically and energetically it can occur during the WR, however, to an extent that will depend on the amount of vibrational energy available.

The potential surfaces of homologues, i.e., the WR of higher α-diazoketones, may differ in detail, but their general features are considered to be quite similar to those derived here. Thus there is experimental evidence that phenyl substituents seem to increase the amount of oxirene participation[10,11] (Table 1), and therefore, the potential energy surface for the compounds Ib-Vb seems to be more favorable for formation of the oxirene IVb.

Hydroxyacetylene is very unlikely to occur in the WR because of high barriers against its formation from either formylmethylene or from ketene. Once formed, however, it should be rather stable, at least monomolecularly. It is surprisingly stable to rearrangement into ketene, of which it is the enol form. Therefore, hydroxyacetylene could very well occur in dense interstellar clouds. The rotational constants as calculated here may very well guide the search of astronomers, and they may also assist in efforts to observe this elusive species terrestrially. Thus we are presently engaged in an effort to generate hydroxyacetylene in an argon matrix from a suitable precursor by either photolysis or by electron impact. Attempts to generate C_2H_2O isomers from vinylidene carbonate by photolysis have so far been unsuccessful,[53] even though the sulfur analogous compounds, namely the trithiocarbonates, have recently been claimed[54] to be a source for the generation of the sulfur analogue of oxirene, namely of thiirene. In an alternate approach we intend to obtain hydroxyacetylene from a peroxide precursor, namely from perpropiolic acid H-C≡C-CO-O-OH. We have previously shown[55] that the photolysis of diacoylperoxides isolated in an argon matrix at 8K yields CO_2 and alkyl radicals, which upon warming to 20K combine or disproportionate to product molecules. We expect the perpropiolic acid to yield hydroxyacetylene according to Scheme 11:

Scheme 11.

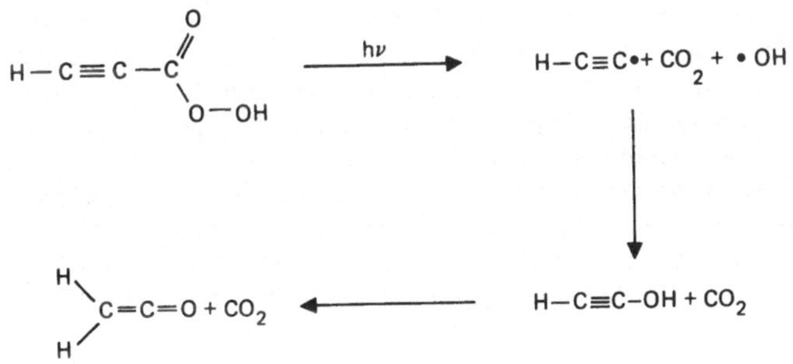

Unfortunately, perpropiolic acid is a dangerous compound which is difficult to purify. We have so far succeeded in obtaining phenol from perbenzoic acid (which can be purified readily by vacuum sublimation[56]) via photolysis in an argon matrix at 8K according to Scheme 12,[57] which is essentially analogous to Scheme 11:

Scheme 12.

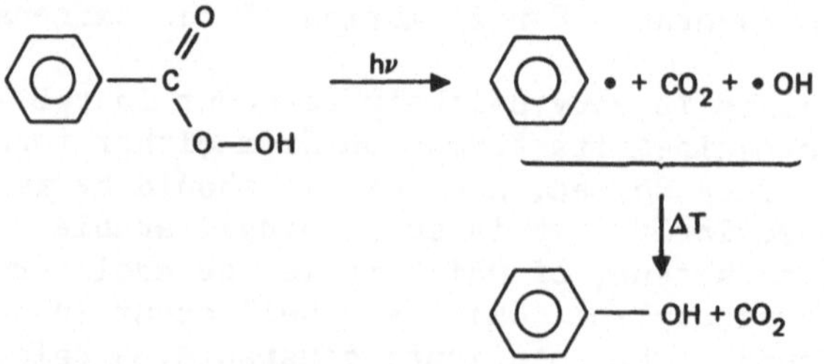

Nevertheless, both oxirene and hydroxyacetylene still remain unknown for the time being.

Stability and Theoretical Properties of the C_2H_2S Isomers

In view of the claims[27,28] that the sulfur-analogous compounds of oxirene, namely thiirene (XVIa), as well as mercaptoacetylene (XVa), which is the sulfur-analogous form of hydroxyacetylene, have been made and characterized, it is interesting to compare the theoretical predictions for the stabilities and decomposition pathways of the C_2H_2O isomers with those of their C_2H_2S relatives.

Early Hückel MO calculations by Zahradnik suggested a very low stability for thiirene.[58] On the basis of thermochemical considerations a much greater stability of thiirene relative to oxirene had been predicted.[28a] Early ab initio studies[32a,59] have been carried out, but without geometry optimization. Extended Hückel[30] and MINDO/3[31] calculations have suffered from the same shortcomings as the corresponding data for oxirene, mentioned earlier. SCF calculations by Rosmus et al.[27g] had confirmed thioketene to be the most stable of the C_2H_2S isomers. More recent ab initio calculations by Strausz et al.,[60] using a STG–4G minimal basis set, have been performed for five isomeric C_2H_2S structures (XIIIa, XIVa, XVa, XVIa and XVIIIa) with full optimization of their geometries for the singlet ground states and for the triplet states. These calculations predict a relative stability thiirene<thioketene≈mercaptoacetylene. Whereas MINDO/3 had predicted the isomer

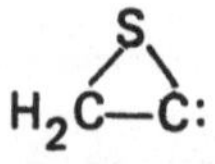

(XVIIa) to be less stable than thiirene by 18.8 kcal/mol,[31] the ab initio calculations[60] assign about the same stability to the isomers XVIa and XVIIIa, the thiirene being actually less stable by 0.3 kcal/mol.

Table 6.   Relative Energies of the C_2H_2O Isomers[60] (in kcal/mol).

ISOMER	SINGLET	TRIPLET
XVa mercaptoacetylene	0	122.7
XIIa thioketene	1	23.8
XIVa thioformylmethylene	15.4	6.3
XVIIIa	30.4	38.4
XVIa thiirene	30.7	48.4

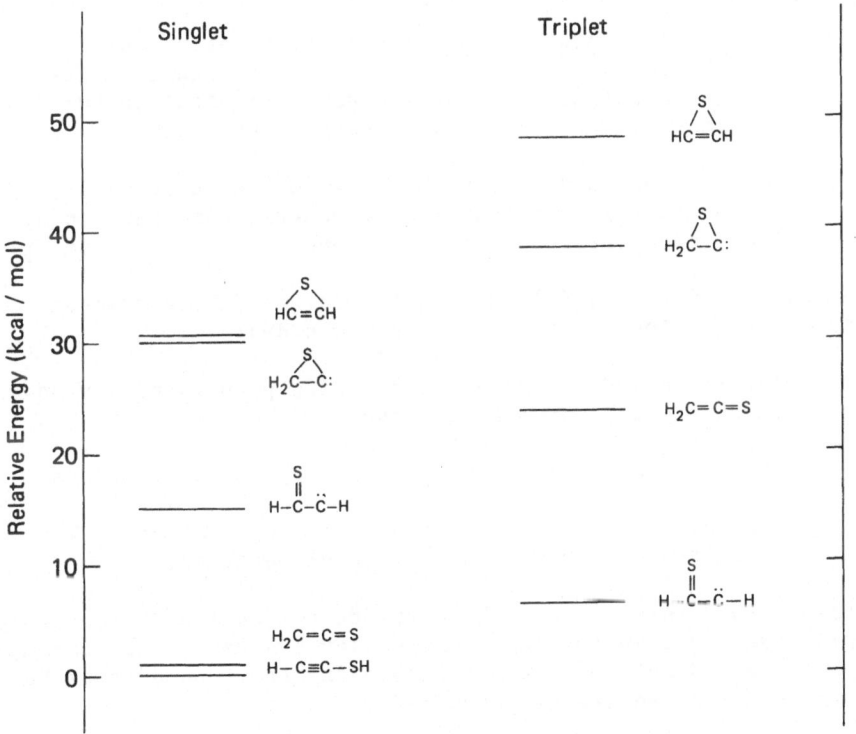

Fig. 7.   Relative energies of C_2H_2S isomers in the singlet ground state and in their triplet state.[60]

Table 6 and Fig. 7 compare the relative energies of all the C_2H_2S isomers as obtained by Strausz et al.[60] Whereas these calculations give some feeling for the relative energetics, they give no clues as to their likelihood of interconversion. The relatively high singlet energy obtained for thiirene makes it difficult to understand why it should be experimentally any easier to observe thiirenes, which have supposedly been characterized directly and

trapped as well, whereas oxirenes remain unknown. Thus more
accurate ab initio data for the C_2H_2S system are needed together
with an evaluation of the energy barriers in the rearrangement
paths of these isomers, before the different experimental behavior
can be explained.

Guidelines for Future Quantum Chemical Studies of Chemical Reactions

The studies conducted here have provided the following insight:

1.   The equilibrium geometries and reaction paths are rather
      insensitive to both polarization functions and electron
      correlation energy corrections.

2.   The relative stabilities, i.e., the relative energies
      and the activation barriers associated with a certain
      path, are rather sensitive to both polarization functions
      and to electron correlation energy corrections.

3.   Correlation energy contributions to the relative
      stabilities are insensitive to polarization functions,
      i.e., the contributions are independent and thus additive.

From these findings the following approach for future quantum
chemical calculation is derived and recommended:

1.   Determine and optimize the reaction path initially on
      the SCF(4-31G) or SCF(DZ) level.

2.   Refine path obtained using SCF(DZ+P) basis functions.

3.   Determine correlation energy contributions via a CI
      calculation with the (4-31G) or DZ basis set.

We believe that this 3-step scheme yields reliable and reasonably
accurate results even for larger molecular systems in an economical
form. It should not only be applicable for the ground state but
also for the excited states potential surfaces.

We have started to put together a new collection of programs,[61]
similar but superior to ALCHEMY,[44] in the sense that this new
collection (ALCHEMY 2) is intended to perform automated searches
for chemical pathways by forming energy gradients. We also plan
to incorporate an improved CI capability for larger systems in
order to compute the potential energy surfaces for systems
containing more carbon and heteroatoms than those outlined here
with comparable accuracy.

REFERENCES

1. L. Wolff, Liebigs Ann. Chem. 325, 129 (1902); 333, 1 (1904); 394, 25 (1912).
2. G. Schroeter, Chem. Ber. 42, 2336 (1909); 49, 2697 (1916).
3. a) H. Meier and K.-P. Zeller, Angew. Chem. 87, 52 (1975); Angew. Chem., Int. Ed. 14, 32 (1975).
   b) C. Wentrup, Topics Current Chem. 62, 173 (1976).
   c) H. Meier and K. P. Zeller, Angew. Chem. 89, 876 (1977), Int. Ed. 16, 835 (1977).
4. a) B. Eistert in W. Foerst, Ed., "Neuere Methoden der Präparativen Chemie," Vol. I, p. 359, Verlag Chemie, Weinheim, Germany (1949).
   b) B. Eistert, Angew, Chem. 54, 124 (1941); 55, 118 (1942).
5. a) J. Pacansky, IBM Journal of Res. and Dev. 23, 42 (1979) and references therein.
   b) J. Pacansky and H. Coufal, J. Am. Chem. Soc. 102, 410 (1980).
6. B. Eistert, Chem. Ber. 68, 208 (1935).
7. A. Krantz, J. Chem. Soc., Chem. Comm. 670 (1973).
8. V. Franzen, Liebigs Ann. Chem. 614, 31 (1958).
9. R. T. Arnold, C. Huggett, and T. I. Taylor, J. Am. Chem. Soc. 64, 3043 (1942).
10. a) G. Czismadia, J. Front, and O. P. Strausz, J. Am. Chem. Soc. 90, 7360 (1968).
    b) D. E. Thornton, R. K. Gosavi, and O. P. Strausz, ibid. 92, 1768 (1970).
    c) G. Frater and O. P. Strausz, ibid. 92, 6654 (1970).
11. K.-P. Zeller, H. Meier, H. Kolshorn, and E. Müller, Chem. Ber. 105, 1875 (1972).
12. K.-P. Zeller, Tetrahedron Lett., 707 (1977).
13. S. A. Martin and P. G. Sammes, J. Chem. Soc., Perkin I, 2623 (1972).
14. R. L. Russell and F. S. Rowland, J. Am. Chem. Soc. 92, 7508 (1970).
15. D. C. Montague and F. S. Rowland, ibid. 93, 5381 (1971).
16. M. Berthelot, C. R. Hebd. Sceances Acad. Sci. 70, 256 (1870); Bull. Soc. Chim. France [2] 14, 113 (1870).
17. W. Madelung and M. E. Oberwegner, Liebigs Ann. Chem. 490, 201 (1931).
18. W. G. Dauben, C. F. Hiskey, and M. A. Muhs, J. Am. Chem. Soc. 74, 2082 (1952).
19. a) K. M. Ibne-Rasa, R. H. Peter, J. Ciabattoni, and J. O. Edwards, J. Am. Chem. Soc. 95, 7894 (1973); P. W. Concannon and J. Ciabattoni, ibid. 95, 3284 (1973).
    b) V. Ogata, Y. Sawaki, and H. Inoue, J. Org. Chem. 38, 1044 (1973).
    c) J. Ciabattoni, R. A. Campbell, C. A. Renner and P. W. Concannon, J. Am. Chem. Soc. 92, 3826 (1970).
    d) J. K. Stille and D. D. Whitehurst, ibid. 86, 4871 (1964).
    e) R. N. McDonald and P. A. Schwab, ibid. 86, 4866 (1964).

f) V. Franzen, Chem. Ber. 88, 717 (1955); 87, 1478 (1954), and earlier papers.

20. a) H. E. Avery and S. J. Heath, Trans. Faraday Soc. 68, 512 (1972).

    b) I. Haller and G. C. Pimentel, J. Am. Chem. Soc. 84, 2855 (1962).

21. H. H. Schlubach and W. Richaud, Liebigs Ann. Chem. 588, 195 (1954).

22. R. Breslow, Angew. Chem., Int. Ed. 7, 565 (1968); Acc. Chem. Res. 6, 393 (1973).

23. R. W. Hoffman and R. Schüttler, Chem. Ber. 108, 844 (1975).

24. H. Hart, J. B.-C. Jiang, and M. Siasaoka, J. Org. Chem. 42, 3840 (1977).

25. A. Padwa, D. Cumrine, R. Hartman, and R. Layton, J. Am. Chem. Soc. 89, 4435 (1967).

26. E. Fahr, Liebigs Ann. Chem. 638, 1 (1960) and references therein.

27. a) E. G. Howard, Jr., Chem. Abstr. 57, 13617f (1962), U.S. Patent 3035030 (1962).

    b) A. Krantz and J. Laureni, J. Am. Chem. Soc. 96, 6768 (1979).

    c) J. Laureni, A. Krantz, and R. A. Hajdn, ibid. 98, 7872 (1976).

    d) A. Krantz and J. Laureni, Ber. Bunsenges. Phys. Chem. 82, 13 (1978).

    e) G. Seibold and C. Heibl, Angew. Chem., Int. Ed. 14, 248 (1975); Chem. Ber. 110, 1225 (1977).

    f) H. Bock, B. Solouki, G. Bert, and P. Rosmus, J. Am. Chem. Soc. 99, 1663 (1977).

    g) P. Rosmus, B. Solouki, and H. Bock, Chem. Phys. 22, 453 (1977).

28. a) O. P. Strausz, J. Front, E. L. Dedio, P. Kebarle, and H. E. Gunning, J. Am. Chem. Soc. 89, 4805 (1967).

    b) J. Front, M. Torres, H. E. Gunning, and O. P. Strausz, J. Org. Chem. 43, 2487 (1978).

    c) M. Torres, A. Clement, J. E. Bertie, H. E. Gunning, and O. P. Strausz, ibid. 43, 2490 (1978), and references therein.

    d) M. Torres, E. M. Lown, and O. P. Strausz, Heterocycles 11, 697 (1978).

    e) M. Torres, I. Safarik, A. Clement, J. E. Berthie, and O. P. Strausz, Nouv. J. Chim. 3, 365 (1979).

29. a) W. Kirmse and L. Horner, Liebigs Ann. Chem. 614, 4 (1958).

    b) K. P. Zeller, H. Meier and E. Müller, Tetrahedron Lett. 537 (1971); Liebigs Am. Chem. 766, 32 (1972).

    c) K. P. Zeller, H. Meier, and E. Müller, Ibid. 749, 178 (1971); Org. Mass. Spec. 5, 373 (1971).

    d) G. E. Castillo and H. E. Bertorello, J. Chem. Soc. Perkin I, 325 (1978) and references therein.

    e) T. Wooldridge and T. D. Roberts, Tetrahedron Lett. 31, 2643 (1977).

    f) L. Benati, P. C. Montevecchi, and G. Zanardi, J. Org. Chem. 42, 575 (1977).

g) U. Timm, H. Bühl, and H. Meier, J. Heterocycl. Chem. 15, 697 (1978).

30. I. G. Czismadia, H. E. Gunning, R. K. Gosavi, and O. P. Strausz, J. Am. Chem. Soc. 95, 133 (1973).

31. M. J. S. Dewar and C. A. Ramsden, Chem. Comm. 688, (1973).

32. a) D. T. Clark, Theor. Chim. Acta 15, 225 (1969).
    b) A. C. Hopkinson, J. Chem. Soc., Perkin II, 794 (1973).
    c) W. A. Lathan, L. Radom, P. C. Hariharan, W. J. Hehre, and J. A. Pople, Topics Current Chem. 40, 1 (1973). and earlier papers.
    d) C. E. Dykstra and H. F. Schaefer, III, J. Am. Chem. Soc. 98, 2689 (1976).
    e) N. C. Baird and K. F. Taylor, ibid. 100, 1333 (1978).
    f) O. P. Strausz, R. K. Gosari, A. S. Denes, and I. G. Csizmadia, ibid. 98, 4784 (1976).
    g) O. P. Strausz, R. K. Gosari, and H. E. Gunning, J. Chem. Phys. 67, 3057 (1977); Chem. Phys. Lett. 54, 510 (1978).
    h) C. Dykstra, J. Chem. Phys. 68, 4244 (1978).

33. E. I. Snyder, J. Am. Chem. Soc. 92, 7529 (1970).

34. The underestimation of strain energy has later been rectified in MINDO/3, cf. M. J. S. Dewar and W.-K. Li, J. Am. Chem. Soc. 96, 5569 (1974).

35. Recent surveys of ab initio quantum chemical calculations include:
    a) H. F. Schaefer, III, "The Electronic Structure of Atoms and Molecules," Addison-Wesley, Reading, Mass (1972).
    b) J. L. Whitten, Acc. Chem. Res. 6, 238 (1973).

36. a) H. Preuss, Z. Naturfschg. 11, 823 (1956).
    b) E. Clementi, C. C. J. Roothaan, and M. Yoshimine, Phys. Rev. 127, 1618 (1962).
    c) E. Clementi and D. Raimondi, J. Chem. Phys. 38, 2686 (1963).
    d) E. Clementi, ibid. 40, 1944 (1964).
    e) S. Hunzinaga, ibid. 42, 1293 (1965).
    f) J. L. Whitten, ibid. 44, 359 (1966).

37. a) C. C. J. Roothaan, Rev. Mod. Phys. 23, 69 (1951); 32, 179 (1960).
    b) R. K. Nesbet, Proc. Roy. Soc. A230, 312 (1955).

38. a) P. O. Löwdin, Advan. Chem. Phys. 2, 207 (1959); Phys. Rev. 97, 1474 (1955).
    b) P. S. Bagus, B. Liu, A. D. McLean, and M. Yoshimine in: "Computational Methods for Large Molecules and Localized States in Solids," E. Herman, A. D. McLean, and R. K. Nesbet, Eds., Plenum Press, New York (1973) p. 87.

39. a) N. Sabelli and J. Hinze, J. Chem. Phys. 50, 684 (1969).
    b) G. Das and A. C. Wahl, Phys. Rev. Lett. 24, 440 (1970); A. C. Wahl and G. Das, Advan. Quantum Chem. 5, 261 (1970).

40. K. Tanaka and M. Yoshimine, to be published.

41. R. Ditchfield, W. J. Hehre, and J. A. Pople, J. Chem. Phys. 54, 724 (1971).

42. A. D. McLean, G. H. Leow, and D. S. Berkowitz, J. Mol. Spec. 64, 184 (1977).

43. F. B. von Duijnevelt, IBM Research Journal, 945 (1971).
44. P. S. Bagus, B. Liu, A. D. McLean, and M. Yoshimine, "The
    Program System ALCHEMY"; A. D. McLean in "Potential Energy
    Surfaces in Chemistry," W. A. Lester, Jr., Ed. (1970).
45. P. Siegbahn in "Proceedings of the SRC Atlas Symposium No. 4,
    "Quantum Chemistry - The State of the Art," (1974).
46. W. J. Hehre, W. A. Lathan, R. Ditchfield, R. F. Stewart, and
    J. A. Pople, "Gaussian 70" QCPE Program No. 236, Quantum
    Chemistry Program Exchange, Indiana University,
    Bloomington, Ind., USA.
47. C. B. Moore and G. C. Pimentel, J. Chem. Phys. $\underline{38}$, 2816 (1963).
48. W. A. Lathan, L. A. Curtiss, W. J. Hehre, J. B. Lisle, and
    J. A. Pople in Prog. in Phys. Org. Chem. Vol. II,
    A. Streitweiser and R. W. Taft, Eds., Wiley, New York
    (1974).
49. J. W. C. Johns and J. M. R. Stone, J. Mol. Spectrosc. $\underline{42}$, 523
    (1972).
50. S. R. Langhoff and E. R. Davidson, Int. J. Quant. Chem. $\underline{8}$, 61
    (1974).
51. D. Feldmann, K. Meier, H. Zacharias, and K. H. Welge, Chem.
    Phys. Lett. $\underline{59}$, 171 (1978).
52. W. A. Goddard, J. Am. Chem. Soc. $\underline{98}$, 6093 (1976).
53. D. C. Brown, J. Pacansky, and J. Bargon, unpublished.
54. M. Torres, A. Clement, H. E. Gunning, and O. P. Strausz, Nouv.
    J. Chim. $\underline{3}$, 149 (1979).
55. J. Pacansky and J. Bargon, J. Am. Chem. Soc. $\underline{97}$, 6896 (1975);
    J. Pacansky, G. P. Gardini, and J. Bargon, ibid. $\underline{98}$, 2665
    (1976); J. Pacansky, D. E. Horne, G. P. Gardini, and
    J. Bargon, J. Phys. Chem. $\underline{81}$, 2149 (1977).
56. F. R. Rodgers and J. Bargon, unpublished results.
57. J. Pacansky, F. Rodgers, and J. Bargon, unpublished results.
58. R. Zahradnik, Advanc. Heterocycl. Chem. $\underline{5}$, 14 (1965);
    R. Zahradnik and J. Kontecky, Collect. Czechoslov. Chem.
    Comm. $\underline{26}$, 156 (1961).
59. a) D. T. Clark, Int. Symposium on the Quantum Aspects of
       Heterocycl. Compounds in Chem. and Biochem., The Israel
       Academy of Sciences and Humanities, Jerusalem (1970)
       p. 238.
    b) D. T. Clark, Intern. J. Sulfur Chem. $\underline{C7}$, 11 (1972).
60. O. P. Strausz, R. K. Gosari, F. Bernardi, P. G. Mezey,
    J. D. Goddard, and I. C. Czismadia, Chem. Phys. Lett. $\underline{53}$,
    211 (1978).
61. B. Liu, A. D. McLean, and M. Yoshimine, unpublished. A similar
    effort is planned for HONDO, see W. A. Lester's
    contribution in this volume.

COMPUTER PROGRAMS FOR THE DEDUCTIVE SOLUTION OF CHEMICAL

PROBLEMS ON THE BASIS OF MATHEMATICAL MODELS -

A SYSTEMATIC BILATERAL APPROACH TO REACTION PATHWAYS

I. Ugi (Speaker), J. Bauer, J. Brandt,
J. Friedrich, J. Gasteiger, C. Jochum,
W. Schubert

Organisch-chemisches Institut der Technischen
Universität München, Lichtenbergstr. 4,
D - 8046 Garching

and J. Dugundji

Department of Mathematics, U.S.C., Los Angeles,
Calif. 90007

1. Trees and Networks of Reaction Pathways in the Solution of Chemical Problems

More than a decade ago it has been recognized that Chemistry is a particularly well-suited substrate for the problem solving capabilities of the computers.

Their first non-numerical use in the solution of chemical problems was in the interpretation of mass spectra[1] and in the design of multistep syntheses [2]. Remarkable progress has been made in the latter field since its beginnings, and many research groups at academic institutions, as well as in industry are active in studying the scope and limitations of computer assistance in the design of syntheses. It has remained the main topic in computer chemistry, although numerous other chemical problems have been found which can be solved with the aid of computers [3].

Initially, the computer programs for the design of syntheses[2] were primarily based on the retrieval and manipulation of filed data on known synthetic reactions. However, soon after the development of these first programs had begun, we proposed an alternative to such techniques, namely an essentially deductive approach[4,5] in which the solution of individual problems is derived from a general theory. A general mathematical theory of constitutional chemistry which could be used as a basis for the deductive design of syntheses was developed soon thereafter[5], and the implementation of a corresponding pilot program, CICLOPS[6] was started.

In the meantime, other investigators have also recognized the advantage of the deductive approach, because it has become that element in computer chemistry which makes fundamental progress still conceivable, although in this field many of the initial goals have already been reached.

AHMOS, a synthesis design program which has been developed by Weise[7], uses the underlying concepts of our mathematical model. Wipke[8] has introduced the "ab initio transforms", as a supplement to his reaction library based synthesis design program SECS[2c]. These "ab initio transforms" correspond directly to the heteroaptic-heterolytic subset of the basis elements of the R-matrices[3,5] of our theory (see Section 2). Hendrickson[9] has proposed an approach to synthesis design whose essential feature is the construction of synthetic reactions from their elementary mechanistic steps. In ASSOR, a mechanistically oriented synthesis design program by Schubert[3,10], such mechanistic steps are found through the combination of three operations, namely (a) supplementing through chemical reasoning the target, or subgoal by conceivable coproducts to form an adequate EM, (b) finding those basis elements of R-matrices which fit mathematically the BE-matrix of the latter EM, and (c) simultaneously taking into account the valence chemical boundary conditions for each of the participating atoms. Thus the combination of an elementary mathematical operation with two straightforward essentially chemical decisions leads to a result which is obtained by Hendrickson[9] through a single, somewhat more complex, chemistry-oriented process. Hendrickson's approach and ASSOR[3,10] generate the result of a mechanistic step by different procedures, but they achieve in essence the same: they both find for a given molecular system those "electron shifting arrows" which convert them into other molecular systems without violating some general chemical valence rules.

It is noteworthy that till recently all synthesis design
programs, e.g. EROS[11], followed the same general
pattern, the reaction library based ones, as well as
those with deductive features. Hitherto, all computer
assisted synthetic design was based on the <u>unilateral
retrosynthetic concept</u>, i.e. from the target of the
synthesis its precursors are generated, then their
precursors, etc., etc., until available starting
materials are reached. The retrosynthetic design of
syntheses, as well as the unilateral solution of any
problems involving sequences of chemical reactions,
produces trees of reaction pathways[11,12]. Despite
sophisticated strategies, such trees may often become
too large to be handled without fairly arbitrary
pruning in order to avoid the "combinatorial explosion"
that may result as the tree grows.

In contrast to the other known approaches, our mathematical model of constitutional chemistry affords also
the <u>bilateral solution of chemical problems</u>, i.e.
"from both ends" of reaction pathways. Thus networks
of reaction pathways may be obtained which are much
smaller than the respective trees. The bilateral
approach also avoids the relative overemphasis of
the last stages of a synthesis, which is characteristic
of unilateral synthesis design.

In this paper the advantages of the bilateral solution
of chemical problems will be discussed, and the essential concepts so well as the nature of the required
algorithms will be pointed out.

2. Brief Outline of Some Essential Features of Our Theory
   of Constitutional Chemistry[5]

Our mathematical model has been described in detail[5],
and it has also been discussed in several review
articles[3]. Accordingly, only a brief outline of our
theory will be presented here. Those particular features
of this theory will be pointed out which are essential
for the bilateral solution of chemical problems.

An EM is any collection of one or more molecules which
may all be different, or of the same kind. For example,
the educts of a chemical reaction may be called an EM.
An extension of the concept of isomerism from molecules
to EM leads to the FIEM, the family of all isomeric EM
which can be obtained from a given set $\underline{A} = \{A_1 \ldots A_n\}$

of n atoms by using each atom belonging to A exactly
once. A chemical reaction, or a sequence of chemical
reactions, is the conversion of an EM into an isomeric
EM. All EM which are chemically interconvertible belong
to the same FIEM. Thus the FIEM(A) contains the whole
chemistry of that EM which consists of collection A of
atoms. Since any collection of atoms may be chosen here,
a model of the logical structure of the FIEM applies to
all of chemistry.

From a chemical viewpoint, atoms consist of a core
(atomic nucleus and the electrons of the inner orbitals)
and the valence electrons (electrons of the outer
orbitals). In molecules, the cores are held together by
the valence electrons. The chemical constitution of a
molecular system is described in terms of its covalently
bonded pairs of cores. The distribution of the "free"
valence electrons which do not belong to covalent bonds
can also be considered as part of the constitution.
Customarily, the chemical constitution is described by
constitutional formulas. In these the atomic cores are
represented by element symbols, and covalent bonds by
lines connecting the element symbols. The placement of
the free valence electrons is generally indicated by
dots next to the element symbols.

In our model, the chemical constitution of an EM is
represented by the combination of its atomic vector,
which corresponds to the atoms in A, with an associated
(bond and electron) BE-matrix, whose rows and columns
belong to the individual atomic cores. The off-diagonal
entries are the formal covalent bond orders, and the
diagonal entries are the numbers of the free valence
electrons[5].

The row/column of a BE-matrix describes the distribution of valence electrons at the respective atomic
core. For a given core, only a limited number of such
distributions (valence schemes) is permissible. These
may be collected in a list of acceptable valence
schemes[13].

The n atoms of an EM can be numbered in up to n!
different ways, thus leading to up to n! distinguishable
but equivalent BE-matrices. By appropriate rules one of
these numberings can be declared canonical[14].

For a chemical reaction the following invariances hold which are based on the conservation laws of matter and charge:

> The atomic cores of an EM are preserved.

> The total number of the valence electrons

> of an EM is preserved.

It follows that the conversion of the starting materials EM(B) into the final products EM(E) through a chemical reaction

$$EM(B) \rightarrow EM(E)$$

is representable by those transformations $B \rightarrow E$ of BE-matrices in which the assignment of the rows/columns to the atomic cores, and the sum over all entries is maintained, since this sum is the total number of participating valence electrons.

We define the R-matrix R through $B + R = E$, the master equation of our theory.

Since the entries in $B = (b_{ij})$ and $E = (e_{ij})$ have the same sum, the sum of the entries $r_{ij} = e_{ij} - b_{ij}$ of the matrix $R = E - B$ must be zero.

The matrix R must be symmetric, because $E = B + R$ is a BE-matrix and must therefore by definition be symmetric, so from $b_{ij} = b_{ji}$ and $e_{ij} = e_{ji}$ follows $r_{ij} = r_{ji}$.

The off-diagonal negative entries $r_{ij} = r_{ji} = -1$ reflect the breaking of a covalent bond $A_i - A_j$; a negative diagonal entry r_{ii} indicates the number of free valence electrons that the atom A_i loses by a chemical reaction. Correspondingly, positive off-diagonal entries show how many bonds between the respective cores are made, and positive diagonal entries show how many free valence electrons the cores gain during the reaction.

An R-matrix R with the entries r_{ij} and the i-th row sum

$$r_i = \sum_j r_{ij}$$

is uniquely represented (up to sequence) as a sum of the R-matrices $U^{in} = P_{ii} - P_{nn}$ and $V^{ij} = P_{ij} + P_{ji} - P_{ii} - P_{jj}$, where P_{ij} is an n x n matrix with 1 in the (i,j) position and zero everywhere else; moreover

$$R = \sum_{i<j} r_{ij}V^{ij} + \sum_{i<n} r_i U^{in}$$

The set of the U^{in} and the V^{ij} forms a basis for the R-matrices. Since the R-matrices may be interpreted as vectors belonging to R^{n^2}, the U^{in} and V^{ij} are also called the basis vectors.

The vector basis of the R-matrices of reactions within the chemistry of valence shells with paired electrons only (with $r_{ii} = 0, \pm 2, \ldots$) consists of R-matrices of the type $(U^{ij} + V^{ij})$ and $2 U^{in}$. The basis vector $(U^{ij} + V^{ij})$ of closed shell chemistry represents hetero-aptic reactions of the types

$$A_i + :A_j \rightarrow A_i^{\ominus} - A_j^{\oplus} \qquad \text{and}$$

$$A_i^{\oplus} + :A_j^{\ominus} \rightarrow A_i - A_j,$$

while the basis vector $2 U^{in}$ corresponds to an elementary two-electron redox reaction

$$A_i^{\oplus} + :A_n^{\ominus} \rightarrow :A_i^{\ominus} + A_n^{\oplus}.$$

Per definitionem, a BE-matrix does not have any negative entries. Thus the transformation of a BE-matrix B according to B + R = E corresponds to a chemical reaction only, if we have for all entries

$$e_{ij} = b_{ij} + r_{ij} \geq 0,$$

i.e. the negative entries of an R-matrix must be chosen to satisfy

$$|r_{ij}| \geq b_{ij} \text{ for all } r_{ij} < 0.$$

This is called the mathematical fitting condition for R-matrices.

Moreover, the rows/columns of the resulting matrix E must agree with the acceptable valence chemical schemes of the chemical elements[13] to which the respective atomic cores belong, i.e. we have also "valence chemical boundary conditions" for the entries of the matrices.

An n x n BE-matrix B can also be represented as a vector b with n^2 components

$$b = (b_{11}, \ldots, b_{1n}; \ b_{21}, \ldots; \ldots b_{n1}, \ldots, b_{nn})$$

This results in an embedding of the n x n BE-matrices of an FIEM into $\mathbb{R}^{n^2}$, i.e. an n^2-dimensional metric space over the field of real numbers. The entries b_{ij} of B can also be interpreted as the cartesian co-ordinates of a point P(B) in $\mathbb{R}^{n^2}$. We call P(B) the BE-point of the BE-matrix B.

Likewise an R-matrix R corresponds to a vector r in $\mathbb{R}^{n^2}$. A chemical reaction λ, as given by B + R = E, is represented by the vector r from P(B) to P(E).

We call

$$d_1(B,E) = \sum_{i,j} |b_{ij} - e_{ij}| = \sum_{i,j} |r_{ij}|$$

$d_1(B,E)$ the <u>chemical distance</u>[5] between B and E, and also between EM(B) and EM(E); $d_1(B,E)$ is a metric defined in the space $\mathbb{R}^{n^2}$ the equivalent to the euclidean metric.

The sum $d_1(B,E)$ over the absolute values of the entries of R is twice the number N of valence electrons which are redistributed during the reaction, since these are taken into account by the negative, as well as the positive entries of R.

For simplicity, the factor two is neglected, and the term "chemical distance" is also used for the minimum number of valence electrons which must be shifted in order to convert EM(B) into EM(E).

Closely related to this distance, is the ordinary euclidean distance

$$d_2(E,B) = \sqrt{\sum_{i,j} (e_{ij} - b_{ij})^2} = \sqrt{\sum_{i,j} r_{ij}^2}$$

which does not have the same chemical meaning as the d_1-distance but there is the relation

$$d_2(E,B) \leq d_1(E,B) \leq n\, d_2(E,B)$$

which is always valid. Tighter bounds can be obtained if some assumption about the r_{ij} is made:

(a) All the $|r_{ij}| = 0$ or 1. Say that there are precisely α entries $|r_{ij}| = 1$. Then

$$d_1(E,B) = \sum_{i,j} |r_{ij}| = \sum_{i,j} |r_{ij}|^2 = \alpha = [d_2(E,B)]^2. \text{ Thus}$$

$d_1(E,B) = \alpha$ and $d_2(E,B) = \sqrt{\alpha}$, so $d_2(E,B) = \sqrt{d_1(E,B)}$

(b) All the $|r_{ij}|$ are either 0, 1 or 2. If there are α entries $|r_{ij}| = 1$ and β entries $|r_{ij}| = 2$, then

$$d_1(E,B) = \sum_{|r_{ij}| = 1} |r_{ij}| + \sum_{|r_{ij}| = 2} |r_{ij}| = \alpha + 2\beta$$

$$d_{2(E,B)}^{2} = \sum_{|r_{ij}|=1} |r_{ij}|^2 + \sum_{|r_{ij}|=2} |r_{ij}|^2$$

$$= \alpha + 4\beta = ! + 2\beta + 2\beta$$

$$d_2(E,B) = \sqrt{d_1(E,B) + 2\beta}$$

(c) All the $|r_{ij}|$ are either 0, 1, 2 or 3. If there are α entries $|r_{ij}|$, and β entries $|r_{ij}| = 2$, and γ entries $|r_{ij}| = 3$, then

$$d_1(E,B) = \sum_{|r_{ij}|=1} |r_{ij}| + \sum_{|r_{ij}|=2} |r_{ij}| + \sum_{|r_{ij}|=3} |r_{ij}|$$

and

$$d_2(E,B)^2 = \alpha + 4\beta + 9\gamma = \alpha + 2\beta + 3\gamma + 2\beta + 6\gamma$$

$$= d_1(E,B) + 2\beta + 6\gamma$$

so

$$d_2(E,B) = d_1(E,B) + 2\beta + 6\gamma$$

Since the cases (a) - (c) cover the chemically meaningful entries of R-matrices, this indicates which values
of d_2 above its minimum must be scanned in order to
cover the chemically meaningful parts of min d_1. We
explicitly remark that a, β, γ depend on the R-matrices
used. An EM consisting of n atoms is representable in
up to n! distinct ways by BE-matrices. Any two of these
differ by row/column permutation which correspond to
the permutation of the atomic core indices. Thus, the
BE-matrices are determined by the EM only up to a
permutation, in that two matrices B, B' represent the
same EM if there is a permutation matrix P such that
$B' = P^t BP$ with P^t being the transposed matrix of P.
We shall call the set $\{P^t BP \mid P$ a permutation matrix$\}$
the cluster determined by B; as we have remarked, all
the matrices in a cluster represent the same EM, but
with different labellings of the atoms in that EM.

Now let us assume that an EM(B) is convertible to an
EM(E); we wish to determine the minimal chemical
distance accomplishing this conversion. Due to the
non-uniqueness of the representation by BE-matrices,
this leads to the question:

Find matrices P^tEP and Q^tBQ as close as possible (by chemical distance); their difference P^tEP-Q^tBQ will be a reaction matrix and will fit Q^tBQ. That this search requires permuting the rows/columns of only one matrix, and that it has at least one solution follows from the

Theorem   Let δ_1 (resp. δ_2) be the d_1 (resp. d_2) distance between the clusters P^tEP and Q^tBQ. Then, given any B, there is at least one $P_i{}^tEP_i$ with $d_i(P^t{}_iEP_i,B) = \delta_i$ i = 1 or 2.

Proof   The sets P^tEP and Q^tBQ are finite, therefore compact. Using any metric for the spaces it is well-known that the distance between two compact sets is attained, so that there are two elements P^tEP and Q^tBQ with

$$\delta = d(P^tEP, Q^tBQ)$$

From the definition of the distance functions d_1 and d_2, they are invariant under row/column permutations, so, recalling that $Q^t = Q^{-1}$ for permutation matrices, we find

$$\delta = d(QP^{-1}EPQ^{-1}, B)$$

Since the product of permutation matrices is a permutation matrix, we have $QP^{-1} = L$ is a permutation matrix; and since

$$L^{-1} = PQ^{-1}$$ this gives

$$\delta = d[LEL^t, B]$$

which proves the theorem.

Thus, one can keep B fixed, and permute E alone, to
find a minimal reaction matrix. The minimal d_1-distance
to B may be obtained from any matrices $P^t EP$, as may
also the minimal d_2-distance.

If the reaction matrices found satisfy the above conditions (a), (b), or (c) which correspond to the
chemically significant cases, then the exact relation
of the d_1- and d_2-distances is given: the $P^t EP$ with
$d_1(P^t EP, B)$ = min will be found among those $P^t EP$ having
$d_2(P^t EP, B)$ = min. This result is of considerable
practical importance, because d_1 can only be minimized
by trial and error, whereas min d_2 can be found through
the maximum of the scalar product of E and $P^t BP$[21,22]
by employing a modified version of a quadratic
assignment algorithm which is used in operations
research[22b].

3. The bilateral Solution of Chemical Problems:

Within the framework of our theory[5] its master
equation B + R = E may be used in three ways for the
solution of chemical problems.

It is possible to start from a given EM(B), and to
generate the R-matrices which fit B mathematically
within the valence chemical boundary conditions. This
is a basis for the design of syntheses and the prediction of products which may be formed from given
starting materials.

The determination of all pairs (B,E) of BE-matrices
which mathematically fit a given R-matrix, under the
valence chemical boundary conditions, enables us to

predict chemical reactions in a systematic manner.

With two given BE-matrices, B and E, belonging to the same FIEM, the decomposition of the R-matrix $R = E - B$ into components R_1, R_2, ..., R_r such that $R = R_1 + R_2 + ... R_r$ yields information on the chemical pathways which lead from EM(B) to EM(E). Computer programs on this basis generate networks of reaction pathways which may consist of conceivable reaction mechanisms, or multistep syntheses, through a bilateral approach.

First, the beginning EM(B), and the end EM(E) for the reaction pathway to be analyzed must be stated. In the elucidation of reaction mechanisms and biochemical processes EM(B) and EM(E) are generally known, whereas in the design of syntheses these EM are not directly known, but must be deduced from the structure of the target and the list of the available starting materials. This can be accomplished by picking from a list of starting materials a subset which is convertible into the target compound of the synthesis, and some co-products. For a given target, the starting materials are found by correlating substructures in the target with substructures which occur in members of the list, by "putting together" the target from substructures belonging to starting materials. Thus, a heuristic association of the substructures in the target compounds and the starting materials is accomplished as a first guidance of the synthesis, providing some "total look-ahead". More than one suitable subset of starting materials may be found for a given target. Then one, or some, may be preferentially selected according to cost, or other considerations.

This requires a new type of substructure analysis. Recently we have implemented a computer program which suits the above purpose[15]. The particular structure of data file is the essential feature in the algorithmic solution of the problems guaranteeing short response times of a few seconds in an interactive system. The file consists of a hierarchically ordered network of sub-structures for each molecule. The networks of the individual molecules are connected by the substructures common to the members of the list, just like a "forest" of interlinking trees may be formed from a set of trees with common nodes[16]. In this list of molecular structures common substructures are generally shared by many members of the list; the smaller the substructures, the more likely they are to be common to large subsets of the list. Since each substructure is only represented once, large parts of the individual networks overlap as parts of the universal network. Thus the required storage and problem solving spaces are reduced considerably from that required in the disjoint treatment of individual molecules, characteristic of the known substructure search algorithms[17]. A particularly advantageous feature of this data structure is that its size increases relatively less and less, the more the new members are added to a given list.

In this program the usually time consuming and repeatedly performed search phase of comparable programs is avoided because we anticipate all steps of a conventional search algorithm once and for all in the initial generation phase of the data file[18]. In the correlation of substructures it suffices to look for the largest and smallest substructures which are common. In the search by substructures one enters the universal network at the relevant substructure nodes and moves toward the full molecular structures via the "son-father relations", until all molecular structures with the given substructure have been reached.

The conversion of EM(B) into EM(E) is only represented by B + R = E, if the assignment of the rows/columns in B to the atomic cores in EM(B) is compatible with that of E to the cores of EM(E), i.e. if any atom A_i belongs to the i-th row column in both BE-matrices, B and E. Since the atom by atom correlation of EM(B) and EM(E) is generally not given a priori, it must be established by a suitable procedure. A computer program for this purpose has been recently implemented on the basis of the <u>principle of minimum chemical distance</u>.

The PMCD was first proposed in 1973 as one of the results of our mathematical theory of constitutional chemistry[5]. In essence it says: In general, chemical reactions proceed along a pathway of minimal chemical distance, i.e. with a redistribution of a minimal number of valence electrons[20].

It is closely related to the principle of minimum
structure change, on which chemists have relied since
the introduction of the concept of chemical constitution[19]. The principle of minimum structure change
is an old and widely used heuristic rule in organic
chemistry. Since it has intuitive appeal, and neither
its content nor its scope have been explicitly defined,
its validity has never been questioned. Some of its
more recent corollaries are, however, still under
critical discussion[20]. The change in structure which
takes place through a chemical reaction involves
breaking and making of covalent bonds and a change in
the placement of the free valence electrons. The total
number

$$\frac{1}{2}d_1(E;B)$$

of valence electrons which must be redistributed in
order to achieve the conversion of an ensemble of
molecules EM(B) into an isomeric ensemble EM(E) is given
by the expression

$$\frac{1}{2}d_1(E,B) = {}^{(-)}c(B) + {}^{(+)}c(E) + \frac{1}{2}\left[{}^{(-)}f(B) + {}^{(+)}f(E)\right].$$

Here ${}^{(-)}c(B)$ and ${}^{(+)}c(E)$ are the numbers of the bonds
broken in E; (B) and bonds made in EM(E), while ${}^{(-)}f(B)$
and ${}^{(+)}f(E)$ are the numbers of free electrons which are
removed from atomic cores in EM(B) and added to atomic
cores in EM(E), respectively. Thus, $d_1(E,B)$ the chemical
distance between B and E (see Section 2) appears to be
well-suited as a quantitative measure for the constitutional changes that occur in a chemical reaction
EM(B) $\rightarrow$ EM(E), and the minimum of $d_1(B,E)$ corresponds to
minimum structure change. The number of valence
electrons which are redistributed during a reaction
whose beginning EM(B) and end EM(E) are given depends on
the correlation of the atoms in the participating EM.
When the atoms in EM(B) and EM(E) are identified by
indices, this correlation involves stating for each
indexed atom in EM(B) its index in EM(E). Thus with
arbitrary indices the atoms in EM(B) and EM(E) may be
correlated in up to $(n!)^2$ different ways. Without guiding
theory the trial-and-error search for that correlation
of the atoms which corresponds to min $d_1(E,B)$, and thus
minimum structure change, would require the determination
of $d(E,B)$ for up to $(n!)^2$ atom-onto-atom bijections of
EM(B) and EM(E). In a subsequent paper[21] it will be
outlined how an algebraic theory of constitutional

chemistry may serve to guide the search for those interconversions of isomeric EM which involve the redistribution of a minimum number of valence electrons (see
also Section 2). This leads to the mathematical justification of a very efficient algorithm[22] for determining those atom-onto-atom bijections of isomeric EM
which correspond to their interconversion by the redistribution of a minimal number of valence electrons.
The above algorithm is based on the fact that in general
the minimum of d_2-distance is found among those having
the minimal d_1-distance (see Section 2). It has been
shown to be useful for the solution of a wide variety
of problems in organic chemistry in fields like substructure correlation, the elucidation of reaction
mechanisms and the design of syntheses[20,21].

In fact, the PMCD has thus become one of the most powerful guidelines in the bilateral solution of chemical
problems.

The applications of the PMCD will be illustrated by the
following examples which were obtained through the MCD-
program[20].

Any atom by atom matching of two isomeric EM has a
characteristic chemical distance. In a matching with
minimal chemical distance we have maximum coincidence
of substructures. Therefore the PMCD may be used to
correlate substructures: The chemical similarity of two
isomeric EM in terms of common substructures is obtained
by an atom-onto-atom mapping under the aspect of minimal
chemical distance. The PMCD leads to a well-defined
procedure for finding substructures (largest set of
largest substructures) which are common to a pair of EM,
as is illustrated by the following example.

Intuitively, a six-membered ring might be seen as the
largest substructure common to molecules 1 and 2.

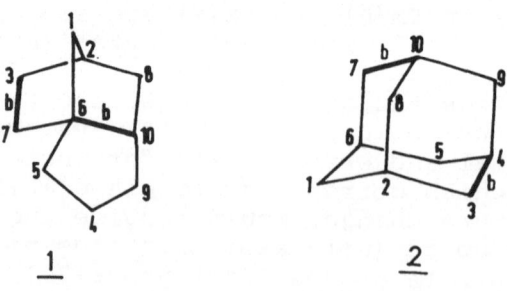

The conversion of 1 and into their largest common substructure according to the PMCD can be achieved by
breaking the bonds labelled by "b". This corresponds to
an atom-onto-atom mapping as shown by the indexings of
structures 1 and 2. As can be seen from this result,
1 and 2 have a disubstituted eight-membered ring in
common (1-2-8-10-9-4-5-6-1) as the largest substructure.
      | |
      3   7

The correlation of the atoms in the educts and products
of a reaction often depends on the reaction mechanism.
Conversely, conclusions about the mechanism can be drawn
if one knows for each atom in the products the corresponding atom in the educts. This is the working basis
for any elucidation of reaction mechanisms by isotopic
labelling. In essence our present computer program
labels all participating atoms and finds a correlation
of labellings which corresponds to a minimal chemical
distance.

In the elucidation of reaction mechanisms the minimal
chemical distance is a measure for its minimal mechanistic complexity, because the minimal chemical distance
is twice the minimal number of arrows which are needed
for describing the reaction by an "electron shifting
pattern". For a given overall reaction there may be
more than one mechanistic pathway corresponding to the
same minimal chemical distance. The preparation of
adamantane required laborious multistep syntheses until
Schleyer's discovery of the conversion of the readily
available tetrahydrodicyclopentadiene 3 into adamantane
4.

        3   4

This conversion proceeds directly from tetrahydrodicyclopentadiene to adamantane through a sequence of
carbonium ions without any identifiable intermediates,
although a multistep sequence of rearrangements must
undoubtedly be involved.

The mechanism of this rearrangement presents an intriguing and not yet completely solved problem.
Whitlock et al.[24] proposed a network of conceivable
pathways, which consists of 1 - 2 shifts only. One of
the pathways was selected by Schleyer et al.[23] as
the likeliest one through molecular mechanics calculations. With the PMCD, new insights in this process can
be obtained. Beginning with an arbitrary indexing of
tetrahydrodicyclopentadiene 3, we find an indexing of
adamantane 4 which correspond to a minimal chemical
distance of eight. If the reaction proceeded from
hydrocarbon to hydrocarbon according to a pathway of
PMCD, the bonds indicated in bold face in 3 would be
broken "b" and those "m" in 4 be made. The numbering
of the atoms in adamantane indicates their origin in
tetrahydrodicyclopentadiene. In comparison, the sum of
the chemical distance of the individual steps of the
pathways suggested by Schleyer[23] is twenty.

The PMCD is not only applicable to chemical reactions
but also to sequences of reactions such as multistep
syntheses. Here the PMCD provides guidance in the design
of multistep syntheses.

Most of the great natural product syntheses seem to have
been designed according to the principle of minimum
structure change. Syntheses following this principle
result from the human mind's striving for straightforward solutions to problems.

Under given conditions the mechanistic pathways for
chemical reactions are determined by the educts and
products, whereas, in synthesis, the pathways for the
conversion of given starting materials into the desired
synthetic target, are chosen by the chemist. We con-
jecture that in the design of syntheses the best
synthetic routes from a given EM(B) to EM(E) will be
found among those for which the characteristic overall
chemical distance is equal, or close, to the minimal
chemical distance. Synthetic pathways with a chemical
distance close to the minimum go directly from the
starting material to the target without significant
detours. However, we wish to emphasize that minimal
chemical distance is just one criterion for the
evaluation of syntheses. A synthesis along a pathway
of minimal chemical distance may not necessarily be
feasible, or best in yield. Sometimes a pathways with
a moderate detour may be more effective in practice.

In order to evaluate the performance and merits of the PMCD-program for synthetic design, we applied the program to the target and its coproducts, using the known starting materials of various published syntheses.

Because of its interesting biological properties and its unusual structure, colchicine has attracted much attention from synthesis chemists. It has been synthesized by various research groups in substantially different ways. From these syntheses we chose five [24-29]. These involve rather complex chemical changes. For none of the investigated sequences of reactions can one tell off-hand the pattern according to which the precursors are incorporated into the target molecule. Within each of these syntheses a sequence of steps was selected which reflects the essential concept of the whole synthesis.

Four of these five most elegant syntheses[24-28] proceed in agreement with the PMCD, whereas the colchicine synthesis of Eschenmoser et al.[29] demonstrates that in synthesis the PMCD is not a rule without exception. The essential overall changes in this synthesis correspond to the breaking and making of bonds according to the scheme below, with a chemical distance of 20.

Our program finds a minimal chemical distance of 14, if the same conversion was accomplished as follows:

In the formulation of the PMCD, we avoided carefully the categoric requirement that the result of each and any sequence of reactions, in particular the in vitro syntheses, must be achieved through a pathway of minimal chemical distance. A sequence of reactions must not necessarily proceed with the displacement of a minimal number of valence electrons. It may be well possible that for a given chemical conversion a pathway with a higher chemical distance is preferable by certain criteria, e.g. yield.

Despite of all the dramatic progress that organic synthesis has made in the past decades, the synthesis of strychnine by the late R.B. Woodward and his group [30] has remained one of the greatest achievements in the art of synthesis. We considered the complex problem of synthesizing strychnine 13 as a suitable test case for studying the scope and limitations of the PMCD-program.

The incorporation of veratrylindole 10 into strychnine is one of the ingenious features of this synthesis. First, we wanted to see how the program would use veratrylindole as a starting material for strychnine. For this, 10 was augmented by six carbon atoms and a nitrogen atom to yield an EM which has the same set of non-hydrogen atoms as strychnine. Two optimal matchings were obtained. One of them corresponds to the utterly novel incorporation of veratrylindole into strychnine by Woodward et al.[30]. This result could hardly have been achieved by any one of the customary substructure search programs.

In Woodward's synthesis, in addition to veratrylindole, two C_2-units, a C_1-unit were used for the constructions of the strychnine skeleton. As a second part of our study, we therefore investigated the matching of veratrylindole and these building blocks onto strychnine. Several matchings corresponding to the minimal chemical distance were found, and among these contained also the pattern of Woodward's synthesis.

This analysis shows (particularly) convincingly the potential of the PMCD in guiding the design of syntheses. It provides the most efficient schemes of bond breaking and making as a basis for synthetic macrostrategies.

The atom-by-atom matchings and "electron shifting" patterns of multistep syntheses are obtained by minimization of chemical distance with complete neglect of the stereochemical aspect. It should, however, be noted that efficient synthetic pathways which also account for the stereochemistry should belong to the set of minimal chemical distance pathways. The examples which we have analyzed confirm this.

Once the atom-by-atom matching of the beginning and end of a chemical conversion, EM(B) → EM(E) has been established, it is representable by an "overall R-matrix" R = E − B. The decomposition of this R-matrix into its components, or basis vectors, $R = R_1 + \ldots\ldots\ldots + R_r$, and the successive additive transformations of B under the mathematical fitting conditions and the valence chemical boundary conditions may be used to generate a network of reaction pathways that lead from EM(B) to EM(E).

This is illustrated by a pathway which was found with the aid of the basis vector oriented program ASSOR within the network between histidine 14 plus 2 water as EM(B) and formyl isoglutamine 15 plus ammonia as EM(E).

This pathway corresponds to an elucidated metabolic process[31].

Our preliminary results indicate that the stepwise bilateral solution of chemical problems will probably become a most efficient, widely usable approach. The three steps, namely

   the definition of the initial and final EM,

   their atom-by-atom matching according to the PMCD,

   and the production of a network of reaction pathways from the components of an overall R-matrix

corresponds to a hierarchic, chemically meaningful order of the conceivable solutions of given chemical problems.

We wish to acknowledge gratefully the financial support of our work in the past decade by Stiftung Volkswagenwerk e.V. the Deutsche Forschungsgemeinschaft, the European Communities and the Bundesministerium für Forschung und Technologie (project administrator: Gesellschaft für Information und Dokumentation), Fonds der Chemischen Industrie, and the Alexander von Humboldt Foundation through a Senior Scientist Award to one of us (J.D.).

Abbreviations:

BE-matrix = bond and electron matrix, also beginning
                 and end matrix.

EM = ensemble of molecules.

FIEM = family of isomeric EM.

PMCD = principle of minimum chemical distance.

R-matrix = reaction matrix.

References

1. J. Lederberg, G.L. Sutherland, B.G. Buchanan,
    E.A. Feigenbaum, A.V. Roberts, A.M. Duffield,
    C. Djerassi, J. Amer. Chem. Soc. 91, 2973 (1969);
    see also: N.J. Nilsson "Problem-Solving Methods in
    Artificial Intelligence", McGraw-Hill, N.Y. (1971).

2. a) E.J. Corey, W.T. Wipke, Science 166, 178 (1969);
    E.J. Corey, W.T. Wipke, R.D. Cramer, W.J. Howe,
    J. Amer. Chem. Soc. 94, 421, 431 (1972);

    b) H. Gelernter, N.S. Sridharan, A.J. Hart,
    S.C. Yen, F.W. Fowler, J.J. Shue, Top. Curr. Chem.
    41, 113 (1973); H. Gelernter et al., Science 197,
    1041 (1977).

    c) W.T. Wipke, "Computer Planning of Research in
    Organic Chemistry", Plenum Press, N.Y. 1976; see
    also: C ∤ EN 57 (6 - 18) 29 (1979);

    d) I. Ugi, G. Kaufhold (unpublished 1966 - 67),
    described in ref. 4a and in I. Ugi, Rec. Chem.Progr.
    30, 289 (1969).

3. I. Ugi, J. Bauer, J. Brandt, J. Friedrich,
    J. Gasteiger, C. Jochum and W. Schubert, Angew.Chem.
    91 , 99 (1979, Angew. Chem. internat. Edit. 18, 111
    (1979).

4. a) I. Ugi, Intra-Sci. Chem. Rep. 5, 229 (1971).

    b) I. Ugi, P.D. Gillespie, Angew. Chem. 83, 982, 990
    (1971); Angew. Chem. internat. Edit. Engl. 10, 914,
    915 (1971).

    c) I. Ugi, P.D. Gillespie, Ann. N.Y. Acad. Sci. 34,
    416 (1972); see also: I. Ugi, IBM-Nachr. 24, 180
    (1974); I. Ugi, J. Gasteiger, J. Brandt, J. Brunnert
    W. Schubert, IBM-Nachr. 24, 185 (1974).

5. J. Dugundji, I. Ugi, Top. Curr. Chem. 39, 19 (1973);
    see also: I. Ugi, J. Bauer, J. Brandt, J. Friedrich,
    J. Gasteiger, C. Jochum, W. Schubert, J. Dugundji,
    Inform. Commun. Math. Chem., in press.

6. J. Blair, J. Gasteiger, C. Gillespie, P. Gillespie, I. Ugi in: "Computer Representation and Manipulation of Chemical Information", eds. W.T. Wipke, S. Heller R. Feldmann, E. Hyde; Wiley, New York, 1974, p. 129.

7. A. Weise, Z. Chem. 15, 33 (1975).

8. W.T. Wipke, H. Braun, G. Smith, F. Chiplin and W. Sieber in "Computer Assisted Organic Synthesis", ACS Symposium Series 61, 97 (1977).

9. J.B. Hendrickson, J. Amer. Chem. Soc. 93, 6854 (1974); Top. Curr. Chem. 62, 49 (1976).

10. W. Schubert, Doctoral Thesis, Technische Universität München, 1978; Inform. Commun. Math. Chem., in press.

11. J. Gasteiger and C. Jochum, Top. Curr. Chem. 74, 93 (1978).

12. J. Dugundji, P. Gillespie, D. Marquarding, I. Ugi and F. Ramirez, in "Chemical Applications of Graph Theory", A. Balaban, ed., Academic Press, N.Y. 1976, p. 107.

13. J. Brandt, J. Friedrich, J. Gasteiger, C. Jochum, W. Schubert, P. Lemmen, I. Ugi, Pure and Appl. Chem. 50, 1303 (1978).

14. W. Schubert and I. Ugi, J. Amer. Chem. Soc. 100, 37 (1978); Chimia 33, 183 (1979).

15. J. Friedrich and I. Ugi, J. Chem. Res., submitted (this paper will include a general description of the procedure, as well as a computer readable full documentation of the program).

16. A.V. Aho, J.E. Hopcraft and J.D. Ullmann, "The Design of Computer Algorithms", Addison-Wesley, Reading, Mass. 1974.

17. J.E. Ash and E. Hyde, "Chemical Information Systems" Wiley, N.Y. 1975.

18. J. Friedrich, Doctoral Thesis, Technische Universität München, 1979.

19. I. Ugi, Giessener Universitätsbl. 11, 68 (1978).

20. C. Jochum, J. Gasteiger and I. Ugi, Angew. Chem.,
    in press.

21. J. Dugundji, C. Jochum, J. Gasteiger and I. Ugi,
    J. Amer. Chem. Soc., in preparation.

22. a) C. Jochum, Doctoral Thesis, Technische Universi-
    tät München, 1978.

    b) R.E. Burkard, "Methoden der ganzzahligen Opti-
    mierung", Springer-Verlag, Wien, 1972.

23. E.M. Engler, M. Farcasiu, A. Sevin, J.M. Cense and
    P. v. R.Schleyer, J. Amer. Chem. Soc., 95, 5768
    (1973).

24. H.W. Whitlock jr. and M.W. Siefken, J. Amer. Chem.
    Soc. 90, 4929 (1968).

25. A.I. Scott, F. McCapra, R.L. Buchanan, A.C. Day and
    D.W. Young, Tetrahedron, 21, 3605 (1965).

26. E.E. van Tamelen, T.A. Spencer, D.S. Allen and
    R.L. Orvis, Tetrahedron, 14, 8 (1961).

27. R.B. Woodward, A.A. Patchett, D.H.R. Barton, D.A.J.
    Ives and R.B. Kelly, J. Chem. Soc., 1957, 1131.

28. D.A. Evans, D.J. Hart and P.M. Koelsch, J. Amer.
    Chem. Soc., 100, 4593 (1978).

29. J. Schreiber, W. Leimgruber, M. Pesaro, P. Schudel,
    T. Threlfall and A. Eschenmoser, Helv. Chim. Acta,
    44, 540 (1961).

30. R.B. Woodward, M.P. Cava, W.D. Ollis, A. Hunger,
    H.U. Daeniker, K. Schenker, J. Amer. Chem. Soc.,
    76, 4749 (1954); Tetrahedron, 19, 247 (1963).

RECENT DEVELOPMENTS IN COMPUTATIONAL CHEMISTRY IN THE U.S.:

THE NRCC (NATIONAL RESOURCE FOR COMPUTATION IN CHEMISTRY)

William A. Lester, Jr.

National Resource for Computation in Chemistry
Lawrence Berkeley Laboratory
University of California
Berkeley, California 94720

BACKGROUND

The National Resource for Computation in Chemistry (NRCC) officially came into being on October 1, 1977, at Lawrence Berkeley Laboratory (LBL). The concept of the NRCC evolved over many years and was the subject of many national committee and workshop studies.[1] The National Science Foundation (NSF) and the U.S. Energy Research and Development Administration (ERDA)[2] joined forces to sponsor the NRCC and to site it at LBL.

MISSION

The NRCC, the nation's first major collective effort in the field of chemistry, provides computational facilities and personnel dedicated to advancing chemistry and related sciences through widespread, innovative, and intensive use of high-speed computational equipment. This mission is accomplished by making appropriate facilities available to a wide group of scientists, by providing and developing software to expedite and upgrade computer use, by encouraging and supporting research efforts to build new and more effective computational methods, and by carrying out an informational and educational program to bring the benefits created through the center to the widest possible scientific public.

INTERNAL STRUCTURE

The NRCC Policy Board is the governing body of the organization, and is broadly representative of the chemical community in the United States. Its members are selected by the LBL Director in consultation with DOE's Division of Basic Energy Sciences and NSF. The Board has responsibility for NRCC policy and for setting scientific goals and priorities. The present members are listed in the Appendix.

The NRCC Program Committee is constituted to have wide representation of important areas in chemical computation. The Committee's principal responsibility is to review the scientific content of all major proposals to the NRCC and to recommend relative priorities among competing programs and proposals, within guidelines established by the Policy Board. The Policy Board is responsible for appointments to the Committee, which is chaired by the NRCC Director. The remaining members of the Committee are listed in the Appendix.

The NRCC Users Association consists of interested members of the user community of the NRCC. Membership is by request. By providing feedback to the Director and the Policy Board, the Association helps ensure that the NRCC is responsive to the needs of the chemical community. Operations of the User Association are governed by its Executive Committee. The members of the Executive Committee are listed in the Appendix.

COMPUTER CAPABILITY

The NRCC's computer capability is based primarily on LBL's CDC 7600 computer front-ended with a CDC 6600 and CDC 6400. In addition, a minicomputer, VAX 11/780, was recently purchased to facilitate the development of portable software for both minicomputers and large mainframes.

The NRCC has a small in-house staff of theoretical chemists who also have considerable experience in computer science. The staff adapts programs to the LBL system, develops and codes new algorithms, collaborates with the user community in code development and application, and performs basic research.

The NRCC has considerable flexibility in arranging computer assistance for users. Computer center consultants are available to assist the user at his or her home laboratory.

ACTIVITIES

Workshops

One of the ways NRCC interacts with the chemical community is through the workshops it sponsors. The workshops are directed towards:

- Important and unsolved computational problems in particular areas.

- Cooperative efforts to develop software.

- Effective usage of major computer programs available at the facility.

The workshops provide guidance to the staff on future directions for the NRCC, help minimize duplicative effort by different groups, and serve to inform chemists on the use of programs designed for theoretical calculations.

Proceedings for the workshops held in 1978 are available upon request. These workshop titles are:

- Computational Methodology in Crystallography: Evaluation and Extension, organized by Professor David Templeton, University of California, Berkeley, and LBL, and Dr. Carroll Johnson, Oak Ridge National Laboratory

- The Minicomputer and Computations in Chemistry, organized by Professor Phillip Certain, University of Wisconsin

- Numerical Algorithms in Chemistry: Algebraic Methods organized by Professor Cleve Moler, University of New Mexico and Dr. Isaiah Shavitt, Batelle Columbus Laboratories

- Post Hartree-Fock: Configuration Interaction, organized by Dr. Charles Bender, Lawrence Livermore Laboratory and Professor John Pople, Carnegie-Mellon University

Workshop activity in 1979 has been vigorous. A two-part workshop on "Algorithms and Computer Codes for Atomic and Molecular Quantum Scattering Theory," co-chaired by Professor John Light, University of Chicago, and Dr. Lowell Thomas, NRCC, was organized. The purpose of this workshop is to encourage the identification, development, testing, and documentation of the best codes in the subject area and to make them available through the NRCC software library. A number of methods have been developed for solving the coupled-channel equations of quantum scattering theory which are not available to the general scientific community. At the first meeting on June 25-27, 1979 at Argonne National Laboratory, key workers in this field came together to discuss testing their algorithms against a fixed set of problems developed by Thomas in consultation with other experts on the subject. The second meeting is planned for the end of this week, September 20-21, 1979 at the NRCC to compare and evaluate the performance of the algorithms. All computations for this workshop are being carried out under NRCC support on the LBL CDC 7600 to avoid ambiguities in accuracy and efficiency that might arise from the use of different hardware.

A workshop on "Stochastic Molecular Dynamics" was held July 9-13, 1979 at Woods Hole, Massachusetts, organized by Dr. John Tully of Bell Laboratories. The workshop brought together a small group of workers active in developing and applying methods which incorporate effects of surrounding media and reduce the number of degrees of freedom in molecular dynamics simulations through the introduction of friction and fluctuating forces into the equations of motion. Prospects for application of this approach for studying condensed phase processes, polymer dynamics, surface chemistry and gas phase chemistry were explored. The proceedings of this workshop is in press and will be available for distribution in a couple of months.

A one and a half day workshop on "Array Processors for Chemical Computations" was arranged by Professor Neil Ostlund, University of Arkansas, to provide additional input for a report that the NRCC had commissioned Ostlund to prepare on the topic. His report is in press and provides a timely pedagogical presentation on array processors, per se, and their present and future use in computational chemistry. As with all NRCC reports, upon completion it will be available upon request.

The NRCC sponsored a workshop on "Software Standards for Chemical Computation at the University of Utah, Salt Lake City, Utah, on July 25-27, 1979. Dr. Frank Harris of the University of Utah was the conference organizer. The conferees considered the broad topics of:

1.   Computer system specifications. Definition of the hardware and system-software configurations to which chemistry software standards are designated to apply. These specifications may differ somewhat in different fields of chemistry, but are important because they will influence the ease with which standardized software will be able to be used with machines of various sizes, architectures, and operating systems.

2.   **Modularization specifications**. Extent to which complex tasks should be broken into modules with standard interface structures.

3.   **Data-list specifications**. Design of generalized data formats permitting data generated by one module to be used easily by other modules.

4.   **Software portability specifications**. Standards to ensure a reasonable level of machine independence in software for chemical computation. Recommended methods for achieving portability.

5.   **Recommend standards for programming practices and documentation.** Level of documentation recommended for chemical computation software. Programming practices which will simplify the attainment of portability and program readability and reliability. Recommended portable utility-libraries.

A workshop joint with the Quantum Chemistry Program Exchange (QCPE) on "Computational Methods for Molecular Structure Determination: Theory and Technique" was held August 13-24, 1979 at the QCPE, Indiana University, Bloomington, Indiana. The workshop provided an introduction to the use of state-of-the-art computer codes for the semi-empirical and ab initio computation of electronic structure and geometry of small and large molecules for 54 participants. Besides lectures on theory embodied in codes, remote RJE and interactive terminal access to the Lawrence Berkeley Laboratory CDC 7600 was provided to give participants "hands on" experience with NRCC hardware and NRCC-QCPE software.

An NRCC workshop entitled "Problem of Long-Range Forces in Computer Simulation of Condensed Media" will be held January 8-11, 1980 in Menlo Park, California, and it is being co-chaired by Professor Harold Friedman, Department of Chemistry, and Professor George Stell, Department of Mechanical Engineering, both of the State University of New York at Stony Brook, New York.

The workshop will consist of seven half-day sessions. Each session will have a review lecture by an invited speaker to be followed by short talks and discussion. The session titles are:

1.   The long-range force problem in simulation of ionic systems.

2.   The long-range force problem in simulations of dipolar systems.

3. Effect of long-range forces on solvation and biopolymer hydration simulations. Consideration of Coulomb and hydration forces.

4. Reaction field methods for long-range force problems.

5. Periodic perturbing fields to handle long-range forces; long-range force problems with polarizable molecules.

6. Long-range force problems in simulations of surface phenomena.

PROPOSALS

Calls for proposals go out twice each year. Proposals are presently sought for an allotment of computer time to be used at the LBL Computer Center or elsewhere, but proposals need not be limited to this form of support. Requests for support in other forms are reviewed by the Policy Board. In general, the Program Committee reviews the proposals, utilizing external reviewers where desirable, and awards funds. A proposal may be from any area of chemistry and should promote the development of chemistry by one of the following modes:

● The advancement of computational methods.

● The innovative application of existing methods.

● The study of problems of significant scientific merit that cannot be carried out in the proposer's institution due to lack of large-scale computational facilities.

● The resolution of intradisciplinary computational methodology issues.

● The development of interdisciplinary approaches to forefront problems in chemistry.

Users of the NRCC may be divided into two classes: clients and customers. Clients are those users whose research is funded by the NRCC. Clients are expected to make available to the other users the programs they develop under NRCC support. Customers are not required to contribute any new software. They pay for services from their own funds, usually derived from individual research grants. Customers are allocated computer time only after client requirements have been satisfied.

SOFTWARE

The NRCC software library has devloped steadily and the present holdings are described in recent issues of the NRCC BULLETIN, the quarterly news organ of the organization.

Codes are installed at the NRCC with particular emphasis on:

- Portability, insofar as this is practicable and consistent with the need to fine-tune codes for efficient CDC 7600 execution.

- Dynamic storage allocation to allow programs to efficiently adjust their run-time storage requirements to the amounts of memory available.

- Modularity and open-endedness.

- Standard data interfaces between program modules.

- Documentation.

- Testing and certification for "robustness".

The NRCC is also interested in the development of specialized software and higher-level languages particularly suited for chemical computations. One example is the use of specialized user interface languages for setting up (staging) the calculation of energies and wave functions of atoms and molecules.

A unique capability that has been discussed for the NRCC is the establishment of a computational data base of the accumulated results of various types of computations, such as molecular wave functions and properties. Such data should be of considerable value to other scientists.

Following is a brief description of software holdings in selected areas.

Chemical Kinetics - L. D. Thomas, W. A. Lester, Jr.

The goal for the chemical kinetics software library in the first year was to gather a representative collection of programs for the most commonly used approximation methods for computing atom-molecule and molecule-molecule collision cross sections. These include the quasiclassical (3), close coupling (4), coupled states (5), effective potential (6), ℓ-dominant (7) and IOS (8) methods.

The library now includes quasiclassical programs for A + BC (rigid - rotor and vibrotor) inelastic collisions and A + BC reactive collisions. Research is now being done under an NRCC award for the development of a quasiclassical program for polyatomic target molecules.

For spherically symmetric potentials, the library now contains programs for the quantum mechanical calculation of scattering phase shifts, collisional time delays, orbiting resonances, shape resonances and bound states.

For inelastic quantum mechanical scattering, Dr. Sheldon Green's program MOLSCAT, is now in the library. This program includes all of the above-mentioned quantum mechanical approximation methods for any of the following collision partners: atom - linear rigid rotor, atom - linear vibrotor, atom - symmetric top rotor, and linear rigid rotor - linear rigid rotor. In addition to the question of which approximate method to use, there remains the question of how to best integrate the resulting set of coupled second-order differential equations. This is an active area of research and many methods have been developed. The program MOLSCAT allows the user the option of either the Gordon (9) or the de Vogelaere (10) methods.

Quantum and Physical Organic Chemistry - M. Dupuis, S. A. Hagstrom D. P. Spangler and J. J. Wendoloski

Much of the in-house program development activity in the area of molecular electronic structure calculations centers around the HONDO system of programs introduced by H. F. King and coworkers[11] and subsequently extended by M. Dupuis and the NRCC staff. This program features the Rys polynomial method[12] for computation of electron repulsion integrals over s, p, d, f, and g-type cartesian Gaussian orbitals. Algorithms for calculation of closed-shell Hartree-Fock (HF), open-shell unrestricted Hartree-Fock (UHF) and open-shell restricted Hartree-Fock are included along with computation of the energy gradiants. Geometry optimization and force constant matrix calculations, as implemented by A. Komornicki[13], are also provided. The program also takes advantage of molecular symmetry, as described by King[14]. In an ongoing process of refinement, a number of improvements have recently been made to HONDO including (1) incorporation of GAUSSIAN 76 integral techniques for s- and p-type orbitals, (2) refinements to the SCF convergence procedures, (3) a perfect pairing approximation (essentially the Generalized Valence Bond approximation)[15] with gradients, and (4) improved eigenvalue methods.

A multi-configuration self-consistent field (MCSCF) capability is planned for HONDO. The exponential transformation method of B. Levy[16], D. Yeager[17], and C.C.J. Roothaan[18] will be used. In this method an exponential transformation is applied to the molecular orbitals as well as to the configuration expansion and a quadradically convergent iterative method, similar to the Newton-Raphson method, is used to optimize the orbitals and CI coefficients variationally without use of Lagrangian multipliers. The second derivative of the energy provides the information needed to determine the stability of the solution.

Finally, a general configuration interaction (CI) capability using the unitary group approach[19] (direct CI) is being added to the HONDO system of programs. The starting point for this effort is the code of Brooks and Schaefer[20] which has recently been installed on the LBL CDC 7600. Future refinements include the direct calculation of energy gradients at the CI level, which will have practical benefits in the study of chemical reactions, transition states, and force constants, as well as in the calculation of potential energy surfaces. A knowledge of both the energy and energy gradients should greatly reduce the number of ab initio points that need to be calculated.

HONDO is a highly portable program. Versions exist for IBM and for CDC machines (small core) as well as a highly optimized large core-small core version for the LBL CDC 7600. It is planned also to make the entire system available eventually on the DEC VAX-11/780 minicomputer.

Macromolecular Science and Statistical Mechanics - D. M. Ceperley

Dr. D. Ceperley has developed a program package CLAMPS (Classical Many Particle Simulator) which can perform many of the simulations of interest to statistical mechanicians. The version available now can handle up to five different types of particles interacting with spherically symmetric pair potentials. In addition, the particles can bond together into chains. The number of particles is only dependent on the computer resources available since all arrays are dimensioned at execution time. A bin-sorting scheme is used to reduce the computer time for simulations of large systems (more than 100 particles). Four different simulation algorithms have been programmed: molecular dynamics (i.e., Newton's equations of motion)[21], Metropolis Monte Carlo[22], polymer reptation (chains of particles are moved in a snake-like motion)[23], and Brownian Dynamics (equivalent to "Smart Monte Carlo")[24]. A single run of CLAMPS can consist of several of these types of algorithms intermixed, thus enabling the user, for example, to bring the system to equilibrium with Monte

Carlo and generate a trajectory with molecular dynamics. A number of analysis programs for the configurations generated in CLAMPS are also available (e.g., time-dependent correlation functions and the pair distribution function). The program is written in FORTRAN and has been tested on a CDC 7600 and a DEC VAX-11/780.

MINICOMPUTERS

Since fairly sophisticated minicomputers have come on the market, some researchers are using these machines extensively for problems in computational chemistry. The NRCC recently obtained a large minicomputer (VAX 11/780) in order to facilitate the development of machine-independent software and to explore the potential of such machines in all areas of computational chemistry. Many of the popular application codes currently running on the large CDC and IBM machines could be installed, at least in limited form, on many of the larger minis available today. For many of these machines, the available operating system software for interactive access and the mode of accommodation of very large overlayed and segmented programs leaves much to be desired. In fact, with the recent advent of inexpensive main memory and disk storage, the chief limitations of minicomputers are the lack of sophisticated operating system software (and this is beginning to be remedied), limited applications software, and slow execution speed. For a large range of smaller problems (10 hours of CDC 7600 time), however, slow execution speed is less important than it might seem since the elapsed times on the minicomputer are still manageable (100 hours) and the cost may be considerably less. Truly, large computations become impractical on minicomputers merely because the elapsed time becomes too great.

It would appear that for the near future, at least, the absence of applications software will be a significant hindrance in the use of minicomputers in computational chemistry. Therefore, the NRCC is undertaking a study to identify those codes that should be adapted to these small machines. The availability of such software will further increase the range of practical applications of minicomputers in computational chemistry.

In addition to its use as a computational tool, the NRCC minicomputer will also serve as a front end to the LBL CDC 7600/6600/6400 complex to provide the resident staff and a limited number of remote users with text editing and interactive program development capabilities. It will also provide RJE access to the CDC 7600 and associated mass storage devices.

A minicomputer project already underway stems from the NRCC Conference on Software Standards held in Salt Lake City in late July. At that meeting a group of crystallographers recommended that the NRCC sponsor the development of crystallographic code that has a standard data format and is portable from computer to computer. In order to demonstrate the feasibility of the idea, the NRCC will hold a workshop in early November 1979 at which 10 scientists will come together for one week in order to develop and code a multiple isomorphous replacement (MIR) program for distribution among crystallographers. The code, which is particularly useful to protein crystallographers, will be developed on the NRCC's VAX 11/780 and will be tested on other computers to ensure portability.

GRAPHICS

A desired component for the NRCC is high performance graphics capability: capability for displaying in spatial terms the two- and three-dimensional results of molecular calculations; capability for demonstrating the time dependent behavior of dynamic simulations; capability for modeling and manipulating molecular structures and assemblies in order to understand the complex interrelationships that govern their properties and interactions; in short -- capability for visualizing and integrating molecular information. We plan to pursue this molecular graphics capability at the NRCC through the acquisition of a commercially available state-of-the-art vector refresh processor and display unit to be interfaced with our VAX-11/780 minicomputer.

The establishment of molecular graphics at the NRCC will serve two broad purposes: 1) the development, standardization and evaluation of graphics tools, and 2) the application of graphics tools to user problems. The development of high performance molecular graphics systems is an active area of research in roughly a dozen laboratories world-wide. Upward of 100 man-years of effort have gone into these developments. Clearly, this effort and experience should be preserved through future hardware generations. One of the long-range goals of the NRCC is the design of a molecular graphics system aimed at achieving a synthesis of existing systems. The realization of this design would not only utilize a body of work already produced, but would also facilitate incorporation of new ideas emerging from research done on any of the existing facilities. A significant component of this project would be the standardization of graphics codes that will be produced with the intent of giving them a reasonable degree of portability.

RESEARCH

National Theoretical Group on Combustion

On June 28, 1979, an informal meeting of primarily quantum chemists and collision dynamicists was held at Argonne National Laboratory under the aegis of the NRCC to explore developing a collaborative arrangement to advance theoretical contributions in combustion. The twelve participants from Basic Energy Sciences-supported groups at Ames, Argonne, Brookhaven, Lawrence Berkeley and Sandia Livermore Laboratories agreed to pursue a course of research jointly to elucidate the electronic structure and collision dynamics of the system of reactions involving one atom each of C, H, and O. This activity is coordinated by the NRCC.

OTHER

In addition to this effort and the large base of NRCC supported external research, the NRCC staff members have been able to initiate a limited number of in-house research investigations. A brief description of somne of these activities follows.

Chemical Kinetics - L. D. Thomas

Research in the chemical kinetics area is currently focused on the computation of atom-molecule collision cross sections and has the following goals. 1) Improve the convergence of the recently developed iterative method[25] for a single column of the S-matrix through the use of a non-diagonal Green function matrix. 2) Extend the method to electron-molecule and reactive collisions. And 3) study the combustion research related problem of
$$O + CH \rightarrow CHO(+) + e(-).$$

Macromolecular Science - D. M. Ceperley

Simulation of polymeric materials. In a collaborative effort with M. Bishop, M. H. Kalos and H. Frisch, more efficient algorithms to generate equilibrium configurations of bulk polymer systems (reptation Monte Carlo)[23] have been developed and are beginning to get results on these systems to test against the theories of Flory, de Gennes[26] and others. In this work, we are interested in studying the variation in shape of a single polymer chain as the concentration of chains increases and the dependence of the correlations on the number of monomers which make up the chain. In addition, we wish to examine the dynamics of a concentrated polymeric system. It is believed by some that because of entanglements, diffusion can only occur with a reptating motion[26]. It should be possible to observe such behavior in a molecular dynamics simulation.

Statistical Mechanics – D. M. Ceperley

Simulations of quantum many body systems. Monte Carlo methods have been used to calculate exact ground state properties of boson systems[27] such as superfluid helium 4. We are extending these methods to fermion systems and have completed a study of the ground state of the electron gas[28]. Ceperley, joint with M. H. Kalos and B. Alder, is currently exploring the application of these methods to quantum chemical problems. Their long-range goal is to be able to simulate a molecular system, such as water, without constructing effective two molecule potentials.

THE FUTURE

The NRCC is now operating in Phase I. In the Spring of 1980, the NRCC will be reviewed by its two funding agencies -- the Department of Energy and the National Science Foundation -- to determine the desired form of operation in Phase II.

FOOTNOTES AND REFERENCES

1. Reports of the National Academy of Sciences studies are contained in "Needs and Opportunities for the National Resource for Computation in Chemistry", 1976 available from the Office of Chemistry and Chemical Technology, National Research Council, 2101 Constitution Avenue, Washington, D.C. 20418

2. The October 1 date for start-up of the NRCC coincided with that for start-up of the U.S. Department of Energy (DOE), into which the U.S. Energy Research and Development Administration (ERDA) was incorporated.

3. R. N. Porter and L. M. Raff, Dynamics of Molecular Collisions, edited by W. H. Miller (Plenum, New York, 1976).

4. A. M. Arthurs and A. Dalgarno, Proc. R. Soc. (London) Ser. A 256, 540 (1960).

5. P. M. McGuire and D. M. Kouri, J. Chem. Phys. 60, 2488 (1974).

6.   H. Rabitz, J. Chem. Phys. <u>57</u>, 1718 (1972).

7.   A. DePristo and M. H. Alexander, J. Chem. Phys. <u>63</u>, 3552 (1975), 63, 5327 (1975, and 64, 3009 (1976).

8.   R. Goldflam, S. Green and D. J. Kouri, J. Chem. Phys. <u>67</u>, 4149 (1977); extensive references can be found here.

9.   R. G. Gordon, J. Chem. Phys. <u>51</u>, 14 (1969).

10. W. A. Lester, Jr., Methods Comput. Phys. <u>10</u>, 211 (1971).

11. M. Dupuis, J. Rys, H. F. King, J. Chem. Phys. <u>65</u>, 111 (1976).

12. H. F. King and M. Dupuis, J. Comp. Phys. <u>21</u>, 144 (1976)

13. J. W. McIver and A. Komornicki, J. Am. Chem. Soc. <u>94</u>, 2625, (1972).

14. M. Dupuis and H. F. King, Inter. J. Quan. Chem. <u>11</u>, 613 (1927).

15. F. W. Bobrowicz and W. A. Goddard, <u>Methods of Electronic Structure Theory</u>, edited by H. F. Schaefer, III (Plenum, New York, 1976).

16. B. Levy, Chem. Phys. Letter <u>18</u>, 59 (1973).

17. D. L. Yeager and P. Jorgenson, J. Chem. Phys. <u>71</u>, 755 (1979).

18. C.C.J. Roothaan, J. Detrich, D. G. Hopper, Inter. J. Quan. Chem. (to be published).

19. I. Shavitt, Inter. J. Quan. Chem. Symposium 11 (1977).

20. B. R. Brooks and H. F. Schaefer, J. Chem. Phys. <u>70</u>, 5091 (1979).

21. L. Verlet, Phys. Rev. <u>159</u>, 98 (1967).

22. M. Metropolis, A. W. Rosenbluth, M. N. Rosenbluth, A. N. Teller, and E. Teller, J. Chem. Phys. <u>21</u>, 1087 (1953)

23. F. T. Well and F. Mandel, J. Chem. Phys. <u>63</u>, 4592 (1975); I. Webman, D. M. Ceperley, M. H. Kalos, and J. L. Lebowitz (to be published).

24. P. J. Rossky, J. D. Doll, and H. L. Friedman, J. Chem Phys. <u>69</u>, 4628 (1978).

25. L. D. Thomas, J. Chem. Phys. 70, 2979 (1979).

26. D. G. de Gennes, J. Chem. Phys. 55, 572 (1971).

27. P. A. Whitlock, D. M. Ceperley, G. V. Chester, and M. H. Kalos
     Phys. Rev. B 19, 5598 (1979).

28. D. Ceperley, Phys. Rev. B 18, 3126 (1978).

APPENDIX

I. <u>NRCC Policy Board</u>

Terms of Appointments
October 1, 1977-December 31, 1980

Professor Bruce J. Berne
Department of Chemistry
Columbia University
New York, NY   10027

Dr. Charles Bender
Lawrence Livermore Laboratory
University of California
P. O. Box 808
Livermore, CA   94550

Professor Mary L. Good
Division of Engineering Research
Louisiana State University
Baton Rouge, LA   70803

Professor William Guillory
Department of Chemistry
University of Utah
Salt Lake City, UT   84112

Professor James A. Ibers,
Chairman, NRCC Policy Board
Department of Chemistry
Northwestern University
Evanston, IL   60201

Dr. Carroll J. Johnson
Oak Ridge National Laboratory
P. O. Box X
Oak Ridge, TN   37830

Professor Martin Karplus
Department of Chemistry
Harvard University
12 Oxford Street
Cambridge, MA   02138

Professor Herbert Keller
Applied Mathematics Department
California Inst. of Technology
1201 East California Boulevard
Pasadena, CA   91125

Professor William A. Miller
Department of Chemistry
University of California
237A Hildebrand Hall
Berkeley, CA   94720

Professor John A. Pople
Department of Chemistry
Carnegie-Mellon University
4400 Fifth Avenue
Pittsburgh, PA   15213

Dr. Anessur Rahman
Argonne National Laboratory
Argonne, IL   60439

Professor Kenneth Wiberg
Department of Chemistry
Yale University
New Haven, CT   06520

II. <u>NRCC Program Committee</u>

Professor Richard Bernstein
Department of Chemistry
Columbia University
New York, NY   10017

Professor Ernest Davidson
Department of Chemistry
University of Washington
Seattle, WA   98195

Dr. William A. Lester, Jr.
Chairman, NRCC Program Committee
National Resource for Computation
   in Chemistry
Lawrence Berkeley Laboratory
University of California
Berkeley, CA   94720

Professor John C. Light
Department of Chemistry
University of Chicago
Chicago, IL   60637

Professor Josef Michl
Department of Chemistry
University of Utah
Salt Lake City, UT   84112

Professor Robert Parr
Department of Chemistry
University of North Carolina
Chapel Hill, NC   27514

Professor Stuart Rice
The James Franck Institute
The University of Chicago
Chicago, IL   60637

Professor David Templeton
Department of Chemistry
University of California
Berkeley, CA   94720

Professor W. Todd Wipke
Department of Chemistry
University of California
Santa Cruz, CA   95064

III.   NRCC User Association Executive Committee

Professor Gerald M. Maggiora   Professor Barbara J. Garrison
Chairperson   Chairperson-Elect
Department of Biochemistry   Department of Chemistry
University of Kansas   Pennsylvania State University
Lawrence, KS   66045   University Park, PA   16802

Term of appointments for the above members: May 1979-February 1980

Professor Stanley A. Hagstrom   Professor Gilda Loew
NRCC   Department of Genetics S009
Lawrence Berkeley Laboratory   Stanford University Medical
Berkeley, CA   94720   Stanford, CA   94305

Dr. David Silver   Professor George Schatz
Johns Hopkins Applied   Department of Chemistry
   Physics Laboratory   Northwestern University
Johns Hopkins University   Evanston, IL   60201
Baltimore, MD   21205

Dr. John J. Wendoloski
Liaison Officer
NRCC
Lawrence Berkeley Laboratory
Berkeley, CA   94720

Term of appointments for the above members: May 1979-February 1981

IV. 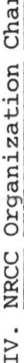 NRCC Organization Chart

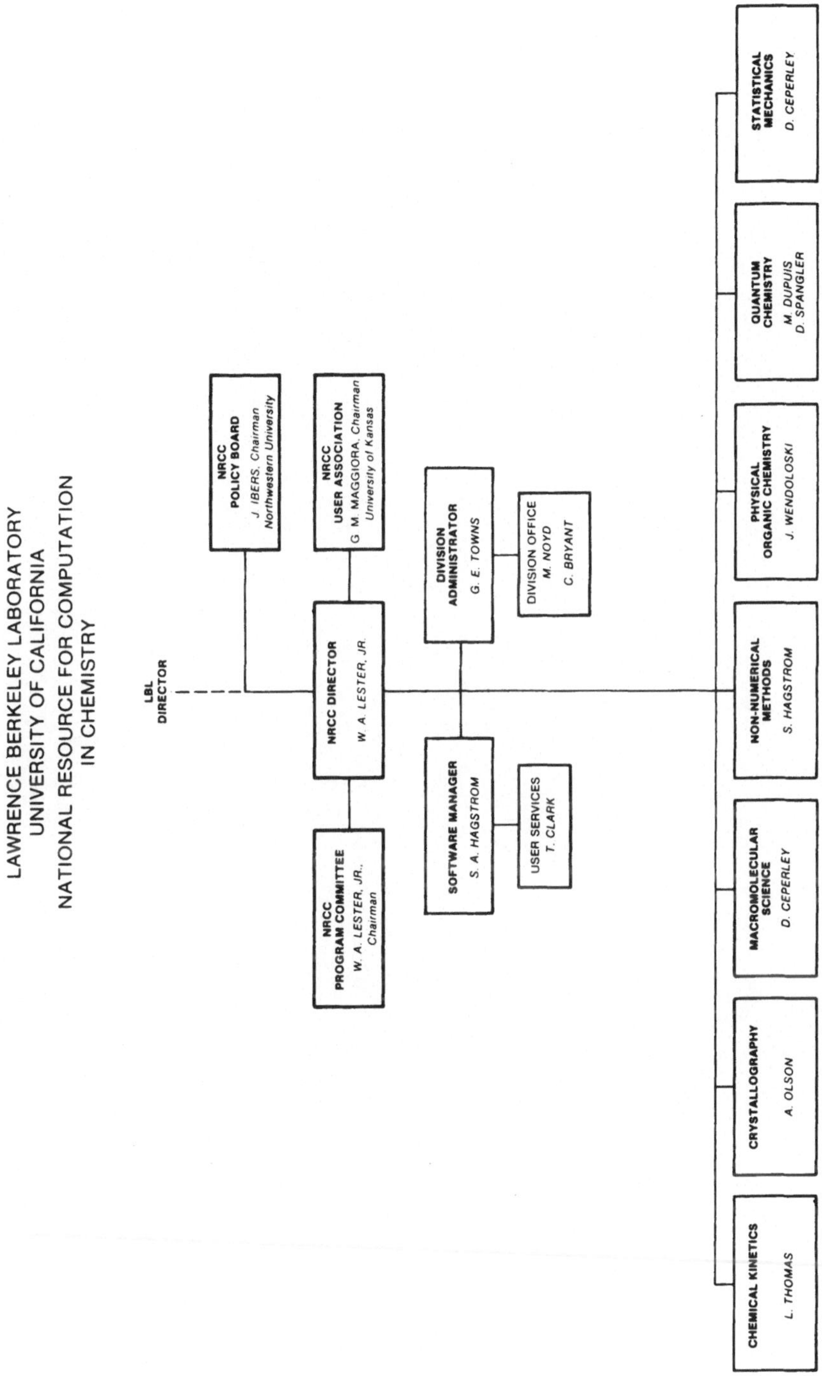

LAWRENCE BERKELEY LABORATORY
UNIVERSITY OF CALIFORNIA
NATIONAL RESOURCE FOR COMPUTATION
IN CHEMISTRY

LBL
DIRECTOR

**NRCC
POLICY BOARD**
J IBERS, Chairman
Northwestern University

**NRCC
USER ASSOCIATION**
G M MAGGIORA, Chairman
University of Kansas

**NRCC
PROGRAM COMMITTEE**
W A. LESTER, JR.,
Chairman

NRCC DIRECTOR
W A. LESTER, JR.

**DIVISION
ADMINISTRATOR**
G. E. TOWNS

DIVISION OFFICE
M. NOYD
C. BRYANT

SOFTWARE MANAGER
S. A. HAGSTROM

USER SERVICES
T. CLARK

CHEMICAL KINETICS
L. THOMAS

CRYSTALLOGRAPHY
A. OLSON

**MACROMOLECULAR
SCIENCE**
D. CEPERLEY

**NON-NUMERICAL
METHODS**
S. HAGSTROM

**PHYSICAL
ORGANIC CHEMISTRY**
J. WENDOLOSKI

**QUANTUM
CHEMISTRY**
M DUPUIS
D SPANGLER

**STATISTICAL
MECHANICS**
D. CEPERLEY